GENETIC ENGINEERING

Kruger Brentt Publishers

GENETIC ENGINEERING

Jordan Karubian
Daniel Cesar Raimunda

Kruger Brentt
Publishers
2025

Kruger Brentt Publishers UK. LTD.
Company Number 9728962

Regd. Office: 68 St Margarets Road, Edgware, Middlesex HA8 9UU

ISBN: 978-1-78715-286-1

For information on all our publications visit our website at http://krugerbrentt.com/

PREFACE

Genetic engineering represents a revolutionary field of science with the potential to transform our understanding of life and revolutionize various industries, from medicine and agriculture to biotechnology and environmental conservation. By manipulating the genetic material of organisms, scientists can engineer novel traits, develop new therapies, and address pressing challenges facing humanity. However, the power of genetic engineering also raises profound ethical, social, and environmental questions that require careful consideration and responsible stewardship.

This book, titled "Genetic Engineering," is a comprehensive exploration of the principles, applications, and implications of genetic engineering. Drawing upon the expertise of geneticists, biotechnologists, ethicists, and policymakers, this volume offers insights, methodologies, and critical perspectives for understanding the promises and pitfalls of genetic engineering.

From the fundamentals of gene editing techniques and genetic modification to the ethical considerations of human genome editing and the regulation of genetically modified organisms (GMOs), "Genetic Engineering" covers a wide range of topics essential for grasping the complexities of this rapidly advancing field. Through clear explanations, case studies, and ethical analyses, readers will gain a deep understanding of the scientific principles underlying genetic engineering and the societal implications of its applications.

Moreover, this book explores the diverse applications of genetic engineering in medicine, agriculture, industry, and beyond. By examining cutting-edge research and real-world examples, we highlight the potential of genetic engineering to address global challenges such as disease eradication, food security, and environmental sustainability.

As we navigate the opportunities and challenges of genetic engineering in the 21st century, it is our hope that "Genetic Engineering" will serve as a valuable resource for scientists, policymakers, educators, and citizens alike. Whether you are a researcher exploring the frontiers of genetic engineering, a policymaker shaping regulatory frameworks, or a concerned citizen grappling with the ethical implications of genetic technologies, this book invites you to engage in informed dialogue and ethical reflection on the future of genetic engineering and its impact on society and the environment.

Jordan Karubian
Daniel Cesar Raimunda

CONTENTS

PREFACE V

1. INDUCTION OF EXOGENOUS GENES IN A HOST ORGANISM NOT NATIVE TO THEM 1

1. Introduction: An Overview 1

2. Mechanisms of gene expression 2

3. Considerations for the expression of exogenous DNA 10

4. Regulatory sequences that must be considered for optimal expression 17

5. Investigation of exogenous expression: A case study on the expression of the bacterial luciferase gene cassette in mammals. 23

6. Concise Summary 31

8. References 32

2. PLANT GENETIC ENGINEERING EMPLOYS GATEWAY VECTORS FOR SEVERAL APPLICATIONS, INCLUDING PLANT VECTORS, BIMOLECULAR FLUORESCENCE COMPLEMENTATION (BIFC), AND MULTIGENE CONSTRUCTION 41

1. Introduction: An Overview 41

2. Basic Ti-binary vector for Agrobacterium-mediated transformation and Gateway cloning 42

3. Binary vectors compatible with Gateway cloning 46

4. Development of Gateway binary vector (pGWB) series 47

5. Application of the pGWB system 54

6. Concise Summary 62

7. References 62

3. ENHANCING THE THERMAL STABILITY OF FIREFLY LUCIFERASES BY THE APPLICATION OF GENETIC ENGINEERING TECHNIQUES 71

1. Introduction: An Overview 71

2. Overview of bioluminescent system of fireflies 72

3. *Stability of firefly luciferases in solution* 77
4. *Thermostabilization of firefly luciferases by random and site-directed mutagenesis* 80
5. *Rational protein design approach to produce the stable and active enzyme* 81
6. *Site-directed mutagenesis of cysteine residues of Luciola mingrelica firefly luciferase* 85
7. *Increase of P. pyralis luciferase thermostability by introduction of disulfide bridges* 92
8. *Thermostabilization of the Luciola mingrelica firefly luciferase by in vivo directed evolution* 93
9. *Concise Summary* 98
10. *References* 99

4. MANIPULATION OF THE PHENYLPROPANOID PATHWAY USING GENETIC ENGINEERING IN LEUCAENA LEUCOCEPHALA 109

1. *Introduction: An Overview* 109
2. *Lignin: Occurrence, structure and function* 110
3. *Lignin as barrier for paper production* 118
4. *Genetic engineering of phenylpropanoid biosynthetic pathway* 119
5. *Concise Summary* 130
6. *References* 130

5. PLANT GENETIC ENGINEERING FOR VIRAL RESISTANCE 145

1. *Introduction: An Overview* 145
2. *Plant transformation* 145
3. *Agrobacterium-mediated transformation* 152
4. *Microprojectile bombardment/ biolistics* 156
5. *Electroporation and PEG-mediated transformation of protoplasts* 158
6. *Electroporation of intact cells and tissues* 158
7. *Electro-transformation* 159
8. *Other methods of plant transformation* 163
9. *Concise Summary* 168
10. *References* 168

6. ENHANCING SOYBEAN REGENERATION VIA SOMATIC EMBRYOGENESIS AND GENETIC TRANSFORMATION: A STRATEGIC APPROACH 177

1. *Introduction: An Overview* 177
2. *Soybean somatic embryogenic process* 178

3. *Genotype-dependent response and screening of highly responsive cultivars* 183

4. *Optimization of soybean somatic embryogenesis protocol* 185

5. *Genetic transformation of soybean somatic embryos via particle bombardment and bombardment/Agrobacterium integrated system* 196

6. *Concise Summary* 199

7. *References* 200

7. GENETIC ENGINEERING AND BIOTECHNOLOGY ARE USED UO MANIPULATE GROWTH HORMONES 213

1. *Introduction: An Overview* 213

2. *The growth hormone (GH) family* 213

3. *Growth hormone of animal origin* 217

4. *Biological potential of GHs* 219

5. *Expression systems for growth hormones* 221

6. *Recombinant growth hormones* 222

7. *Technological platform for the production of recombinant GHs* 223

8. *Construction of GHs producing Pichia pastoris strains* 225

9. *Production and analysis of recombinant hormones in the flask* 227

10. *Production scaling in the bioreactor* 228

11. *Semipurification of recombinant hormones* 231

12. *Testing the biological activity of recombinant hormones* 231

13. *Contributions* 233

14. *Perspectives: At a Glance* 233

15. *References* 234

8. ASSESSING THE ENVIRONMENTAL RISKS AND MANAGING CHALLENGES OF GENETICALLY ENGINEERED VIRUS-VECTORED VACCINES 239

1. *Introduction: An Overview* 239

2. *Creation and applications of GMVs* 242

3. *Vaccines and vaccination (modified from Traavik, 1999)* 243

4. *Viruses that are used as GMVs* 244

5. *Features by GMVs that are relevant for risk assessment and management frameworks* 252

6. *The connection between risk assessment and risk management* 255

7. *Biosafety Implications for environmental risk assessment and management frameworks* 258

8. The normative challenge by the concepts of harm and unwanted ecological consequences 262
9. Precautionary principle and precautionary approaches 265
10. Concise Summary 267
11. References 267

9. THE ETHICAL IMPLICATIONS OF GENETIC ENGINEERING 273

1. Introduction: An Overview 273
2. The rise of genetic science 275
3. Developments in genetic science and moral questions 276
4. Rights and responsibilities of science in society 297
5. Concise Summary 299
6. References 301

INDEX 311

01 INDUCTION OF EXOGENOUS GENES IN A HOST ORGANISM NOT NATIVE TO THEM

1. INTRODUCTION: AN OVERVIEW

Genetic engineering can be employed to enhance the efficiency of certain metabolic and functional processes in a target organism. Nevertheless, it is frequently desired to bestow a particular host organism with supplementary functionality and/or novel phenotypic traits. In such situations, the concepts of genetic engineering can be employed to introduce foreign genes into the host organism, resulting in the production of protein products that were previously not present. Although this approach has greatly contributed to the advancement of fundamental scientific research and biotechnology in the last five decades, it has become evident that there are numerous elements that need to be taken into account in order to accurately express foreign genetic constructions.

The primary aspects to be addressed are mostly attributed to the variations in how different species have evolved to duplicate, mend, and manifest their inherent genetic structures with a notable degree of effectiveness. Therefore, when considering the expression of exogenous genes in a surrogate host, it is important to take into account how well the replication and expression machinery can identify and interact with the gene of interest. This chapter will primarily focus on the disparities in gene expression machinery and tactics between bacterial and eukaryotic species. When expressing a prokaryotic gene in a eukaryotic organism, or vice versa, we will take into account factors such as the presence or absence of exons, the functionality of polycistronic expression systems, and differences in ribosomal interaction with the gene sequence. These factors will help us understand and address the discrepancies that may arise.

There are other problems that apply regardless of the degree of relatedness between the surrogate host and the natural organism. In order to adequately train researchers in the expression of genes in different organisms, we will take into account factors such as variations in codon usage bias between the surrogate and native host, as well as the impact of differences in overall GC content on gene expression efficiency and the long-term stability of the construct. We will also consider how the host organism recognizes and eliminates foreign DNA. This will offer a fundamental comprehension of the biochemical mechanisms accountable for genetic replication and expression, as well as their potential application in expressing non-native structures.

Furthermore, this document will provide a comprehensive description of the primary regulatory signals that govern gene expression, including their presence, position, and function. Special attention will be given to the necessary modifications required for exogenous expression. This section will specifically address the existence, position, and makeup of typical promoter elements, the purpose and position of the Kozak sequence, and the significance of restriction and other regulatory sites in relation to expression in various host types. Factors regarding the probable physical impacts of introducing foreign genes into an organism will also be examined, particularly in relation to how they may interact with the organism's metabolism or affect the accumulation of the protein produced by the foreign gene. This will give readers a fundamental comprehension of how common sequences can be utilized to either amplify or moderate the production of a target gene within a surrogate host in order to achieve efficient expression.

To demonstrate the need for using both mechanisms together to activate foreign genes in a different host organism, we will show a case study involving the production of the complete bacterial luciferase gene cassette in a human kidney cell host. This example illustrates a rare situation when several simultaneous factors were taken into account to demonstrate the functionality of a set of six genes, which were first thought to only work in prokaryotic species, in a eukaryotic substitute. The culmination of over two decades of research by multiple groups, the final manifestation of the complete bacterial luciferase gene cassette effectively showcases the necessity of each major topic discussed in this chapter in order to achieve self-sustaining bioluminescence in a diverse surrogate host. This text will provide a concise summary of the introduced concerns and offer the reader a comprehensive review of how these principles might be used in laboratory settings to accomplish a certain objective.

2. MECHANISMS OF GENE EXPRESSION

Prior to the external expression of a gene in a host organism from another species, it is crucial to have a comprehensive understanding of the fundamental principles governing gene expression and maintenance. By comprehending the inherent genetic mechanism, we can gain a deeper insight into the alterations that contribute to the improved expression of genes that are not naturally present. Fortunately, regardless of whether genes are prokaryotic or eukaryotic in origin, they are all susceptible to the same fundamental processes: replication, transcription, and translation. The main distinctions between eukaryotic and prokaryotic gene expression arise from the involvement of specific proteins in each of these processes. Ultimately, the goal remains consistent: to convert DNA into messenger RNA (mRNA), then translate the mRNA into protein, and last, enable the protein to perform a certain function. The sequence of events shown here is sometimes referred to as the central dogma of biology (Fig. 1). To gain a deeper comprehension of why specific adjustments are necessary when expressing a prokaryotic gene in a eukaryotic host, and vice versa, it is important to comprehend the disparities in the genetic machinery utilized by eukaryotes and prokaryotes.

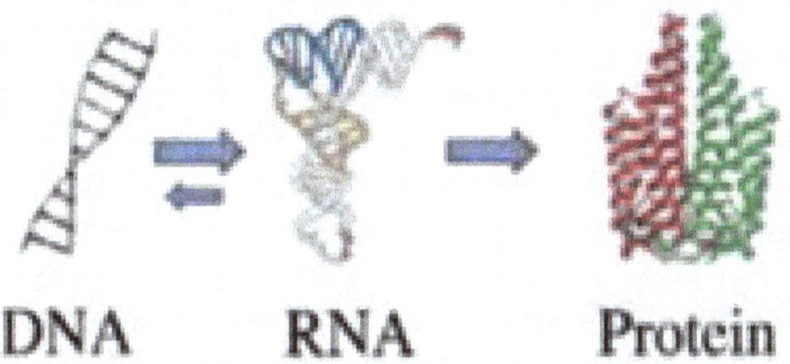

Fig. 1. The central dogma of biology shown in schematic form. DNA is transcribed to RNA and the RNA is then translated into protein. This process is the fundamental platform of our understanding of life. Adapted from (Schreiber, 2005)

2.1 Replication

The ultimate objective of the replication process is identical for all species, regardless of whether they are eukaryotic or prokaryotic: to duplicate genetic information for transmission to the succeeding generation. Replication is a crucial stage in the gene expression process, as it allows for the transmission of genetic information. Any mistakes that happen during this phase might change the genetic code and pass on that change to future generations. The primary distinctions in replication between prokaryotes and eukaryotes arise from the specific site of replication and the organization of the genome. Prokaryotic organisms normally store their DNA in a circular chromosome, which is found in the unenclosed cytoplasm of the cell. In eukaryotic creatures, the DNA is organized into linear chromosomes and kept within the cell's nucleus. The process of DNA replication is comparable in both prokaryotes and eukaryotes. An origin of replication is a specific site on the DNA where the interaction of DNA helicase enzyme causes the DNA to unwind. This unwinding exposes both strands of the DNA, allowing them to be used as templates for replication (Keck & Berger, 2000; So & Downey, 1992). After being unraveled, an RNA primer is attached to the 5' end of the DNA, and the DNA polymerase enzyme initiates the addition of complementary nucleotides in the 5' to 3' direction. Due to its antiparallel structure, the unwinding of DNA results in the production of both a leading strand and a lagging strand. The leading strand facilitates continuous replication and requires just one primer. In contrast, the trailing strand is exposed in the 3' to 5' direction, necessitating discontinuous replication. The lagging strand necessitates the use of several primers to enable the polymerase to produce numerous small DNA fragments, known as Okazaki fragments. These fragments are subsequently joined together to form a continuous strand (Falaschi, 2000; So & Downey, 1992). Prokaryotic DNA is located in a circular chromosome, which enables bidirectional replication. Replication terminates when the two replication forks meet at a termination sequence (Keck & Berger, 2000), as mentioned earlier. Nevertheless, due to the linear nature of eukaryotic chromosomes, termination occurs when the chromosome's end is reached. At this point, a telomerase enzyme elongates the 3' end of the chromosome, allowing the template DNA to finish the replication process (Zvereva et al., 2010).

2.2 Transcription

2.2.1 Transcription initiation

Transcription is the procedure by which an mRNA message is generated from a DNA template. This process occurs in three fundamental stages, namely initiation, elongation, and termination, in both eukaryotic and prokaryotic species. A significant distinction lies in the fact that prokaryotes possess a sole coding area for genetic material, whereas eukaryotes possess both coding and non-coding sections known as exons and introns, respectively. Exons contain the necessary genetic material for transcription and translation, while introns interrupt sequences of exons with non-coding genetic sequences (Watson et al., 2008). The initiation phase commences with the attachment of an RNA polymerase enzyme to a particular DNA sequence that contains the gene or genes being expressed. The difference in this step is small between prokaryotic and eukaryotic species. Prokaryotes possess just one RNA polymerase, whereas eukaryotes possess three RNA polymerases. The bacterial RNA polymerase utilizes a distinct characteristic known as a sigma (σ) factor to identify an upstream initiation location referred to as a promoter. The region consists of a minimum of two DNA sequences situated at -35 and -10 base pairs (bp) upstream from the transcription start site (Murakami & Darst, 2003). Furthermore, an additional DNA component known as a UP-element is occasionally situated even higher upstream in the promoter region. This UP-element facilitates a more robust interaction between the DNA template and the RNA polymerase upon their engagement. Upon the RNA polymerase interaction, the DNA experiences a conformational shift that causes it to unwind and expose the single template strand necessary for the transcription process to advance to the elongation stage. The process of DNA separation often takes place inside the -11 and +3 bp locations in relation to the transcription start site. While the fundamental method of transcription initiation is similar throughout eukaryotes, several enzymes are employed to execute the aforementioned steps. Eukaryotic organisms possess three distinct RNA polymerase enzymes, namely Pol I, Pol II, and Pol III, which is in contrast to prokaryotes. Among these three enzymes, Pol II is the most prevalent during regular transcription. Prokaryotes have only one initiation factor, the σ factor, while Pol II collaborates with numerous general transcription factors (GTFs). Irrespective of these variations, the process of polymerase binding remains consistent, since initiation factors identify certain locations on the promoter and facilitate the binding of Pol II (Ebright, 2000). The TRIIB site, the TATA box, the initiator, and downstream promoter elements are the most often observed recognition sites in eukaryotes (Boeger et al., 2005). Upon binding to the DNA, Pol II and the GTFs facilitate the unwinding of the DNA, thereby prepping for the elongation process and the initiation of mRNA message assembly synthesis.

2.2.2 Elongation during transcription

During the elongation step, a conformational shift occurs that enables the RNA polymerase to detach from the promoter and start synthesizing an mRNA message while moving along the template sequence. During transcription in prokaryotes, the DNA template binds to the polymerase-promoter complex. This binding results in the formation of a short transcript made up of mRNA nucleotides that are complementary to the DNA sequence. During the repetition of this process, the recently created mRNA nucleotide cannot remain inside the polymerase and must depart through a specific exit channel. This leads to the dissociation of the σ factor from the polymerase and, in the same way, the dissociation of the polymerase from the template. This enables the ongoing elongation of the nascent mRNA message. As the polymerase elongates the mRNA by sequentially adding mRNA nucleotides, the DNA undergoes winding and unwinding to maintain a consistent size of the transcription bubble formed on the DNA template. In eukaryotes, the process of escaping the promoter involves two processes to separate the GTFs from the polymerase and the polymerase from the promoter. The initial stage involves the intake of energy obtained by the hydrolysis of ATP. In the absence of the energy obtained from the breakdown of ATP, a time of interruption would take place, which might bring an end to the elongation phase and thus halt transcription completely (Dvir et al., 1996, 2001). Phosphorylation of Pol II is the second necessary step. When phosphates are introduced to the polymerase tail, it releases the linked GTFs and separates from the promoter region (Boeger et al., 2005). After the polymerase is released from the GTFs, elongation factors can attach and enhance the incorporation of nucleotides into the mRNA strand being synthesized.

2.3 Termination of transcription

Transcription concludes with the termination step once the entire mRNA molecule has been produced. The termination phase, as implied by its name, serves the aim of halting the creation of mRNA once the template gene has been transcribed. Prokaryotes employ two distinct ways for termination, namely Rho-dependent and Rho-independent. Rho binding sequences are specific DNA sequences that serve as signals for the termination of elongation and facilitate the detachment of the polymerase from the DNA. The Rho protein consists of six identical subunits that exhibit a strong attraction to RNA sequences rich in C nucleotides. Transcription termination occurs when the ribosome slows down translation to a level where it can attach to the RNA located between the RNA polymerase and the ribosome (Richardson, 2003). When a Rho binding region is present, the appropriate Rho protein can attach to the RNA molecule after it has been released from the polymerase. The inherent ATPase activity of the Rho protein subsequently halts elongation, hence ceasing the synthesis of RNA (Richardson, 2003). Rho-independent terminators can trigger termination of RNA synthesis without the need for the Rho protein to bind. Conversely, the DNA template sequence contains a reversed duplicate and a sequence of AT base pairs that, when transcribed into RNA, create a hairpin structure, which is then followed by a sequence of AU base pairs. The creation of this secondary structure leads to the cessation of RNA synthesis and the liberation of the newly formed mRNA molecule from the polymerase (Abe & Aiba, 1996). Termination in eukaryotes differs from that in prokaryotes due to the occurrence of three RNA processing events: capping, splicing, and polyadenylation. After the mRNA message leaves the polymerase, capping takes place by adding a methylation guanine to the 5' end of the newly formed mRNA (Wahle, 1995). Subsequently, splicing takes place, during which the non-coding sections of the mRNA are excised. Finally, the 3' end of the mRNA gets polyadenylated, enabling it to detach from the polymerase and conclude transcription. Table 1 provides a concise summary of the key distinctions in the transcription process between prokaryotes and eukaryotes.

Table 1. Comparison of the transcriptional process in prokaryotes and eukaryotes

Prokaryotes	Eukaryotes
Occurs in cytoplasm	Occurs in nucleus
Single polymerase	Pol I, Pol II, and Pol III
-10, -35, and UP recognition elements	TATA box and TRIIB recognition elements
Single coding region	Multiple coding regions: exons and introns
Rho dependent and independent termination	RNA processing 5' capping, splicing, and 3' polyadenylation

2.3 Translation

Once transcription is successfully finished, the mRNA is prepared for translation, which is the process of using the mRNA message to generate a sequence of amino acids, referred to as a protein. Similar to the transcriptional process, there exist nuanced yet significant distinctions in the execution of this process between prokaryotes and eukaryotes. In eukaryotes, transcription occurs in the nucleus, while translation occurs in the cytoplasm. Consequently, the mRNA that was generated earlier must traverse the nuclear membrane in order for translation to take place in the cytoplasm. Transcription in prokaryotes takes place in the cytoplasm, which is not divided into compartments. As a result, there is no need for an intermediate step, and translation can begin immediately after the mRNA leaves the polymerase during transcription. Irrespective of whether this process takes place in a prokaryote or eukaryote, it involves four key components: mRNA, transfer RNA (tRNA), aminoacyl-tRNA synthetases, and ribosomes. The mRNA component consists of codons, which are three-nucleotide units linked together in a linear manner to create open reading frames (ORFs). Eukaryotic genes typically have only one open reading frame (ORF) per mRNA sequence, but prokaryotes may have two or more ORFs per mRNA sequence (Watson et al., 2008). These mRNA sequences containing several open reading frames (ORFs) are known as polycistronic mRNAs. They have the ability to code for many proteins using a single mRNA sequence. TRNA serves as an intermediary to facilitate the recognition and binding of amino acids to the mRNA template. tRNAs bind to specific codons through their anti-codons and, when they recognize their designated codon, they add the relevant amino acid that matches that codon (Kolitz & Lorsch, 2010). After the amino acid is attached to the tRNA, the resulting complex is called an aminoacyl-tRNA synthetase. This complex subsequently attaches to the complementary mRNA, enabling the addition of the correct amino acid to the peptide chain. The ribosome, as the ultimate element of the translational process, functions as the catalyst for the fusion of mRNA and tRNA, resulting in the creation of the polypeptide chain. Ribosomes consist of two distinct subunits, namely the small and large subunits, and possess three binding sites: the A site, the P site, and the E site (Ramakrishnan, 2002). The coordination of these three binding sites facilitates the process of protein synthesis. Just like in transcription, these components collaborate to carry out the initiation, elongation, and termination stages of translation.

2.3.1 Initiation of translation

The translational initiation stage in prokaryotes and eukaryotes follows similar processes, although they employ distinct enzymes for executing these steps. In prokaryotes, the initiation stage of protein synthesis involves the ribosome being recruited to the mRNA by a ribosomal binding site positioned right before the start codon on the mRNA that was made earlier. The process described can take place immediately after the newly formed mRNA has left the polymerase. During this process, three translation initiation factors (IF1, IF2, IF3) attach to certain locations (A, E, and P) on the ribosome. These factors guide the positioning of the initiator tRNA at the start codon of the mRNA (Ramakrishnan, 2002). After binding, the initiation factor that was attached to the E site is released, enabling the large ribosomal subunit to join with the small subunit, resulting in the formation of a 70S initiation complex. This interaction induces the hydrolysis of GTP and subsequent liberation of all additional initiation components. After the initiation factors dissociate, the ribosome/mRNA complex is prepared to enter the elongation phase.

Translation and transcription are distinct processes in eukaryotic species due to the inherent compartmentalization. The nuclear membrane acts as a barrier, preventing the mRNA from interacting with the ribosome until it is released into the cytoplasm. Upon entering the cytoplasm, the eukaryotic mRNA forms a binding with the ribosome through the attachment of a 5' methylated guanine cap, initiating the process. The eukaryotic ribosome shares similarities with its prokaryotic cousin. It also possesses A, P, and E binding sites and employs initiation factors to ensure proper attachment of related tRNA (Figure 2). Contrary to the bacterial ribosome, the small subunit of the eukaryotic ribosome has to attach to the initiator tRNA before it can interact with mRNA (Watson et al., 2008). Once the tRNA is attached, the ribosome identifies the mRNA template and initiates a search for an AUG start codon. After being detected, the initiator tRNA attaches to the mRNA by breaking down GTP, which leads to the removal of the first group of initiation factors and the introduction of a second group (Acker et al., 2009). This enables the binding of the big subunit, triggering another GTP hydrolysis event that separates the remaining initiation components and forms an 80S initiation complex. Once the entire ribosome initiation complex is assembled, the ribosome/ mRNA complex is prepared to proceed into the elongation phase of translation.

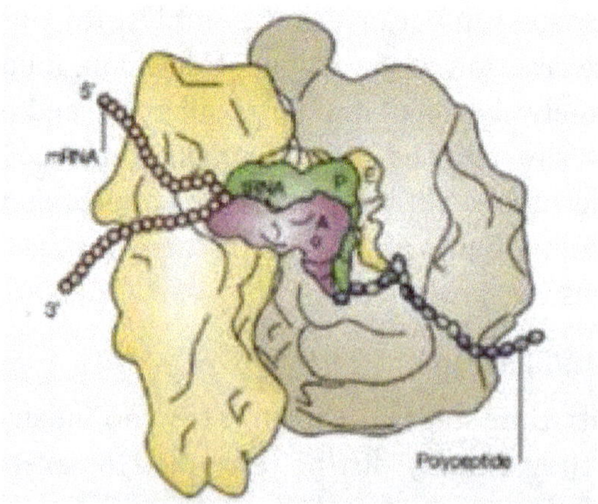

Fig. 2. The ribosome is responsible for translating mRNA into protein. Used with permission from (Lafontaine & Tollervey, 2001)

2.3.2 Elongation during translation

Elongation refers to the process in which the protein synthesized from a particular gene starts to assume its final structure. During elongation, every tRNA codon forms a connection with the corresponding amino acid via a 3′ ester bond. After the attachment of the amino acid, the aminoacyl-tRNA, which carries that specific amino acid, binds to the A site of the ribosome. The ribosome catalyzes the formation of a peptide bond between the amino acid carried by the incoming tRNA and the peptide chain connected to the peptidyl-tRNA in the P site. The interaction between the amino acid and the peptide chain results in the conversion of the aminoacyl-tRNA into a peptidyl-tRNA and induces the movement of this tRNA from the A site to the P site. Subsequently, this translocation compels the peptidyl-tRNA, which formerly occupied the P site, to depart via the E site, resulting in the assembly of a developing polypeptide chain that will ultimately constitute the protein initially specified by the expressed gene. This process is facilitated by elongation factors. Prokaryotes have three elongation factors: EF-Tu, EF-G, and EF-T. On the other hand, eukaryotes only use two elongation factors: eEF-1 and eEF-2. These findings were reported by Lavergne et al. in 1992, Nilsson and Nissen in 2005, and Oldfield and Proud in 1993. The bacterial elongation factor EF-Tu and eukaryotic elongation factor eEF-1 function in a comparable manner to attach to aminoacyl-tRNAs and guide them to the A site of the ribosome (Nilsson & Nissen, 2005; Oldfield & Proud, 1993). After the aminoacyl-tRNA enters the A site, the peptide chain from the peptidyl-tRNA binds to the amino acid on the aminoacyl-tRNA, forming a complex that is prepared for translocation. Translocation occurs through the use of either the EF-G factor in prokaryotic systems or the eEF-2 factor in eukaryotic systems. Both of these factors can bind to the peptidyl-tRNA at the P site after the peptide chain has been transferred to the aminoacyl-tRNA at the A site. This binding leads to the hydrolysis of GTP, which enables the peptidyl-tRNA at the A site to move to the P site, while the peptidyl-tRNA at the P site exits through the E site (Nilsson & Nissen, 2005; Riis et al., 1990; Watson et al., 2008). The final elongation factor, EF-T, present in prokaryotes and lacking a eukaryotic counterpart, is accountable for the displacement of EF-Tu and EF-G from the ribosome, enabling the A site to once again associate with a fresh aminoacyl-tRNA and resume the elongation process (Nilsson & Nissen, 2005). The process of adding amino acids to the mRNA codons continues until all of them have been translated into protein.

2.3.3 Termination of translation

Upon the successful culmination of the protein synthesis process, the elongation phase must be ceased, so concluding the expansion of the polypeptide chain and signifying the creation of a fully-formed protein product. The extension of the polypeptide product will persist until a termination codon is encountered in the mRNA template. Both prokaryotes and eukaryotes use three stop codons, UAG, UGA, or UAA, to halt translation. After the ribosome identifies a stop codon at the A site, a group of release factors (RFs) are activated to facilitate the release of the produced protein. Prokaryotes have two Class I release factors, RF1 and RF2, which specifically identify the UAG and UGA stop codons, respectively, and universally recognize the UAA stop codon. Additionally, there is one Class II release factor, RF3, which facilitates the separation of the Class I release factors from the ribosome once the protein has been released (Moreira et al., 2002). On the other hand, eukaryotes possess a single Class I release factor, eRF1, that can identify all three stop codons, as well as a Class II release factor, eRF3, which is responsible for dissociation (Moreira et al., 2002). Irrespective of the release factor employed, once the stop codon is identified, the process of breaking down the peptide chain commences, resulting in the liberation of the newly formed protein and all termination components from the ribosome. Table 2 provides a concise overview of the host protein machinery that is involved in translation.

Table 2. Host proteins active during translation

	Prokaryotes	Eukaryotes	Function
Initiation	IF-1	eIF-1	Blocks the A site from initiation t-RNA
	IF-2	eIF-2	Binds to initiator t-RNA
	IF-3	eIF-3	Blocks the E site
	N/A	eIF-4	Ribosomal recognition of mRNA
	N/A	eIF-5	Blocks the E site
Elongation	EF-Tu	eEF-1	Binds aminoacyl-tRNA to the A site
	EF-G	eEF-2	Translocation
	EF-T	N/A	Releases elongation factors
Termination	RF-1	eRF-1	Recognizes the UAA and UAG stop codons
	RF-2		Recognizes the UAA and UGA stop codons
	RF-3	eRF-2	Releases all translation factors

3. CONSIDERATIONS FOR THE EXPRESSION OF EXOGENOUS DNA

While nucleic acids function as the universal genetic material and the fundamental dogma applies to all organisms, introducing foreign genes into host cells and expecting them to be expressed is not a simple process. This is due to the evolutionary adaptation of the gene expression machinery in specific species, which enables it to control its own genetic material more effectively than genetic material from other species. This is particularly evident when the foreign genetic material comes from a species that is extremely distantly related. The presence of differences, such as variations in the genomic properties of GC content and codon use patterns, between the original and substitute hosts will significantly impact the effectiveness of expressing foreign genes. Furthermore, certain organisms have developed the ability to identify and eliminate or suppress foreign genetic sequences as a means of safeguarding themselves against the harmful consequences of foreign DNA activation. Efficient expression of a foreign gene in a surrogate host can only be achieved by mimicking, bypassing, or inhibiting these mechanisms. Hence, comprehending the functioning of these processes enhances the probability of devising a strategy for successful exogenous gene expression.

3.1 GC content

GC content is the proportion of guanine (G) and cytosine (C) nucleotides in a DNA sequence. The term can be applied to denote a gene, a chromosome, a genome, or any specific segment of a DNA sequence. Organisms can exhibit substantial variation in their genomic GC content. Plasmodium falciparum has a genome that is very low in GC content, with just around 20% of its genetic material being composed of GC base pairs. In contrast, Streptomyces coelicolor has a high GC content, with as much as 72% of its genome being made up of GC base pairs. Table 3 provides the GC contents of frequently utilized laboratory organisms.

Table 3. GC content varies among common organisms

Species	Genomic GC content (%)
Escherichia coli	51
Saccharomyces cerevisiae	38
Arabidopsis thaliana	36
Caenorhabditis elegans	36
Drosophila melanogaster	33
Homo sapiens	41

The disparity in thermodynamic stability between the GC and AT bonding pairs leads to an impact of GC content on the formation and stability of DNA and RNA secondary structures. These structures play a crucial role in the regulation of gene expression (Kubo & Imanaka, 1989; Kudla et al., 2009). The Shine-Dalgarno ribosome binding site, found in the 5' untranslated region of mRNA in bacteria, is relatively rich in AU nucleotides. The high abundance of AT and low stability of secondary structure at the 5' end of a coding region have been discovered to greatly contribute to the achievement of high translation efficiency in bacteria (Allert et al., 2010; Desmit & Vanduin, 1990). In addition, Kudla et al. (2009) have shown that including AU-rich leader sequences into the 5' untranslated region of mRNAs can enhance the expression levels of proteins that are typically produced at low levels. Gu and colleagues conducted a comprehensive study on 340 genomes from different groups of organisms, such as bacteria, archaea, fungi, plants, insects, fishes, birds, and mammals. They found that there is a consistent decrease in mRNA stability near the start codon in most organisms, except for birds and mammals. This decrease in stability is associated with changes in genomic GC content (Gu et al., 2010).

However, in birds and mammals, there is no reported decrease in mRNA stability around the translation start site over the entire genome, despite the fact that the GC content in these organisms is not considerably different from the species where this tendency was first noticed (Gu et al., 2010). The authors hypothesize that this disparity can be attributed to the isochore-type organization in the genomes of these creatures. An isochore is a consequence of a significant fluctuation in GC content across extensive regions within a genome (Bernardi, 1995). In an isochore structure, the GC content remains consistent throughout, even if the rest of the genome is diverse (Eyre-Walker & Hurst, 2001) (Figure 3). It should be emphasized that, in contrast to E. coli, a high GC content in the coding region generally enhances expression in mammalian cells (Kudla et al., 2006). Kudla and colleagues discovered that genes with a high GC content in mammalian cells were transcribed more effectively compared to genes with a low GC content, resulting in increased protein synthesis. Typically, eukaryotic genes have a GC-rich composition in their 5' untranslated region, which includes the 5' cap and Kozak consensus sequence (Kozak, 1987).

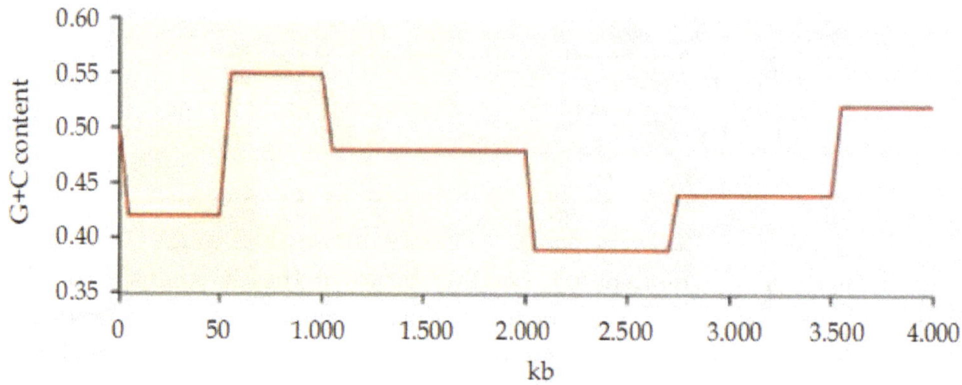

Fig. 3. The classic isochore model of genomic GC content. Used with permission from (Eyre- Walker & Hurst, 2001)

Genomic GC content is commonly acknowledged to have co-evolved with the gene expression machinery in order to achieve optimal expression for the host's fitness (Andersson & Kurland, 1990; Kudla et al., 2009). Thus, the expression level of foreign genes can be influenced by the disparity in GC contents, particularly at the 5' end, between the target genes and the host organism. The challenge of producing

Plasmodium falciparum genes in E. coli is thought to be due to their exceptionally low GC content and the potential destruction of mRNA by ribonuclease E (McDowall et al., 1994; Plotkin & Kudla, 2011). Plotkin and Kudla (2011) have estimated that over 40% of human genes will have low expression in E. coli without modification. This is because the 5' end of mRNA has a high GC content, which leads to low folding energy at the 5' end.

3.2 Codon usage bias

GC content not only affects mRNA stability and secondary structure organization, but also influences the codon usage profile of every genome. The 20 amino acids that are frequently present in protein sequences are all derived from a set of 61 distinct nucleotide triplets. The redundancy inherent in this coding method enables multiple codons to encode the same amino acid. As an illustration, the amino acids alanine and serine can be represented by either four or six codons, respectively (Table 4). The inherent propensity for genetic code to undergo degeneration has developed as a mechanism to safeguard DNA sequences from harmful mutations. This mechanism ensures that the resulting protein sequences remain intact despite the presence of genetic mutations, effectively suppressing these mutations. Nevertheless, the utilization of synonymous codons varies in frequency among different species, regions within the same genome, and even within the same gene (Andersson & Kurland, 1990; Kurland, 1991). The disparity in codon usage profiles is most pronounced amongst distantly related species, whereas closely related species are more inclined to have comparable codon preferences. The precise mechanisms by which an organism develops a specific codon bias have not yet been fully understood (Chamary et al., 2006; Hershberg & Petrov, 2008). However, it is widely believed that the GC content of the preferred codon is the primary factor influencing codon usage biases across genomes (Plotkin & Kudla, 2011).

The genetic code's redundancy enables multiple codons to encode a specific amino acid. Initially, it was thought that synonymous codon substitutions were random silent mutations. However, recent studies have shown that codon usage patterns can directly impact crucial cellular processes, including transcription and translation efficiency, protein translation accuracy, and protein folding (Angov, 2011; Zhang et al., 2009). It is possible to imagine that the particular pattern of codon usage in an organism has evolved together with other cellular machinery to ensure the best possible expression of genes and functioning of proteins in their natural environment (Grantham et al., 1981). For instance, in prokaryotes, the likelihood of a codon being utilized is directly related to the amount of its corresponding tRNA present within the cell (Bulmer, 1987; Dong et al., 1996). Consequently, the presence of differences in codon use patterns between non-native genes and the host organism hinders the expression of these genes.

First Position	Second Position: U		C		A		G		Third Position
	Codon	Amino Acid	Codon	Amino Acid	Codon	Amino Acid	Codon	Amino Acid	
U	UUU	Phe	UCU	Ser	UAU	Tyr	UGU	Cys	U
	UUC		UCC		UAC		UGC		C
	UUA	Leu	UCA		UAA	STOP	UGA	STOP	A
	UUG		UCG		UAG		UGG	Trp	G
C	CUU		CCU	Pro	CAU	His	CGU	Arg	U
	CUC		CCC		CAC		CGC		C
	CUA		CCA		CAA	Gln	CGA		A
	CUG		CCG		CAG		CGG		G
A	AUU	Ile	ACU	Thr	AAU	Asn	AGU	Ser	U
	AUC		ACC		AAC		AGC		C
	AUA		ACA		AAA	Lys	AGA	Arg	A
	AUG	Met	ACG		AAG		AGG		G
G	GUU	Val	GCU	Ala	GAC	Asp	GGU	Gly	U
	GUC		GCC		GAC		GGC		C
	GUA		GCA		GAA	Glu	GGA		A
	GUG		GCG		GAG		GGG		G

The concept that the same DNA sequence can be expressed at various efficiencies in different organisms has been consistently supported in the extensive history of exogenous gene expression (Gustafsson et al., 2004). The reason for this is that the foreign DNA sequence contains codons that are not commonly used in the host organism. As a result, there is a decrease in the efficiency of translation and protein expression. This is caused by a slower rate of translation elongation, which is a result of an imbalance between the codons used in the target gene sequence and the available pool of charged tRNA in the host organism (Kane, 1995; Kim & Lee, 2006; Rosano & Ceccarelli, 2009). These issues with expressing genes are made worse by any mismatch between the host translation machinery and the mRNA secondary structure caused by variations in GC content resulting from different codon usage patterns (Kim & Lee, 2006; Wu et al., 2004).

Codon optimization is a frequently used approach to improve the expression of non-native genes in a surrogate host, with the goal of overcoming these issues. This procedure involves substituting uncommon codons in the DNA sequence to closely align with the host's codon usage preference, while maintaining an exact match to the original amino acid sequence. Codon optimization enables the simultaneous adjustment of expected mRNA secondary structures that may arise from changes in the GC content. This procedure is particularly advantageous for removing structures located at the 5' end of coding sequences, where they are more likely to disrupt downstream protein expression (Wu et al., 2004). In order to enhance expression, the elimination of cis-acting negative regulatory components within the coding sequence is carried out to minimize the likelihood of suppression (Graf et al., 2000). The method of codon optimization can be accomplished experimentally either by performing numerous stages of site-directed mutagenesis on directly cloned DNA or by synthesizing the target gene from scratch. If there are only a few codons that need to be modified, the former method may be favored. However, the latter method has become increasingly feasible because advancements in gene synthesis have made it faster and cheaper to create synthetic DNA sequences. Typically, the process of codon optimization has been demonstrated to enhance the expression of a normal mammalian gene by a factor of five to fifteen when it is expressed in an E. coli host (Burgess-Brown et al., 2008; Gustafsson et al., 2004). In addition, the approach described in this study has been shown to greatly enhance the expression of prokaryotic genes in eukaryotic cells, as demonstrated by previous research conducted by Patterson et al. (2005), Zolotukhin et al. (1996), and Zur Megede et al. (2000).

3.3 Mechanisms for removal and silencing of exogenous genes

In order for a gene from outside to be activated in a different organism, the foreign DNA needs to be physically transported into the organism's cell and then correctly incorporated into the network responsible for gene expression and control within the organism. Extensive research in molecular and cellular biotechnology has yielded numerous efficient methods for introducing genetic material into both prokaryotic and eukaryotic hosts. However, once the gene has been transferred into the host cell, it must undergo recognition and processing by the host cell's replication, transcription, and translation machinery in order to be expressed as a functional protein. Nevertheless, as the expression of a foreign gene frequently harms the survival of the host organism in its natural state, numerous species have developed defense mechanisms to eliminate or suppress foreign DNA. This is done to save themselves from the potentially harmful consequences of this process. In bacteria, the foreign DNA that enters the cell can be cut by restriction endonucleases, which are enzymes that identify and target certain nucleotide sequences that are not part of the organism's own DNA. This process is known as restriction. During this process, methylase enzymes frequently methylate certain places on the original genetic material. This methylation prevents recognition and breakdown by restriction endonucleases, while also assuring the preservation and activation of native DNA sequences. The restriction modification system was initially identified in the 1960s and has subsequently been found to be prevalent in numerous bacterial species (Wilson & Murray, 1991). Nevertheless, the restriction system is not the sole defense mechanism devised to safeguard the host against the manifestation of foreign genetic material. Research has shown that Gram-negative bacteria can selectively suppress horizontally acquired genes by interacting with a histone-like nucleoid structuring (H-NS) protein. The occurrence known as xenogeneic silencing was initially observed in 2006 by Navarre, Lucchini, Oshim, and their colleagues (Lucchini et al., 2006; Navarre et al., 2006; Oshima et al., 2006). The H-NS protein, which is responsible for xenogeneic silencing, is a member of a group of nucleoid-associated proteins that have a tendency to bind to DNA sequences that are rich in AT base pairs, with a relatively low level of sequence specificity. Xenogeneic silencing occurs when the H-NS protein specifically targets and represses the expression of laterally acquired DNA sequences that have a lower GC content compared to the host genome.

While prokaryotic methods for silencing exogenous DNA sequences exist, there is currently no known mechanism for directly removing foreign genetic information in eukaryotic species. However, the introduction of foreign DNA into plants and mammalian cells is frequently hindered by a lack of effectiveness caused by epigenetic alteration. These alterations result in an unstable expression and, in severe instances, the suppression of the transgene over a period of time. Silencing can occur either at the transcriptional or post-transcriptional level by altering the methylation status of the sequence, modifying histones, or through RNA interference (Pal-Bhadra et al., 2002; Pikaart et al., 1998; Riu et al., 2007).

Irrespective of the precautionary actions implemented, the host utilizes these mechanisms to control the expression of foreign genes and safeguard them from harmful consequences. A remaining issue that cannot currently be regulated is that the expression of the transgene might vary significantly depending on where it is inserted into the host genome, as a result of random integration. The expression level between individual efforts often varies greatly because to position effects and epigenetic events, which are influenced by the integration location. Although it is often not possible to consistently manipulate the genomic insertion place of foreign genes, various factors have been suggested to mitigate the resulting position effects and achieve long-lasting expression of the transgene. The discussion of these elements can be found in section 4.4.

4. REGULATORY SEQUENCES THAT MUST BE CONSIDERED FOR OPTIMAL EXPRESSION

To optimize the expression of DNA sequences in a different organism, researchers can gain a deep understanding of the mechanisms that control gene expression. They should also consider factors like GC content and codon usage bias, which affect protein expression. By doing so, investigators can develop theoretical guidelines for designing DNA sequences that are best suited for expression in their desired target organism. This strategy is particularly appealing because to the decreased time and cost associated with gene synthesis, which enables the creation of new genes and expression cassettes from scratch. This means that one can simply design a gene sequence and start working on it. However, there are other considerations that need to be resolved before the successful production of an alien gene sequence may occur. In addition to optimizing the coding region, it is important to examine regulatory sequences that are neither transcribed or translated in order to ensure optimal expression. While these elements may not be present in the final protein product, they play a crucial role in the transcription, translation, and long-term maintenance of target genes in the surrogate host. Therefore, it is equally vital to optimize these elements as it is to optimize the coding sequence itself.

4.1 Regulatory elements involved in transcription

The process of converting a gene into a functional protein begins with the transcription carried out by RNA polymerase. Transcription initiation is frequently a critical regulatory point for the production of foreign proteins. The promoter region is essential for engaging the host's transcription machinery and binding the polymerase responsible for transcribing DNA into mRNA. While the promoter itself does not undergo transcription or translation, selecting a promoter that can be effectively processed by the host's machinery greatly influences the effectiveness of the design strategy. Typically, powerful promoters that are commonly utilized to control the activity of genes that are naturally present in the host organism are selected to achieve a high degree of expression for genes that come from outside sources. The T7, alcohol dehydrogenase 1 (ADH1), and human elongation factor 1 α (EF1α) promoters are frequently used to express foreign proteins in E. coli, S. cerevisiae, and mammalian cells, respectively. The cytomegalovirus immediate early (CMV IE) promoter and the Simian virus 40 (SV40) regulatory sequence are often employed viral promoters to induce transgenic expression in mammalian cells. It is worth mentioning that the strength of the promoter utilized can influence the level of transgenic expression to some extent. However, it is vital to note that different promoters can exhibit varying rates of transcription in different cell lines. Therefore, the choice of a suitable promoter should be made individually for each specific situation. A series of recent studies have conducted comprehensive comparisons of the frequently utilized promoters across different types of cells (Norrman et al., 2010; Qin et al., 2010) (Figure 4). These references are highly valuable for obtaining information when designing constructions that require specific expressions.

It is also important to remember that promoter sequences can be designed *de novo* similar to gene sequences, and that designing a specific primer upstream of a gene construct may be beneficial if no native alternative promoter sequences are available. Analysis of a large number of prokaryotic and eukaryotic promoters has revealed that many promoters contain a conserved core sequence that is essential for recognition and binding of RNA polymerase and its cofactors. Through incorporation of these conserved sequences, it may be possible to specifically design a promoter sequence, allowing one to tailor expression of their genetic rRNA in the ribosome.

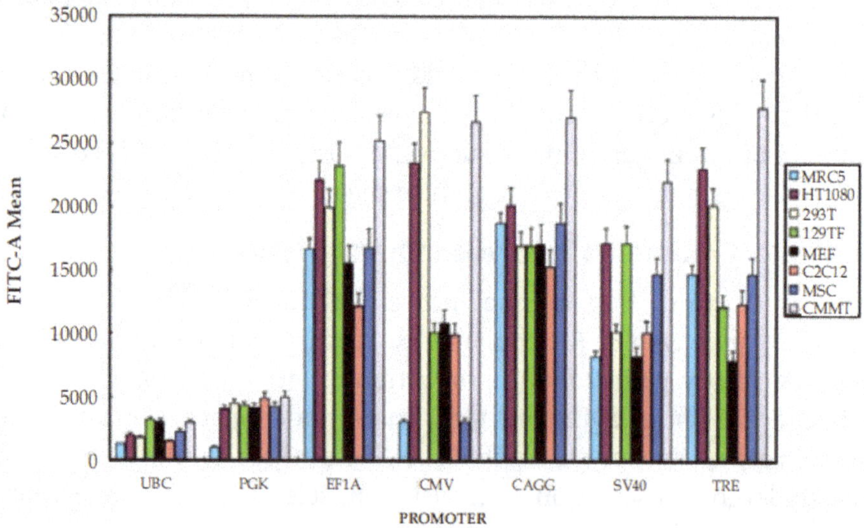

Fig. 4. Systematic comparison of different promoters in different mammalian cell types. Originally published in (Qin et al., 2010)

In prokaryotes, this conserved sequence is known as the Pribnow box, and consists of a consensus sequence of six nucleotides, TATAAT (Pribnow, 1975). In addition, there is another conserved element often found 17 bp upstream of the Pribnow box. This upstream region has a consensus TTGACAT sequence that has been shown to be crucial for transcription initiation (Rosenberg & Court, 1979). In eukaryotic organisms, the counterpart to the Pribnow box is the TATA box with a consensus sequence of TATAAA. Besides recruiting the associated transcription machinery, these core promoter elements are also crucial in defining where RNA synthesis starts. In prokaryotes, RNA synthesis usually begins 10 bp downstream of the Pribnow box, whereas the first transcribed nucleotide is located approximately 25 bp downstream of the TATA box in eukaryotes. Therefore, in addition to the use of an appropriate core promoter sequence, the location of that promoter sequence relative to the coding region should also be carefully considered to ensure complete transcription of the target genes.

It is crucial to acknowledge that while this basic core promoter is necessary for transcription, it is often insufficient on its own to achieve significant protein production. Eukaryotes utilize DNA enhancers in conjunction with the central promoter to augment gene expression by attracting extra transcription factors. These enhancers are located in different positions, such as upstream of the core promoter, within the introns of the gene controlled by the core promoter, and downstream of the genes it regulates (Levine & Tjian, 2003). While the exact mechanism of action for most enhancers is not fully understood, certain viral enhancer elements that have been extensively investigated are frequently incorporated into standard expression vectors to enhance the transcription efficiency of foreign sequences. The CMV IE enhancer, when used in conjunction with many weak promoters, has demonstrated the ability to raise gene expression levels by a range of 8 to 67 times in lung epithelial cells, as evidenced by Yew et al. (1997). Li and colleagues have shown that incorporating an SV40 enhancer into the CMV IE enhancer/promoter or the polyadenylation site's 3' end will significantly improve exogenous gene expression in mouse muscle cells, resulting in a 20-fold increase (Li et al., 2001).

4.2 Regulatory elements involved in translation

Similar to the necessity of a core promoter sequence for transcription initiation, the existence of certain, conserved regions in the 5' untranslated region of mRNA sequences is crucial for translation initiation. The Shine-Dalgarno region on the translated mRNA in prokaryotic organisms functions as the ribosome binding site (RBS). The consensus sequence consists of six nucleotides, AGGAGG, which are the complement of the anti-Shine-Dalgarno sequence found at the 3' end of the 16S During the commencement of translation, the ribosome is brought to the mRNA by the complementary base pairing between the ribosome binding site (RBS) and the 16S ribosomal RNA (rRNA). Hence, the conventional RBS is incorporated as a fundamental component in the Registry of Standard Biological Parts (http://partsregistry.org/). The registry also contains a compilation of constitutive prokaryotic ribosome binding sites (RBSs) that include the Shine-Dalgarno sequence, along with adjacent sequences that are known to impact translation. These sequences are extremely important in the design of promoter and gene sequences, since their inclusion is necessary for the effective production of the synthetic construct.

In eukaryotes, the 40S ribosomal subunit attaches to initiation factors to assist in scanning the mRNA during translation. The Kozak sequence plays a crucial role as the major initiator for translation (Kozak, 1986, 1987). The process of translation typically initiates at the AUG codon nearest to the 5' end of the mRNA, however there are exceptions to this rule. Kozak et al. have shown that the position of the actual initiation site is influenced by the distance from the 5' end, the sequence surrounding the initial AUG codon, and its steric interaction with the 40S ribosomal subunit. Nevertheless, it has been consistently shown that positioning the promoter and Kozak region before the starting codon leads to enhanced expression of target gene sequences (Morita et al., 2000).

In addition to optimizing the codon usage pattern in the coding area, it is necessary to consider other factors when expressing bacterial genes in eukaryotic hosts or vice versa. Genes obtained directly from the genomic library of a eukaryotic organism typically cannot be effectively produced in a prokaryotic host because of the existence of non-coding sections that separate the coding regions within the gene sequence. Prokaryotes do not possess the RNA splicing machinery necessary to eliminate intron regions and generate a fully developed mRNA. Hence, it is necessary to remove any introns that exist in the expression construct before introducing it into the prokaryotic host.

4.3 Elements for simultaneous expression of multiple genes in eukaryotes

On the other hand, a major hindrance to the expression of bacterial genes that have been cloned from the genome in a eukaryotic host is the host's inability to produce many proteins from a single mRNA. In contrast to prokaryotes, where it is common for multiple adjacent genes to be translated from one promoter, translation in eukaryotic cells typically necessitates the presence of a methyl-7-G(5')pppN cap at the 5' end of the mRNA. This cap is required for recognition by the translation initiation complex at the beginning of peptide synthesis (Pestova et al., 2001). Nevertheless, there exist ways that enable the simultaneous expression of two or more genes in eukaryotic cells. At a fundamental level, it is feasible to express each gene autonomously using its own promoter. This can be achieved by either introducing several vectors or by introducing a single vector that has many promoters. A different method involves the expression of several genes utilizing a polycistronic expression vector that utilizes either IRES (Internal Ribosomal Entry Site) or 2A elements. The IRES element, derived from a viral linker sequence, facilitates ribosome binding and translation initiation at the start codon of the downstream gene, independent of the 5'-cap. This enables for the translation of numerous ORFs from a single mRNA (Jackson, 1988; Jang et al., 1988). While the length and sequence of known IRES sequences may differ, specific secondary structures have been found to be preserved and play a crucial role in the function of these elements (Baird et al., 2006). The encephalomyocarditis virus (EMCV) derived IRES sequence is the most commonly utilized for expression in mammalian cells (de Felipe, 2002). 2A elements, like IRES elements, are viral sequences that can serve as a concise linker region to facilitate translation of several genes controlled by a single promoter. The process of translating the 2A element leads to an interaction between the recently produced sequence and the exit tunnel of the ribosome. This interaction results in the omission of the last peptide bond at the C-terminus of the 2A sequence. Notwithstanding the absence of this connection, the ribosome can still proceed with translation, resulting in the production of a separate protein. In order to maintain uninterrupted translation, it is important to modify the stop codon of the open reading frame (ORF) located before the 2A element, so as to prevent superfluous termination. Researchers have shown that it is possible to express five genes at the same time from a single promoter in mammalian cells by using a combination of different IRES and 2A elements. This demonstrates how these elements can be used to mimic the simultaneous expression of multiple genes seen in bacteria.

4.4 Elements for sustained maintenance and expression

The incorporation of foreign DNA sequences into a host chromosome is typically necessary to maintain the expression of a transgene in mammalian cells. Due to the random nature of the insertion event that occurs before the expression, the level of the integrated gene can be significantly influenced by the surrounding sequences and chromatin structure. The two primary challenges related to transgenic expression are unstable expression and substantial diversity among individual clones. Furthermore, if the incorporation of foreign genes takes place inside or near an essential host gene, it can have a detrimental effect on the health or ability of the host to survive. In order to mitigate this form of adverse regulation, a number of DNA components that can avoid such position effects and enhance the stability of transgenic expression have been identified (Table 5). These DNA elements are inherent in mammalian genomes and play a vital role in controlling the appropriate expression of native genes. The locus control regions (LCRs) have the ability to increase the transcription of genes that are connected to them and also allow for gene expression that depends on the number of copies present (Li et al., 2002). However, their application is limited in many types of mammalian cells due to their enormous size and tissue-specific characteristics (Kwaks & Otte, 2006). Insulators, sometimes referred to as barriers or enhancer-blocking elements, are specific DNA sequences that shield genes from the transcriptionally inactive heterochromatin or the influence of enhancers and repressors (Recillas-Targa et al., 2004). For instance, the insulator cHS4 (chicken β-globin hypersensitive site 4) is well-known for its ability to maintain transgene expression for an extended duration (Pikaart et al., 1998) and promote the effective integration of produced sequences (Recillas-Targa et al., 2004). STARs, like insulators, are employed to precisely inhibit repression. The ubiquitous chromatin opening element (UCOE) is a DNA sequence that originates from promoters of genes that are expressed in all cell types. These components have been demonstrated to enhance and maintain the expression of a foreign gene in a way that is not specific to any particular tissue, most likely by preserving an active chromatin structure (Williams et al., 2005). Matrix attachment regions (MAR) are DNA elements that facilitate the binding of the chromosome to the nuclear matrix. They are commonly utilized to ensure long-term expression of transgenes. These elements have demonstrated the ability to reverse the negative effects of position dependent insertion and inhibit the silence of transgenes in many types of cells and transgenic animals (as discussed by Harraghy and colleagues (Harraghy et al., 2008)).

Table 5. Many different elements can be used to enhance and stabilize transgene expression in mammalian cells. Modified from (Kwaks & Otte, 2006) and (Harraghy et al., 2008)

Element	Size	Increased expression	Stability of expression	Cell type-specific	Copy number-dependent	Position-independent
LCR	16 kb	Unknown	Yes	Yes	Yes	Yes, if powerful enough
Insulator	1.2-2.4 kb	Unknown	Yes	Unknown	No	Majority Yes
UCOE	2.5-8 kb	Yes	Yes	No	Unknown	Yes
MAR	~3 kb	Yes	Yes	No	No	Majority Yes
STAR	0.5-2 kb	Yes	Yes	No	Yes	Yes

5. INVESTIGATION OF EXOGENOUS EXPRESSION: A CASE STUDY

ON THE EXPRESSION OF THE BACTERIAL LUCIFERASE GENE CASSETTE IN MAMMALS.

Throughout the years, there have been numerous instances of genes that have been externally introduced and expressed. An illustrative instance that exemplifies the factors covered here is the incorporation of the bacterial luciferase gene cassette to operate independently in a human cell line. The lux cassette, which contains the bacterial luciferase gene, was used in prokaryotic systems for nearly two decades before it was first successfully expressed in a eukaryotic cell. It took almost another ten years before it was successfully expressed in a human cell line. By tracing the evolution of the lux system from a bacterial genetic system to a eukaryotic reporter cassette, we can examine both the genetic changes necessary for expressing foreign genes and the reasoning behind these modifications as adopted by researchers.

5.1 Bacterial luciferase background

The lux gene cassette consists of five genes that collaborate to generate a luminous signal at 490 nm in the blue region of the visible spectrum (Close et al., 2009). Out of the total of five genes, two genes (luxA and luxB) combine to create the luciferase protein, which is made up of two different subunits. The other three genes (luxC, luxD, and luxE) are in charge of producing a co-substrate called long chain aliphatic aldehyde. This co-substrate is essential for the luciferase protein to function properly. (Meighen, 1991). The enzyme may directly scavenge the remaining co-substrates, FMNH2 and O2, which are naturally present within the host. When the substrate complex binds to the luciferase dimer, it undergoes oxidation and emits a photon at a wavelength of 490 nm (Figure 5). The rate of this reaction is exceedingly sluggish, with the reaction proceeding for a duration of up to 20 seconds at a temperature of 20°C (Hastings & Nealson, 1977).

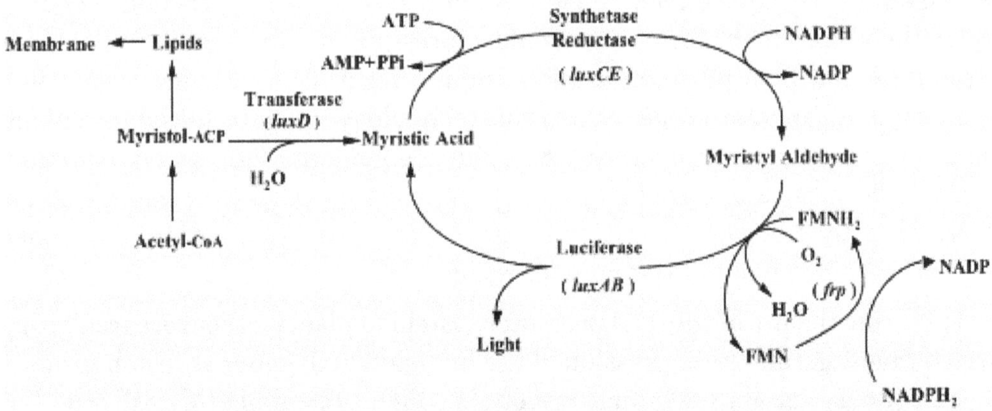

Fig. 5. The bioluminescent reaction catalyzed by the bacterial luciferase gene cassette. Reproduced with permission from (Close et al., 2009)

While these genes are widely distributed in prokaryotic organisms, the bioluminescent system they encode for is quite distinct from those commonly found in eukaryotes, such as the firefly or *Renilla* luciferase systems. Unlike these eukaryotic bioluminescence systems, the *lux* system is organized as a single operon, with all of the genes required for bioluminescent production driven from a single promoter. In addition, its prokaryotic origin means that it is optimized for function in a cellular background that is free from extensive compartmentalization. It is therefore not surprising that extensive genetic modifications were required prior to successful expression in the distantly related human cellular background. These modifications present an interesting case study of the considerations that must be made when exogenously expressing any gene in a non-native host organism.

5.2 Initial attempts at exogenous expression

The initial endeavors to articulate the lux system outside bacteria commenced in the 1980s. Upon recognizing the advantages of using a self-sufficient light-producing system as a bioluminescent reporter in bacteria, there has been a growing interest in adapting this system to work in a greater diversity of organisms, in order to exploit its benefits in a wider range of situations. These original efforts concentrated on expressing only the luxA and luxB genes, rather than expressing the entire cassette. The goal was to understand how to make the luciferase function and then use this knowledge to express the other lux genes.

Due to the inability of eukaryotic organisms to perform polycistronic expression, the initial alteration made to enable the production of the luxA and luxB genes was to place them under the control of separate promoters (Koncz et al., 1987). This technique facilitated the independent transcription of each mRNA sequence. However, given they were all located on the same plasmid, their physical expression in the host should be close to each other. This expression technique avoids the requirement for polycistronic expression and increases the likelihood of the luxA and luxB protein products coming together in a living organism to form a functional heterodimer. When this technology was introduced into plants, the cellular extracts demonstrated the ability to generate light when exposed to an aldehyde substrate. Although this experiment showed that it was possible to externally activate a part of the lux cassette, it was still not feasible for independent and self-sustained bioluminescent expression.

Following the use of the dual promoter system in plants, other research groups started investigating the expression of the luxA and luxB genes as fusion products in yeast (Boylan et al., 1989; Kirchner et al., 1989), Drosophila (Almashanu et al., 1990), and murine cell lines (Pazzagli et al., 1992). Irrespective of the source of the host, the outcomes of these tests were typically consistent. When the aldehyde substrate was applied to yeast cells, the bioluminescent expression was observed, although it was not as common as the bioluminescence shown in other prokaryotic systems studied under the same conditions (Boylan et al., 1989). When attempting to express utilizing this approach in higher eukaryotic hosts like Drosophila and mouse cell lines, a notable issue arose: bioluminescence was observable but found to be extremely sensitive to temperature.

Due to the elevated temperatures necessary for the development of the murine Ltk- cell line, the lux luciferase proteins exhibited limited stability after gene expression. As a consequence, the bioluminescent production from Ltk- cells transfected with the luxA and luxB genes was significantly reduced when cultivated at their ideal temperature of 37°C. When the growing temperature was reduced to an acceptable, if suboptimal, temperature of 30°C, the bioluminescent detection rose by a factor of 10 (Pazzagli et al., 1992). The temperature-dependent decrease in bioluminescence was confirmed by conducting additional tests in E. coli. It was found that when hosts expressed LuxA-LuxB fusion proteins, there was a more than 50,000-fold increase in bioluminescent production when grown at 23°C compared to growth at 37°C (Escher et al., 1989). This emphasizes the need of not only assessing the possible genetic obstacles to introducing a desired gene from an external source, but also taking into account the physiological restrictions that may limit the production of the protein encoded by that gene. The limitation was a substantial obstacle in the process of establishing regular eukaryotic production of these genes. It took a further ten years before this obstacle was successfully surmounted, resulting in the expression of the complete lux cassette in a yeast cell type.

5.3 Autonomous bioluminescent expression from the lux cassette in yeast

Building on the knowledge gained from the previous experiments involving the dual-promoter and fusion-based expression of the luxA and luxB genes, further efforts were made to express the complete lux cassette in a eukaryotic organism. A significant advancement occurred when researchers chose to utilize lux genes from the bacteria Photorhabdus luminescens instead of the traditional lux model organism, Vibrio harveyi (Gupta et al., 2003). Contrary to the V. harveyi template organism employed in prior attempts, P. luminescens is a bacterium that inhabits land rather than the sea. Consequently, it exhibits a greater inherent growth temperature, resulting in the ability of its protein products to remain stable at higher temperatures compared to those produced by V. harveyi, even though they serve the same role in living organisms. The alteration in the choice of the exogenous gene source exemplifies the significance of the selection procedure in gene expression within a foreign host. The P. luminescens proteins provide essential structural stability, without which no combination of genetic changes could induce high-level expression in a eukaryotic host at its optimum growing temperature.

After successfully addressing the inherent challenges associated with gene expression at the yeast's normal growth temperature, there were still other genetic alterations that needed to be taken into account before the complete lux cassette could be produced independently. An initial crucial factor was determining the most effective method for inducing consistent and abundant expression of the genes. This was achieved by integrating yeast-specific promoter sequences that had previously been proven to induce high-level expression in most growth conditions. The glyceraldehyde 3' phosphate dehydrogenase (GPD) and alcohol dehydrogenase 1 (ADH1) promoters were substituted for the original upstream areas found in the wild-type bacterial species. These original regions either had an inducer binding site or an AT rich region (Meighen, 1991). By substituting the AT rich promoter region with well-known promoters that can be produced by the host, it was guaranteed that there would be a significant increase in transcription levels when the genes were expressed in the yeast surrogate.

Subsequently, the researchers needed to devise a technique to express all five lux cassette genes concurrently in the chosen host organism. Due to its eukaryotic nature, S. cerevisiae lacks the ability to naturally express the cassette in a polycistronic manner, as would happen in a prokaryotic host under normal conditions. In order to overcome this obstacle, the researchers replicated the polycistronic expression system by include IRES sites (Gupta et al., 2003). The IRES sites serve as connectors between the lux genes, enabling the transcription of several ORFs into a single mRNA molecule. However, these ORFs are then translated independently through the recruitment of ribosomes without the need for a cap structure during translation (Lupez-Lastra et al., 2005). The researchers utilized an IRES sequence derived from S. cerevisiae to assure optimal functionality within this system, however it is acknowledged that several species can host these IRES components (Gupta et al., 2003).

Despite incorporating IRES linker sections and several promoters, the large number of genes required for autonomous light production utilizing the lux cassette remained a major challenge for exogenous expression. In order to address this issue, it was concluded that the most effective approach to expressing the lux cassette was to split its expression between two separate expression vectors (Gupta et al., 2003). This established a mechanism in which the luxA and luxB genes were separately expressed from two promoters on a single vector, while the luxC, luxD, and luxE genes were expressed from a second vector and connected using IRES sequences (Figure 6). Although the vectors utilized in this example can express genes in yeast without integrating into the chromosome, it is crucial to acknowledge that in typical eukaryotic expression, the gene sequences are usually integrated into the chromosome before expression begins. Due to the lack of control over the integration location of the gene sequences, employing a dual vector expression strategy may result in distal integration of the gene sequences. This, in turn, could increase the likelihood of different gene groups being expressed with varying efficiencies, despite the use of identical promoter sequences.

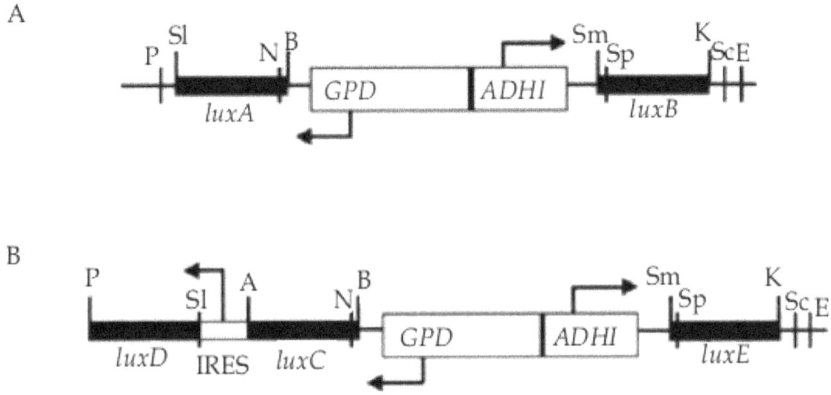

Fig. 6. Expression of the *lux* gene cassette in *S. cerevisiae* was made possible through A) independent expression of the *luxA* and *luxB* genes on one plasmid and B) expression of the remaining *lux* genes using a combination of multiple promoters and IRES linker regions from a second plasmid. Adapted from (Gupta et al., 2003)

The lux cassette genes underwent significant changes, resulting in their ability to produce a distinct and measurable bioluminescent signal when produced in S. cerevisiae (Gupta et al., 2003). This event represented the initial successful exhibition of self-sustained bioluminescent output through the use of lux genes in a eukaryotic host organism. Although the yeast host's eukaryotic nature contributed to its success, it was found that the compartmentalization of the host was restricting the luciferase's ability to reach its FMNH2 co-substrate. In contrast to prokaryotes, eukaryotes lack significant amounts of cytosolically accessible FMNH2. An further modification was made to the lux expression method by including a flavin reductase gene (frp) into the lux cassette. This was achieved by placing the frp gene downstream of the luxE gene, utilizing the IRES linker region as previously reported, and regulating its expression with the ADH1 promoter. This led to a higher concentration of FMHN2 being accessible to the luciferase enzyme in the nearby area. The final change achieved both the stabilization of bioluminescent production and an increase in light output by more than five times (Gupta et al., 2003). Although it is not frequently taken into account when external expression is being discussed, this inclusion serves as a remarkable illustration of the need of considering the expression environment together with overall genetic alterations. Regarding lux expression, the introduction of the frp gene was enough to modify the surroundings into a more advantageous state. Nevertheless, this outcome may not be consistent and should be evaluated individually for each situation.

5.4 Modification of the lux cassette for expression in mammalian cells

After the successful exhibition of self-sustained light emission from the lux genes in S. cerevisiae, investigations were initiated to examine its manifestation in human cell lines. Initially, it was thought that the alterations made during yeast expression development would be enough for expression in human cells. If it had been established that this was the situation, it would have been feasible to directly introduce genetic material into human cells using the vectors that had been created earlier and observe the emission of bioluminescent light. Regrettably, it was found that this statement is false. Despite the inclusion of human-specific, powerful promoters, the expression of the genes could not be observed at levels significantly higher than the background (Close et al., 2010; Patterson et al., 2005). Consequently, it was imperative to further tweak the lux expression system to enhance expression in a human host cell line.

Similar to prior modification methods, this study commenced by specifically targeting the expression of a limited number of genes from the lux cassette, namely luxA and luxB. The knowledge gained from S. cerevisiae expression was applied to place the luxA and luxB genes under the regulation of a potent, unchanging human promoter. These genes were connected using a specialized IRES linker region that is unique to humans. Although this did result in the capacity to identify bioluminescence in cell extracts after adding substrates, it did not represent a substantial advancement compared to expressing it in a yeast host. Having exhausted other possibilities for improving expression through genetic organization and increased promoter sequences, the researchers resorted to codon-optimization as a means of enhancing transcriptional and translational efficiency, with the goal of increasing light output. The codon usage patterns of the P. luminescens lux genes were compared to the codon usage patterns of each amino acid in all known expressed human genes. Subsequently, the codon usage patterns of the lux genes were modified to better align with the preferred codons used by humans. During this period, the gene sequences were also examined for the existence of restriction and other regulatory sequences, such as possible hairpins or terminator sequences. Subsequently, these sequences were removed by replacing the DNA sequence with a sequence that had a 100% identical amino acid output. However, the new sequence was computationally favored because it closely matched human codon preferences and did not contain any regulatory sequences (Table 6) (Patterson et al., 2005). The procedure of codon optimization, in addition to the alterations mentioned earlier, increased the bioluminescent output by 54 times compared to the expression of gene sequences that were not codon-optimized. This notable alteration emphasizes the crucial role of codon optimization in the exogenous expression of genes in a genetically distant organism.

Table 6. Comparison of the *luxA* and *luxB* genes in their wild-type (wt) and codon-optimized (co) forms. The probability of recognition as an exon was determined *in silico* using the genescan algorithm (http://genes.mit.edu). Adapted from (Patterson et al., 2005)

Gene	Predicted Start Position	Length (bp)	% GC	Number of Nucleotide Substitutions	Probability of Recognition as an Exon
wt*luxA*	61	1023	40%	N/A	0.70
co*luxA*	1	1083	54%	190	0.88
wt*luxB*	1	984	35%	N/A	0.97
co*luxB*	1	984	52%	188	0.99

Based on the success of the codon-optimization process for expression of the *luxA* and *luxB* genes in a human host cell, work then immediately began on implementing expression of the full *lux* cassette for autonomous bioluminescent production from a human host. For this process, the vector that was developed for expression of *luxA* and *luxB* was maintained, and the additional *lux* genes were placed into a second vector, mimicking the strategy employed for full *lux* cassette expression in *S. cerevisiae*. One important change that was incorporated, however, was the replacement of the yeast specific glyceraldehyde 3' phosphate dehydrogenase and alcohol dehydrogenase 1 promoters with CMV and EF1-α promoters (Close et al., 2010). These promoters allowed for strong constitutive expression of the remaining *lux* genes in a way that would not be possible if the original bacterial AT rich regions or yeast promoters were used. The benefits of the codon-optimization process were again highlighted during optimization of the remaining *lux* genes. The removal of regulatory sequences had a dramatic effect on the expression of the *luxE* gene, where their presence would have moved the predicted translational start point back to the 102nd nucleotide of the DNA sequence. In addition, the GC content of each of the genes was significantly altered to more closely match that of human coding regions, aiding in the recognition, expression and stability of each of the gene sequences following transfection into the human cellular genome (Table 7). As before, the *frp* flavin reductase gene was included in these constructs as well to compensate for the diminished cytosolic availability of FMNH2 in the highly compartmentalized eukaryotic host.

Table 7. Codon-optimization of the remainder of the *lux* genes was responsible for significant changes in both transcriptional start sites and the overall GC content. Each of these changes contributed significantly to the probability of the sequence being recognized as a coding sequence in the human host as determined *in silico* using the genescan algorithm (http://genes.mit.edu). Reproduced with permission from (Close et al., 2010)

Gene	Predicted Start Position	Length (bp)	% GC	Number of Nucleotide Substitutions	Probability of Recognition as an Exon
wt*luxC*	1	1443	37%	N/A	0.921
co*luxC*	1	1443	60%	449	0.999
wt*luxD*	1	924	38%	N/A	0.875
co*luxD*	1	924	59%	294	0.999
wt*luxE*	102	1087	38%	N/A	0.443
co*luxE*	1	1113	60%	331	0.999
wt*frp*	1	613	47%	N/A	0.715
co*frp*	1	723	64%	249	0.999

While the changes required to induce bioluminescent production from the *lux* cassette genes in the human cellular background were extensive, they were all necessary for proper function. The failure of even a single modification would lead to cells that may be capable of expressing the genes but not maintaining expression at a high enough level to be useful as a reporter system (Figure 7).

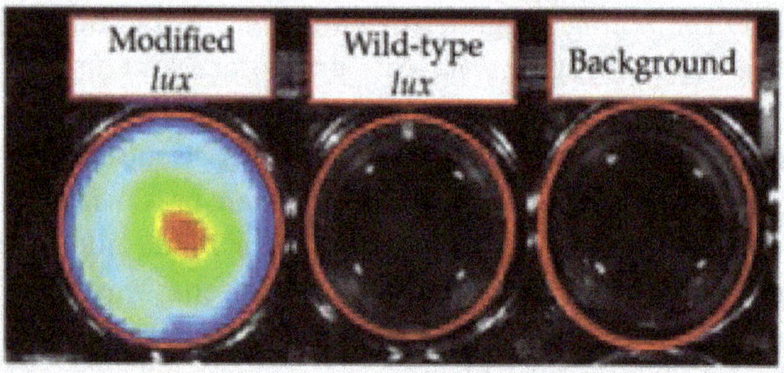

Fig. 7. Comparison of the bioluminescent expression from the *lux* genes expressed in a human host cell either following the modifications described above (modified *lux*) or without the aforementioned modifications (wild-type *lux*), and background light detection from host cells without *lux* genes (background). Adapted from (Close et al., 2010)

By applying the techniques and considerations described in this chapter, it was possible to create a set of six gene sequences from a reporter system that was previously thought to only work in prokaryotic organisms. This new bioluminescent reporter system can now be expressed in human cells and produces a strong signal that can be detected through tissue, similar to natural eukaryotic genes like firefly luciferase (Figure 8).

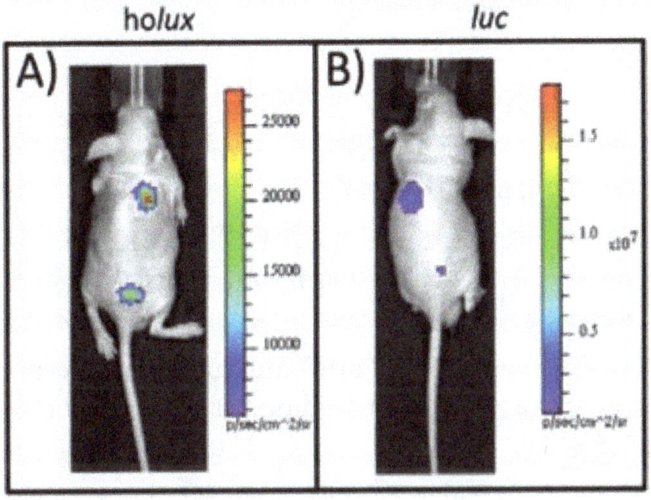

Fig. 8. Following modification of the full *lux* cassette, it was capable of being expressed in a human cell line host and producing bioluminescence at levels comparable to detection patterns of the native eukaryotic bioluminescent firefly luciferase (*luc*) gene. Adapted from (Close et al., 2011)

6. CONCISE SUMMARY

This chapter has extensively discussed the various factors that need to be taken into account when trying to externally express a gene of interest in a different organism. A comprehensive knowledge of the transcriptional, translational, and regulatory mechanisms that govern the preservation and manifestation of all genes is essential for comprehending the necessity of specific modifications to achieve significant levels of exogenous expression. However, the illustrations presented here should suffice for the average researcher to initiate the development of an adequate expression protocol. Applying all of the modifications presented in this chapter to every gene is not mandatory. However, having a comprehensive knowledge of the potential alterations can offer a diverse range of techniques for expressing difficult gene sequences. Similar to the lux cassette method, it is frequently required to make multiple modifications to achieve satisfactory levels of gene expression in a distant host organism when foreign genes are produced. Frequently, following a systematic approach will provide indications of the necessary adjustments and the phases that can be skipped to optimize time and cost efficiency when creating a novel expression platform for a gene sequence that has not been expressed before. It is important to mention that the techniques described in this chapter do not cover all possible scenarios. Occasionally, the host environment may be unsuitable for expressing the desired gene sequence, and it may not be feasible to modify the environment by adding or removing genes. As we discover new technologies and gain a better knowledge of biological processes that maintain and express genes, the number of genes that cannot be expressed will decrease.

8. REFERENCES

Abe, H., & Aiba, H. (1996). Differential contributions of two elements of rho-independent terminator to transcription termination and mRNA stabilization. *Biochimie,* 78, 11- 12, pp. 1035-1042.

Acker, M. G., Shin, B.-S., Nanda, J. S., Saini, A. K., Dever, T. E., & Lorsch, J. R. (2009). Kinetic analysis of late steps of eukaryotic translation initiation. *Journal of Molecular Biology,* 385, 2, pp. 491-506.

Allert, M., Cox, J. C., & Hellinga, H. W. (2010). Multifactorial determinants of protein expression in prokaryotic open reading frames. *Journal of Molecular Biology,* 402, 5, pp. 905-918.

Almashanu, S., Musafia, B., Hadar, R., Suissa, M., & Kuhn, J. (1990). Fusion of *luxA* and *luxB* and its expression in *Escherichia coli, Saccharomyces cerevisiae* and *Drosophila melanogaster. Journal of Bioluminescence and Chemiluminescence,* 5, 1, pp. 89-97.

Andersson, S. G. E., & Kurland, C. G. (1990). Codon preferences in free-living microorganisms. *Microbiological Reviews,* 54, 2, pp. 198-210.

Angov, E. (2011). Codon usage: Nature's roadmap to expression and folding of proteins. *Biotechnology Journal,* 6, 6, pp. 650-659.

Baird, S. D., Turcotte, M., Korneluk, R. G., & Holcik, M. (2006). Searching for IRES. *RNA - A Publication of the RNA Society,* 12, 10, pp. 1755-1785.

Bernardi, G. (1995). The human genome: Organization and evolutionary history. *Annual Review of Genetics,* 29, 445-476.

Boeger, H., Bushnell, D. A., Davis, R., Griesenbeck, J., Lorch, Y., Strattan, J. S., et al. (2005). Structural basis of eukaryotic gene transcription. *FEBS Letters,* 579, 4, pp. 899-903.

Boylan, M., Pelletier, J., & Meighen, E. A. (1989). Fused bacterial luciferase subunits catalyze light emission in eukaryotes and prokaryotes. *Journal of Biological Chemistry,* 264, 4, pp. 1915-1918.

Bulmer, M. (1987). Coevolution of codon usage and transfer RNA abundance. *Nature,* 325, 6106, pp. 728-730.

Burgess-Brown, N. A., Sharma, S., Sobott, F., Loenarz, C., Oppermann, U., & Gileadi, O. (2008). Codon optimization can improve expression of human genes in *Escherichia coli*: A multi-gene study. *Protein Expression and Purification,* 59, 1, pp. 94-102.

Chamary, J. V., Parmley, J. L., & Hurst, L. D. (2006). Hearing silence: non-neutral evolution at synonymous sites in mammals. *Nature Reviews Genetics,* 7, 2, pp. 98-108.

Close, D. M., Hahn, R., Patterson, S. S., Ripp, S., & Sayler, G. S. (2011). Comparison of human optimized bacterial luciferase, firefly luciferase, and green fluorescent protein for continuous imaging of cell culture and animal models. *Journal of Biomedical Optics,* 16, 4, pp. e12441.

Close, D. M., Patterson, S. S., Ripp, S., Baek, S. J., Sanseverino, J., & Sayler, G. S. (2010). Autonomous bioluminescent expression of the bacterial luciferase gene cassette (*lux*) in a mammalian cell line. *PLoS ONE,* 5, 8, pp. e047003.

Close, D. M., Ripp, S., & Sayler, G. S. (2009). Reporter proteins in whole-cell optical bioreporter detection systems, biosensor integrations, and biosensing applications. *Sensors,* 9, 11, pp. 9147-9174.

de Felipe, P. (2002). Polycistronic viral vectors. *Current Gene Therapy,* 2, 3, pp. 355-378.

Desmit, M. H., & Vanduin, J. (1990). Secondary structure of the ribosome binding site determines translation efficiency - A quantitative analysis. *Proceedings of the National Academy of Sciences of the United States of America,* 87, 19, pp. 7668-7672.

Dong, H. J., Nilsson, L., & Kurland, C. G. (1996). Co-variation of tRNA abundance and codon usage in *Escherichia coli* at different growth rates. *Journal of Molecular Biology,* 260, 5, pp. 649-663.

Dvir, A., Conaway, J. W., & Conaway, R. C. (2001). Mechanism of transcription initiation and promoter escape by RNA polymerase II. *Current Opinion in Genetics & Development,* 11, 2, pp. 209-214.

Dvir, A., Conaway, R. C., & Conaway, J. W. (1996). Promoter escape by RNA polymerase II - A role for an ATP cofactor in suppression of arrest by polymerase at promoter- proximal sites. *Journal of Biological Chemistry,* 271, 38, pp. 23352-23356.

Ebright, R. H. (2000). RNA polymerase: Structural similarities between bacterial RNA polymerase and eukaryotic RNA polymerase II. *Journal of Molecular Biology,* 304, 5, pp. 687-698.

Escher, A., Okane, D. J., Lee, J., & Szalay, A. A. (1989). Bacterial luciferase alpha-beta fusion protein is fully active as a monomer and highly sensitive *in vivo* to elevated temperature. *Proceedings of the National Academy of Sciences of the United States of America,* 86, 17, pp. 6528-6532.

Eyre-Walker, A., & Hurst, L. D. (2001). The evolution of isochores. *Nature Reviews Genetics,* 2, 7, pp. 549-555.

Falaschi, A. (2000). Eukaryotic DNA replication: a model for a fixed double replisome. *Trends in Genetics,* 16, 2, pp. 88-92.

Graf, M., Bojak, A., Deml, L., Bieler, K., Wolf, H., & Wagner, R. (2000). Concerted action of multiple *cis*-acting sequences is required for Rev dependence of late human immunodeficiency virus type 1 gene expression. *Journal of Virology,* 74, 22, pp. 10822-10826.

Grantham, R., Gautier, C., Gouy, M., Jacobzone, M., & Mercier, R. (1981). Codon catalog usage is a genome strategy modulated for gene expressivity. *Nucleic Acids Research,* 9, 1, pp. R43-R74.

Gu, W. J., Zhou, T., & Wilke, C. O. (2010). A universal trend of reduced mRNA stability near the translation-initiation site in prokaryotes and eukaryotes. *PLoS Computational Biology,* 6, 2, pp. e1000664.

Gupta, R. K., Patterson, S. S., Ripp, S., & Sayler, G. S. (2003). Expression of the *Photorhabdus luminescens lux* genes (*luxA, B, C, D,* and *E*) in *Saccharomyces cerevisiae. FEMS Yeast Research,* 4, 3, pp. 305-313.

Gustafsson, C., Govindarajan, S., & Minshull, J. (2004). Codon bias and heterologous protein expression. *Trends in Biotechnology,* 22, 7, pp. 346-353.

Harraghy, N., Gaussin, A., & Mermod, N. (2008). Sustained transgene expression using MAR elements. *Current Gene Therapy,* 8, 5, pp. 353-366.

Hastings, J., & Nealson, K. (1977). Bacterial bioluminescence. *Annual Reviews in Microbiology,* 31, 1, pp. 549-595.

Hershberg, R., & Petrov, D. A. (2008). Selection on codon bias. *Annual Review of Genetics,* 42, 1, pp. 287-299.

Jackson, R. J. (1988). RNA translation - Picornaviruses break the rules. *Nature,* 334, 6180, pp. 292-293.

Jang, S.K., Krausslich, H.G., Nicklin, M.J.H., Duke, G.M., Palmenberg, A.C., & Wimmer, E. (1988). A segment of the 5' nontranslated region of encephalomyocarditis virus RNA directs internal entry of ribosomes during in vitro translation. *Journal of Virology,* 62, 8, pp. 2636-2643.

Kane, J. F. (1995). Effects of rare codon clusters on high-level expression of heterologous proteins in *Escherichia coli. Current Opinion in Biotechnology,* 6, 5, pp. 494-500.

Keck, J. L., & Berger, J. M. (2000). DNA replication at high resolution. *Chemistry & Biology,* 7, 3, pp. R63-71.

Kim, S., & Lee, S. B. (2006). Rare codon clusters at 5'-end influence heterologous expression of archaeal gene in *Escherichia coli*. *Protein Expression and Purification*, 50, 1, pp. 49- 57.

Kirchner, G., Roberts, J. L., Gustafson, G. D., & Ingolia, T. D. (1989). Active bacterial luciferase from a fused gene: Expression of a *Vibrio harveyi luxAB* translational fusion in bacteria, yeast and plant cells. *Gene*, 81, 2, pp. 349-354.

Kolitz, S. E., & Lorsch, J. R. (2010). Eukaryotic initiator tRNA: Finely tuned and ready for action. *FEBS Letters*, 584, 2, pp. 396-404.

Koncz, C., Olsson, O., Langridge, W. H. R., Schell, J., & Szalay, A. A. (1987). Expression and assembly of functional bacterial luciferase in plants. *Proceedings of the National Academy of Sciences USA*, 84, 1, pp. 131-135.

Kozak, M. (1986). Point mutations define a sequence flanking the AUG initiator codon that modulates translation by eukaryotic ribosomes. *Cell*, 44, 2, pp. 283-292.

Kozak, M. (1987). An analysis of 5'-noncoding sequences from 699 vertebrate messenger RNAs. *Nucleic Acids Research*, 15, 20, pp. 8125-8148.

Kubo, M., & Imanaka, T. (1989). mRNA secondary structure in an open reading frame reduces translation efficiency in *Bacillus subtilis*. *Journal of Bacteriology*, 171, 7, pp. 4080-4082.

Kudla, G., Lipinski, L., Caffin, F., Helwak, A., & Zylicz, M. (2006). High guanine and cytosine content increases mRNA levels in mammalian cells. *PLoS Biology*, 4, 6, pp. 933-942.

Kudla, G., Murray, A. W., Tollervey, D., & Plotkin, J. B. (2009). Coding-sequence determinants of gene expression in *Escherichia coli*. *Science*, 324, 5924, pp. 255-258.

Kurland, C. G. (1991). Codon bias and gene expression. *FEBS Letters*, 285, 2, pp. 165-169.

Kwaks, T. H. J., & Otte, A. P. (2006). Employing epigenetics to augment the expression of therapeutic proteins in mammalian cells. *Trends in Biotechnology*, 24, 3, pp. 137-142.

Lafontaine, D. L. J., & Tollervey, D. (2001). The function and synthesis of ribosomes. *Nature Reviews Molecular Cell Biology*, 2, 7, pp. 514-520.

Lavergne, J. P., Reboud, A. M., Sontag, B., Guillot, D., & Reboud, J. P. (1992). Binding of GDP to a ribosomal protein after elongation factor-2 dependent GTP hydrolysis. *Biochimica et Biophysica Acta - Gene Structure and Expression*, 1132, 3, pp. 284-289.

Levine, M., & Tjian, R. (2003). Transcription regulation and animal diversity. *Nature,* 424, 6945, pp. 147-151.

Li, Q. L., Peterson, K. R., Fang, X. D., & Stamatoyannopoulos, G. (2002). Locus control regions. *Blood,* 100, 9, pp. 3077-3086.

Li, S., MacLaughlin, F. C., Fewell, J. G., Gondo, M., Wang, J., Nicol, F., et al. (2001). Muscle- specific enhancement of gene expression by incorporation of SV/40 enhancer in the expression plasmid. *Gene Therapy,* 8, 6, pp. 494-497.

Lucchini, S., Rowley, G., Goldberg, M. D., Hurd, D., Harrison, M., & Hinton, J. C. D. (2006). H-NS mediates the silencing of laterally acquired genes in bacteria. *PLoS Pathogens,* 2, 8, pp. 746-752.

Lupez-Lastra, M., Rivas, A., & Barrìa, M. (2005). Protein synthesis in eukaryotes: the growing biological relevance of cap-independent translation initiation. *Biological Research,* 38, 121-146.

McDowall, K. J., Linchao, S., & Cohen, S. N. (1994). A+U content rather than a particular nucleotide order determines the specificity of RNase E cleavage. *Journal of Biological Chemistry,* 269, 14, pp. 10790-10796.

Meighen, E. A. (1991). Molecular biology of bacterial bioluminescence. *Microbiological Reviews,* 55, 1, pp. 123-142.

Moreira, D., Kervestin, S., Jean-Jean, O., & Philippe, H. (2002). Evolution of eukaryotic translation elongation and termination factors: Variations of evolutionary rate and genetic code deviations. *Molecular Biology and Evolution,* 19, 2, pp. 189-200.

Morita, S., Kojima, T., & Kitamura, T. (2000). Plat-E: An efficient and stable system for transient packaging of retroviruses. *Gene Therapy,* 7, 12, pp. 1063-1066.

Murakami, K. S., & Darst, S. A. (2003). Bacterial RNA polymerases: The whole story. *Current Opinion in Structural Biology,* 13, 1, pp. 31-39.

Navarre, W. W., Porwollik, S., Wang, Y. P., McClelland, M., Rosen, H., Libby, S. J., et al. (2006). Selective silencing of foreign DNA with low GC content by the H-NS protein in *Salmonella. Science,* 313, 5784, pp. 236-238.

Nilsson, J., & Nissen, P. (2005). Elongation factors on the ribosome. *Current Opinion in Structural Biology,* 15, 3, pp. 349-354.

Norrman, K., Fischer, Y., Bonnamy, B., Sand, F. W., Ravassard, P., & Semb, H. (2010). Quantitative comparison of constitutive promoters in human ES cells. *PLoS ONE,* 5, 8, pp. e12413.

Oldfield, S., & Proud, C. G. (1993). Phosphorylation of elongation factor-2 from the lepidopteran insect, spodoptera frugiperda. *FEBS Letters,* 327, 1, pp. 71-74.

Oshima, T., Ishikawa, S., Kurokawa, K., Aiba, H., & Ogasawara, N. (2006). *Escherichia coli* histone-like protein H-NS preferentially binds to horizontally acquired DNA in association with RNA polymerase. *DNA Research,* 13, 4, pp. 141-153.

Pal-Bhadra, M., Bhadra, U., & Birchler, J. A. (2002). RNAi related mechanisms affect both transcriptional and posttranscriptional transgene silencing in *Drosophila*. *Molecular Cell,* 9, 2, pp. 315-327.

Patterson, S. S., Dionisi, H. M., Gupta, R. K., & Sayler, G. S. (2005). Codon optimization of bacterial luciferase (*lux*) for expression in mammalian cells. *Journal of Industrial Microbiology & Biotechnology,* 32, 3, pp. 115-123.

Pazzagli, M., Devine, J. H., Peterson, D. O., & Baldwin, T. O. (1992). Use of bacterial and firefly luciferases as reporter genes in DEAE-dextran mediated transfection of mammalian cells. *Analytical Biochemistry,* 204, 2, pp. 315-323.

Pestova, T. V., Kolupaeva, V. G., Lomakin, I. B., Pilipenko, E. V., Shatsky, I. N., Agol, V. I., et al. (2001). Molecular mechanisms of translation initiation in eukaryotes. *Proceedings of the National Academy of Sciences of the United States of America,* 98, 13, pp. 7029- 7036.

Pikaart, M. I., Recillas-Targa, F., & Felsenfeld, G. (1998). Loss of transcriptional activity of a transgene is accompanied by DNA methylation and histone deacetylation and is prevented by insulators. *Genes & Development,* 12, 18, pp. 2852-2862.

Plotkin, J. B., & Kudla, G. (2011). Synonymous but not the same: the causes and consequences of codon bias. *Nature Reviews Genetics,* 12, 1, pp. 32-42.

Pribnow, D. (1975). Nucleotide sequence of an RNA polymerase binding site at an early T7 promoter. *Proceedings of the National Academy of Sciences of the United States of America,* 72, 3, pp. 784-788.

Qin, J., Zhang, L., Clift, K., Hulur, I., Xiang, A., Ren, B., et al. (2010). Systematic comparison of constitutive promoters and the doxycycline-inducible promoter. *PLoS One,* 5, 5, pp. e10611.

Ramakrishnan, V. (2002). Ribosome structure and the mechanism of translation. *Cell,* 108, 4, pp. 557-572.

Recillas-Targa, F., Valadez-Graham, V., & Farre, C. M. (2004). Prospects and implications of using chromatin insulators in gene therapy and transgenesis. *Bioessays,* 26, 7, pp. 796-807.

Richardson, J. P. (2003). Loading Rho to terminate transcription. *Cell,* 114, 2, pp. 157-159.

Riis, B., Rattan, S. I. S., Clark, B. F. C., & Merrick, W. C. (1990). Eukaryotic protein elongation factors. *Trends in Biochemical Sciences,* 15, 11, pp. 420-424.

Riu, E. R., Chen, Z. Y., Xu, H., He, C. Y., & Kay, M. A. (2007). Histone modifications are associated with the persistence or silencing of vector-mediated transgene expression *in vivo. Molecular Therapy,* 15, 7, pp. 1348-1355.

Rosano, G. L., & Ceccarelli, E. A. (2009). Rare codon content affects the solubility of recombinant proteins in a codon bias-adjusted *Escherichia coli* strain. *Microbial Cell Factories,* 8, 1, pp. 41.

Rosenberg, M., & Court, D. (1979). Regulatory sequences involved in the promotion and termination of RNA transcription. *Annual Review of Genetics,* 13, 1, pp. 319-353.

Schreiber, S. L. (2005). Small molecules: The missing link in the central dogma. *Nature Chemical Biology,* 1, 2, pp. 64-66.

So, A. G., & Downey, K. M. (1992). Eukaryotic DNA replication. *Critical Reviews in Biochemistry and Molecular Biology,* 27, 1-2, pp. 129-155.

Szymczak, A. L., & Vignali, D. A. A. (2005). Development of 2A peptide-based strategies in the design of multicistronic vectors. *Expert Opinion on Biological Therapy,* 5, 5, pp. 627-638.

Wahle, E. (1995). 3'-End cleavage and polyadenylation of mRNA precursors. *Biochimica et Biophysica Acta - Gene Structure and Expression,* 1261, 2, pp. 183-194.

Watson, J., Baker, T., Bell, S., Gann, A., Levine, M., & Losick, R. (2008). *Molecular Biology of the Gene* (6 ed.). Cold Spring Harbor: Cold Spring Harbor Laboratory Press.

Williams, S., Mustoe, T., Mulcahy, T., Griffiths, M., Simpson, D., Antoniou, M., et al. (2005). CpG-island fragments from the *HNRPA2B1/CBX3* genomic locus reduce silencing and enhance transgene expression from the hCMV promoter/enhancer in mammalian cells. *BMC Biotechnology,* 5, 1, pp. 17.

Wilson, G. G., & Murray, N. E. (1991). Restriction and modification systems. *Annual Review of Genetics,* 25, 1, pp. 585-627.

Wu, X. Q., Jornvall, H., Berndt, K. D., & Oppermann, U. (2004). Codon optimization reveals critical factors for high level expression of two rare codon genes in *Escherichia coli*: RNA stability and secondary structure but not tRNA abundance. *Biochemical and Biophysical Research Communications,* 313, 1, pp. 89-96.

Yew, N. S., Wysokenski, D. M., Wang, K. X., Ziegler, R. J., Marshall, J., McNeilly, D., et al. (1997). Optimization of plasmid vectors for high-level expression in lung epithelial cells. *Human Gene Therapy,* 8, 5, pp. 575-584.

Zhang, G., Hubalewska, M., & Ignatova, Z. (2009). Transient ribosomal attenuation coordinates protein synthesis and co-translational folding. *Nature Structural & Molecular Biology,* 16, 3, pp. 274-280.

Zolotukhin, S., Potter, M., Hauswirth, W., Guy, J., & Muzyczka, N. (1996). A "humanized" green fluorescent protein cDNA adapted for high-level expression in mammalian cells. *Journal of Virology,* 70, 7, pp. 4646-4654.

Zur Megede, J., Chen, M. C., Doe, B., Schaefer, M., Greer, C. E., Selby, M., et al. (2000). Increased expression and immunogenicity of sequence-modified human immunodeficiency virus type 1 *gag* gene. *Journal of Virology,* 74, 6, pp. 2628.

Zvereva, M., Shcherbakova, D., & Dontsova, O. (2010). Telomerase: Structure, functions, and activity regulation. *Biochemistry,* 73, 13, pp. 1563-1583.

02 PLANT GENETIC ENGINEERING EMPLOYS GATEWAY VECTORS FOR SEVERAL APPLICATIONS, INCLUDING PLANT VECTORS, BIMOLECULAR FLUORESCENCE COMPLEMENTATION (BIFC), AND MULTIGENE CONSTRUCTION

1. INTRODUCTION: AN OVERVIEW

Transgenic technologies play a crucial role in the genetic modification of plants, serving as a vital tool for both fundamental plant study and advancements in biotechnology. For instance, in fundamental plant research, promoter analysis is commonly employed to ascertain the expression pattern of genes of interest using a reporter like green fluorescent protein (GFP). Furthermore, the function of these genes can be determined by conducting downregulation or controlled expression experiments of target genes. The transgenic approach is commonly employed in plant biotechnology to enhance the expression of foreign genes, hence improving the productivity of economically significant agricultural plants. In recent times, genome projects that concentrate on different types of higher plants have yielded a substantial amount of sequence data. Consequently, comprehensive investigations of gene function and gene regulation across the entire genome are currently underway. Transgenic analyses utilizing genetically modified plants will increasingly be indispensable in various scientific domains. An example of a method that can be used to investigate the regulation of gene expression, the localization of gene products, and the activities of different genes is high-throughput promoter analysis. This technique involves examining the temporal and spatial regulation of gene expression, determining the subcellular localization of gene products using reporter genes, and studying the effects of ectopic production of cDNA clones and RNAi. The binary system of Agrobacterium-mediated transformation is the most often employed method for gene manipulation in plants. This system comprises two plasmids that are derived from Ti plasmids, specifically disarmed Ti plasmids and binary vectors (Bevan, 1984). The former consists of the majority of genes responsible for transferring T-DNA from Agrobacterium tumefaciens to plants, while the latter is made up of a functioning T-DNA and essential elements for replication in both Escherichia coli and A. tumefaciens. The majority of the commonly employed binary vectors developed in the 1990s were created using a conventional process that relies on restriction endonucleases. Consequently, the process of creating altered genes on binary vectors was arduous and time-consuming due to the scarcity of restriction sites and their huge size, as well as the presence of several restriction sites outside their cloning sites. In order to address this drawback and achieve effective and speedy synthesis of changed genes on binary vectors, there was a need for a novel cloning technology that could facilitate high-throughput study of plant genes. One of the available alternatives is the Gateway cloning system offered by Invitrogen, a company based in Carlsbad, CA, USA. We have developed a range of Gateway-compatible Ti binary vectors specifically designed for plant transgenic research.

2. BASIC TI-BINARY VECTOR FOR *AGROBACTERIUM*-MEDIATED TRANSFORMATION AND GATEWAY CLONING

The process of gene manipulation in plants is commonly achieved by the use of A. tumefaciens, a soil bacterium. This bacterium possesses massive Ti-plasmids, which exceed 200 kb in size, and has the capability to transfer the T-DNA portion of the Ti-plasmid for the purpose of infecting plant chromosomes. The Ti-mediated transformation pathway, which occurs naturally, can be utilized to introduce new genes into the genome of a plant. Binary vectors including the T-DNA region were produced to facilitate gene modification. The vectors must have a plant selection marker gene, a bacterial antibiotic resistance gene, a site for cloning foreign genes, T-DNA border sequences for transferring genes to the plant genome, an origin of replication (ori) for a wide range of host plasmids, and an ori for E. coli. While binary vectors are smaller than native Ti-plasmids, they nonetheless pose challenges in gene cloning using standard methods due to their relatively large size. Gateway Technology, which may be obtained from Invitrogen, utilizes the site-specific recombination system between phage lambda and E. coli DNA. The technique underwent modifications to enhance its specificity and efficiency, enabling its utilization as a universal cloning technology. Gateway cloning offers several advantages. Firstly, it eliminates the requirement for restriction endonucleases and DNA ligase. Secondly, it utilizes a straightforward and standardized technique. Additionally, it provides highly efficient and dependable cloning, as well as convenient manipulation of fusion structures. Hence, the creation of a diverse range of Gateway cloning compatible vectors for various objectives will enhance the utility of this approach in plant research.

2.1 Ti-binary vector for Agrobacterium-mediated plant transformation

A. tumefaciens harboring a Ti-plasmid can transfer a specific segment of the plasmid, the T-DNA region, which is bounded by a right border (RB) and a left border (LB) sequence, to the genome of an infected plant (Figure 1). Expression of the T-DNA genes causes the overproduction of phytohormones in the infected cells, which causes crown gall tumors. Although T-DNA genes are required for crown gall tumor formation, other genes called the *vir* genes outside of the T-DNA region are essential for transfer of T-DNA into the host plant genome. These *vir* genes work even when they reside on another plasmid in *A. tumefaciens*. Based on these findings, a Ti-binary vector system was developed to overcome the difficulty of manipulating the original Ti plasmids *in vitro* by recombinant DNA methods due to their huge size (Bevan, 1984).

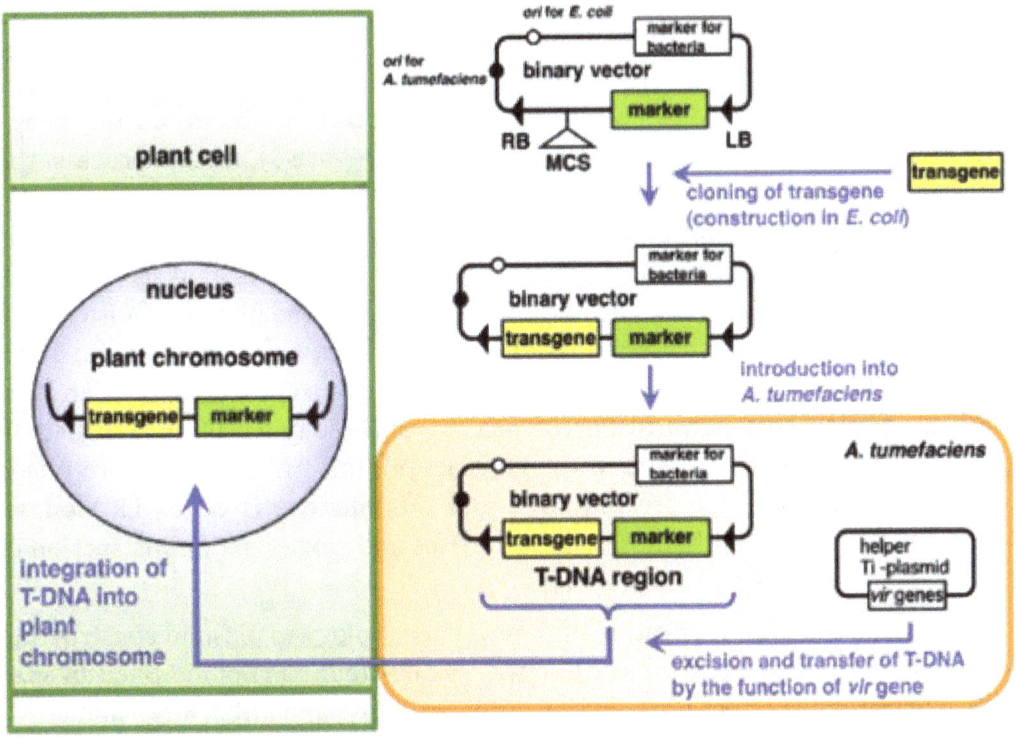

Fig. 1. Ti-binary vector system for *Agrobacterium*-mediated plant transformation. A binary vector, in which a target gene and plant selection marker gene are cloned between the two border sequences (RB and LB), is transformed into *A. tumefaciens* harboring a disarmed Ti-plasmid without the T-DNA region. Plant cells are infected by the transformed *A. tumefaciens* and then the target gene and marker gene are transferred into a plant chromosome by the *vir* genes on Ti-plasmid

A wide range of shuttle vectors for *E. coli* and *A. tumefaciens* was constructed that contain T-DNA border sequences flanking multiple restriction sites for foreign DNA cloning and marker genes for selection in plant cells. Using this vector system, DNA manipulation and vector construction can be done in *E. coli*; the vector is then transferred to *A. tumefaciens* harboring an artificial Ti-plasmid in which the T-DNA has been deleted. The vector is maintained stably in *A. tumefaciens,* and the cloned foreign DNA and marker gene between RB and LB can be transferred to the host plant genome by the transformation system encoded by *vir* genes on the T-DNA deletion Ti-plasmid. In early studies, several dicot plants were transformed by an *Agrobacterium* method. However, various dicot and monocot plants can now be transformed by co-cultivation of leaf slices or cultured calli with chemicals inducing expression of *vir* genes. Transformed cells are selected by marker gene phenotype such as antibiotic resistance and regenerated to transgenic plants. The most important model plant, *Arabidopsis thaliana*, can be easily transformed by *A. tumefaciens* using a floral dip procedure.

2.2 Outline of Gateway cloning

Gateway cloning technology is based on the lambda phage infection system, in which site- specific reversible recombination reactions occur during phage integration into and excision from *E. coli* genome (Figure 2). In this process, the *att*P site (242 bp) of lambda phage and the *att*B site (25 bp) of *E. coli* recombine (in a BP reaction) and the lambda phage genome is integrated into the *E. coli* genome. After the recombination reaction, the lambda phage genome is flanked by the *att*L (100 bp) and *att*R (168 bp) sites. In the reverse reaction, the phage DNA is excised from the *E. coli* genome by recombination between the *att*L and *att*R sites (in an LR reaction). The BP reaction needs two proteins, the phage integrase (Int) and the *E. coli* integration host factor (IHF). The mixture of these two proteins is called BP clonase in the Gateway system. In the LR reaction, Int, IHF and one more phage protein, excisionase (Xis), are required, and this mixture is called LR clonase. The Gateway cloning method uses these *att* sites and clonases for construction of recombinant DNA *in vitro*.

Basic strategies for application of Gateway technology to plasmid construction are shown in Figure 3. For the basic Gateway system, four pairs of modified *att* sites were generated for directional cloning. They are *att*B1 and *att*B2, *att*P1 and *att*P2, *att*L1 and *att*L2, and *att*R1 and *att*R2; a recombination reaction can occur only in the combinations of *att*B1 and *att*P1, *att*B2 and *att*P2, *att*L1 and *att*R1, or *att*L2 and *att*R2, since recombination strictly depends on *att* sequences (Hartley et al., 2000; Walhout et al., 2000). In addition to these *att* sites, the negative selection marker *ccd*B, the protein product of which inhibits DNA gyrase, and a chloramphenicol-resistance (Cmr) marker are used for selection and maintenance of Gateway vectors. Usually, *att*1 is located at the 5‘ end of the open reading frame (ORF) and *att*2 is located at the 3‘ end. This orientation is maintained in all cloning steps.

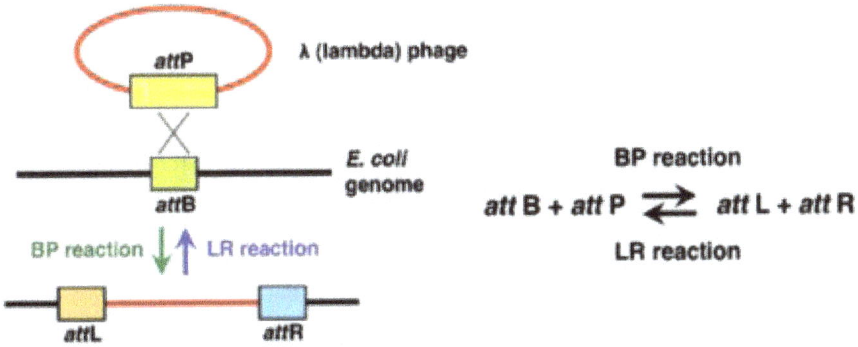

Fig. 2. BP and LR reactions in lambda phage infection of *E. coli*. **The site-specific reversible BP and LR recombination reactions occur during lambda phage integration into and excision from the** *E. coli* **genome**

First, the gene of interest should be cloned in an entry vector by TOPO cloning (pENTR/D-TOPO), a BP reaction (pDONR221), or restriction endonuclease and ligase (pENTR1A). Each vector is available from Invitrogen. To make an entry clone by a BP reaction, the *att*B1 and *att*B2 sequences are added to the 5‘ and 3‘ ends, respectively, of the ORF by adapter PCR. The product (*att*B1-ORF-*att*B2) is subjected to a BP reaction with a donor vector, pDONR221, which possesses an *att*P1-*ccd*B-Cmr-*att*P2 cassette. Because of the negative selection marker *ccd*B between *att*P1 and *att*P2, only transformants harboring the recombined vectors carrying *att*L1-ORF-*att*L2 (the entry clone) can grow on the selection plate. Once the entry clone is in hand, the ORF is transferred to a destination vector that possesses an *att*R1-Cmr-*ccd*B-*att*R2 cassette. Since destination vectors also contain *ccd*B between *att*R1 and *att*R2, and have a selection marker gene that is different from the entry clone, only the recombined destination vectors carrying *att*B1-ORF-*att*B2 will be selected. Gateway cloning is designed so that the smallest *att* sequence, *att*B (25 bp), appears in the final product to minimize the length of cloning junctions after the clonase reaction. In N- or C-terminal fusion constructs, the ORF is linked to a tag with eight or more amino acids encoded by the *att*B1 or *att*B2 sites. Because the reading frame of *att*B1 and *att*B2 is unified in the Gateway system, any entry clone incorporated into a destination vector is correctly fused to the tag sequence.

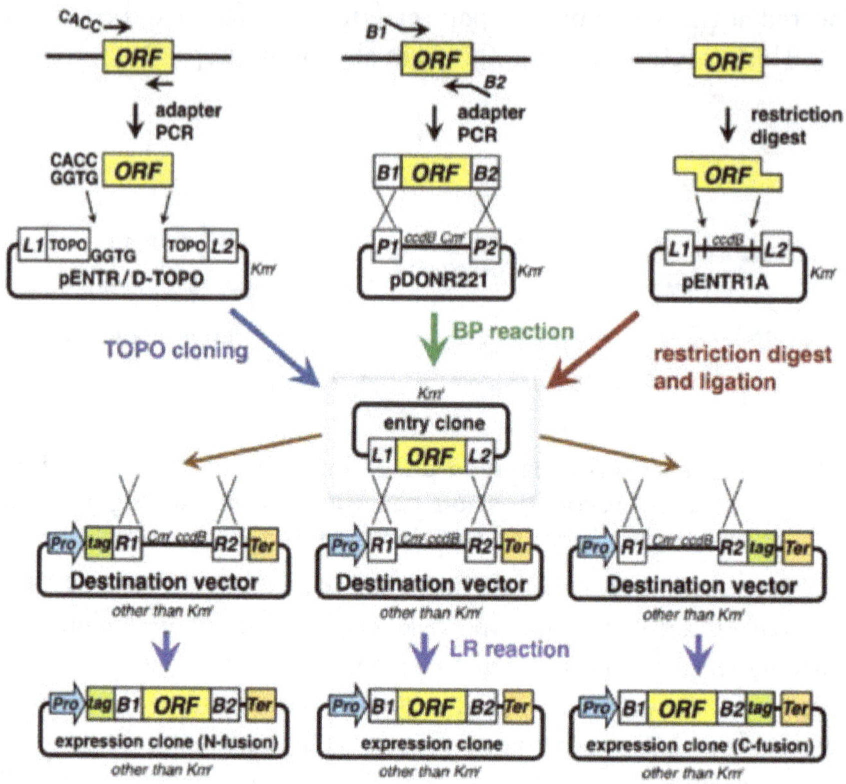

Fig. 3. Schematic illustration of Gateway cloning. An entry clone is constructed by TOPO directional cloning, a BP reaction or restriction digestion and ligation. For construction using the BP reaction, the ORF region is amplified by adapter PCR and the resulting *att*B1-ORF-*att*B2 fragment is cloned into pDONR221 by a BP reaction to generate an entry clone containing *att*L1-ORF-*att*L2. Subsequently, the ORF is cloned into destination vectors by an LR reaction to generate expression clones including tagged fusion constructs. For D-TOPO cloning, CACC is added to the ORF by adapter PCR, and the resulting CACC-ORF fragment is cloned into pENTR/D TOPO. *B1*, *att*B1; *B2*, *att*B2; *P1*, *att*P1; *P2*, *att*P2; *L1*, *att*L1; *L2*, *att*L2; *R1*, *att*R1; *R2*, *att*R2; *Pro*, promoter; *Ter*, terminator; *Cmr*, chloramphenicol resistance marker; *ccd*B, negative selection marker in *E. coli.*; *Kmr*, kanamycin-resistance marker

3. BINARY VECTORS COMPATIBLE WITH GATEWAY CLONING

A large number of binary vectors compatible with Gateway cloning, known as destination vectors, have been developed and are summarized in a review (Karimi et al., 2007b). Gateway compatible binary vectors for promoter analysis have the general structure *att*R1-Cmr-*ccd*B-*att*R2-tag-terminator, and after an LR reaction with an *att*L1-promoter-*att*L2 entry clone, they yield an *att*B1-promoter-*att*B2-tag-terminator binary construct. Gateway compatible binary vectors for expression of tagged fusion proteins have the general structure promoter-tag-*att*R1-Cmr-*ccd*B-*att*R2-terminator (for N-terminal fusions) or promoter-*att*R1-Cmr-*ccd*B-*att*R2-tag-terminator (for C-terminal fusions). After an LR reaction with an *att*L1-ORF-*att*L2 entry clone, they respectively yield promoter-tag-*att*B1-ORF-*att*B2- terminator or promoter-*att*B1-ORF-*att*B2-tag-terminator. The tag added to the N-terminus of the ORF is linked by the peptide encoded by the *att*B1 sequence (XSLYKKAGX), and the tag added to the C-terminus is linked by the peptide encoded by the *att*B2 sequence (XPAFLYKVX). Gateway compatible binary vectors for RNAi analysis (Helliwell & Waterhouse, 2003; Hilson et al., 2004; Karimi et al., 2002; Miki & Shimamoto, 2004) generally have the inverted structure of cassettes: promoter-*att*R1-*ccd*B-*att*R2-linker-*att*R2-*ccd*B-*att*R1- terminator. By an LR reaction with an *att*L1-trigger-*att*L2 entry clone, the trigger sequence is incorporated into both sites in opposite orientations, yielding a promoter-*att*B1-trigger- *att*B2-linker-*att*B2-(complementary trigger)-*att*B1-terminator construct. When the construct is introduced into plants, hairpin RNA is expressed and processed into small interfering RNA that functions in gene silencing.

Among many Gateway compatible binary vector series, the pW (Karimi et al., 2002), pMDC (Brand et al., 2006; Curtis & Grossniklaus, 2003) and pEarleyGate (Earley et al., 2006) series contain vectors available for many kinds of experiments in plants. The pW series consists of vectors for overexpression or antisense repression by the cauliflower mosaic virus 35S promoter (P35S), for promoter analysis using luciferase (LUC), β-glucuronidase (GUS), or GFP-GUS as reporters, and for construction of gene fusions with GFP, cyan fluorescent protein (CFP), yellow fluorescent protein (YFP) or red fluorescent protein (RFP). The pMDC series consists of vectors for cloning, for overexpression by P35S, for inducible expression by heat shock or estrogen treatment, for promoter analysis using GFP-6xHis or GUS as reporter, and for gene fusions with GFP, GFP-6xHis, or GUS. The pEarleyGate is a BASTA®- resistance binary vector series consisting of vectors for overexpression by P35S, for promoter analysis using HA, FLAG, Myc, or AcV5, and for gene fusions with YFP, HA, FLAG, Myc, AcV5, tandem affinity purification (TAP) tags, YFP-HA, or GFP-HA.

The vectors described above are useful tools; however, sometimes it is necessary to use a different series if an existing one does not have a vector of the required type. In order to carry out most experiments within the same series (having a unified backbone and a unified junction sequence), we constructed a comprehensive Gateway compatible binary vector system carrying many reporters and tags based on the same backbone, as mentioned in next section.

4. DEVELOPMENT OF GATEWAY BINARY VECTOR (PGWB) SERIES

To make Gateway compatible binary vectors efficiently, we first tried to establish a systematic method for construction of a vector series. For this purpose, we designed a construction method for introducing a tag sequence by blunt end ligation to save time and labor caused by restriction sites in the tag sequence. Based on this notion, platform vectors pUGW0 and pUGW2 (Nakagawa et al., 2007a) were made using pUC119 as the backbone. As described below, many Gateway binary vector (pGWB) series were constructed from intermediate plasmid pUGWs, which were made with pUGW0 or pUGW2. The characteristics and accession nos. of each pGWB are summarized in Information of Gaeway Binary Vectors (pGWBs) (http://shimane-u.org/nakagawa/gbv.htm).

4.1 Platform vectors pUGW0 and pUGW2 for construction of pGWB series

The platform vectors pUGW0 and pUGW2 include P35S and the nopaline synthase terminator (Tnos), as shown in Figure 4. A pUGW0 was the starting vector for N-terminal fusions, with the structure *Hin*dIII-P35S-*Xba*I-ATG-*Aor*51HI-*att*R1-Cmr-*ccd*B-*att*R2-*Sac*I-Tnos. A tag (reporter or epitope tag) sequence amplified by blunt-end PCR was introduced into the *Aor*51HI site (blunt end) to yield *Hin*dIII-P35S-*Xba*I-ATG-*tag*-*att*R1-Cmr-*ccd*B-*att*R2-*Sac*I- Tnos. In the case of a small epitope tag, an oligonucleotide could be introduced directly into the *Aor*51HI site. Translation is initiated at the ATG just upstream of the *Aor*51HI site. pUGW2 was the starting vector for C-terminal fusions, with the structure *Hin*dIII-*Xba*I- *Hin*dIII-P35S-*Xba*I-*att*R1-Cmr-*ccd*B-*att*R2-*Aor*51HI-*Sac*I-Tnos.

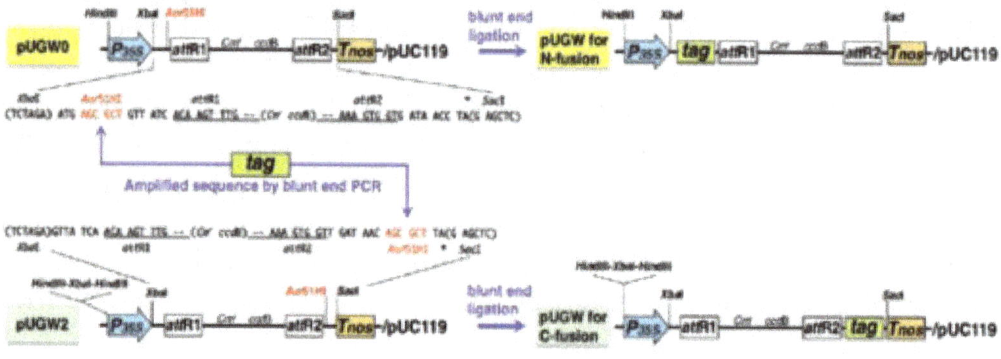

Fig. 4. Procedure for construction of pUGWs. pUGW0 and pUGW2 are the starting vectors for construction of new pUGW derivatives. The tag sequence amplified by blunt-end PCR is introduced into the *Aor*51HI site of pUGW0 or pUGW2, which yields pUGWs for N-fusion or C-fusion. The region between P35S and Tnos is indicated. The nucleotide sequence corresponding to the region from *att*R1 to *att*R2 is underlined. *Cmr*, chloramphenicol resistance marker; *ccd*B, negative selection marker in *E. coli*.; *P*35S, 35S promoter

Tag sequences were introduced by the same method used for pUGW0. The P35S region could be easily removed by digestion with *Xba*I followed by self-ligation for construction of promoter-less pUGWs. Because there is no need to digest the tag fragment with restriction enzymes to introduce it into the *Aor*51HI site of pUGW0 and pUGW2, any tag fragment can be cloned by the same method. With these simple procedures, a pUGW series containing a variety of tags was efficiently generated. They were sources of Gateway cassettes including tag sequences, and were used for construction of a Gateway binary vector (pGWB). Moreover, the pUGWs are Gateway compatible plant vectors useful for transient expression analysis after particle bombardment or protoplast transformation. Because of their small size and high copy number in *E. coli*, preparation and handling of pUGW plasmids are very easy.

4.2 The pGWB series (pGWBxx and pGWB2xx) based on the pBI plasmid

Initially, pGWB was constructed on the backbone of modified pBI carrying a nopaline synthase promoter (Pnos) driven neomycin phosphotransferase II (NPTII) and P35S-driven hygromycin phosphotransferase (HPT), which confer kanamycin-resistance (Kmr) and hygromycin-resistance (Hygr), respectively, to plants (Mita et al., 1995).

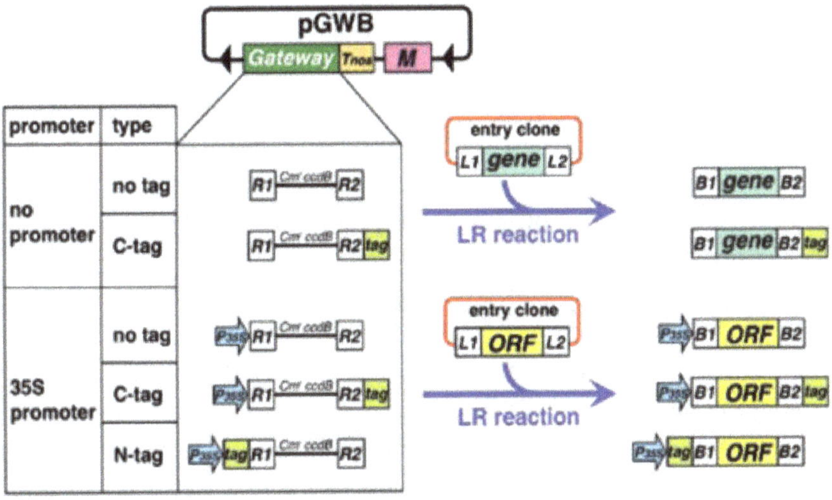

Fig. 5. Cloning into pGWB by LR reaction. The Gateway region in pGWB (top of the figure) represents a variety of acceptor sites (R1-R2) described in the box. The pGWB series includes plasmids with no promoter and no tag, or with no promoter and a C-tag. These are used for expression controlled by a gene's own promoter. The pGWB plasmids also include the following types: a 35S promoter and no tag, a 35S promoter and a C-tag, and a 35S promoter and an N-tag. These are used for constitutive expression using the 35S promoter. After an LR reaction with the entry clone, the expression clones indicated in the right panel are obtained. The tag is fused via the *att*B sequence. *B1*, *att*B1; *B2*, *att*B2; *L1*, *att*L1; *L2*, *att*L2; *R1*, *att*R1; *R2*, *att*R2; *Tnos*, nopaline synthase terminator; *M*, selection marker for plant; *Cmr*, chloramphenicol-resistance marker; *ccd*B, negative selection marker in *E. coli*.; *P35S*, 35S promoter

The initial pGWB series (pGWBxx) consists of 36 vectors designed for simple cloning of genes (pGWB1), for overexpression of ORF clones (pGWB2), and for fusion with a variety of tags (pGWB3 through pGWB45) as shown in the Complete List of pGWB (http://shimane- u.org/nakagawa/gbv.htm). GUS, TAP and LUC are available for C-fusion, and 10 other tags, sGFP, 6xHis, FLAG, 3xHA, 4xMyc, 10xMyc, GST, T7, enhanced yellow fluorescent protein (EYFP), and enhanced cyan fluorescent protein (ECFP), are available for both N- and C-fusion. The promoter-less C-fusion vectors can be used for promoter analysis. By an LR reaction with a promoter entry clone, a binary construct of promoter:tag is created. The remaining N- and C-fusion vectors contain P35S for constitutive expression. By an LR reaction with an ORF entry clone, binary constructs expressing tag-ORF or ORF-tag are easily obtained (Figure 5). With the pGWBs, promoter activity, detection of tagged proteins, and subcellular localization of proteins can be analyzed effectively (Nakagawa et al., 2007a).

We also constructed pGWBs carrying the *Pnos:HPT:Tnos* marker instead of *P35S:HPT:Tnos* (pGWB1-45) to avoid a possible effect of the P35S sequence on the expression pattern and strength of the cloned gene (Zheng et al., 2007). These vectors are named pGWB203, 204, 228 and 235, and their characters are shown at the bottom of the Complete List of pGWB (http://shimane-u.org/nakagawa/gbv.htm). In early experiments, when the phosphate transporter PHT1 promoter was used for promoter analysis in *A. thaliana*, GUS activity in plant extracts was 5-fold higher with pGWB3 than with pGWB203 (Nakagawa et al., 2007a).

4.3 Improved Gateway binary vector (ImpGWB) series (pGWB4xx, pGWB5xx, pGWB6xx and pGWB7xx) based on the pPZP plasmid

We next constructed improved Gateway binary vectors (ImpGWBs) using pPZP as a backbone (Hajdukiewicz et al., 1994). In the ImpGWB system, handling of plasmid is largely improved, transformation efficiency in *E. coli* is drastically increased and much larger amount of plasmid DNA was recovered. The structures and characters of pGWBs (pBI backbone) and ImpGWBs (pPZP backbone) are summarized in Figure 6.

At present, four kinds of ImpGWB, the Kmr subseries (pGWB4xx) (Nakagawa et al., 2007b), Hygr subseries (pGWB5xx) (Nakagawa et al., 2007b), BASTA®-resistance subseries (pGWB6xx) (Nakamura et al., 2010) and tunicamycin-resistance subseries (pGWB7xx) (Tanaka et al., 2011), are available, and they are useful for introducing multiple transgenes into plants by repetitive transformation. Each subseries is composed of 46 vectors as summarized in the Complete List of ImpGWB (http://shimane-u.org/nakagawa/gbv.htm). A set of 16 tags, sGFP, GUS, LUC, EYFP, ECFP, G3 green fluorescent protein (G3GFP), monomeric red fluorescent protein (mRFP), TagRFP, 6xHis, FLAG, 3xHA, 4xMyc, 10xMyc, GST, T7, and TAP, is available in ImpGWB. Because ImpGWB is highly efficient in transformation of *E. coli*, this series was used for development of a new cloning system using multiple LR reactions as described below.

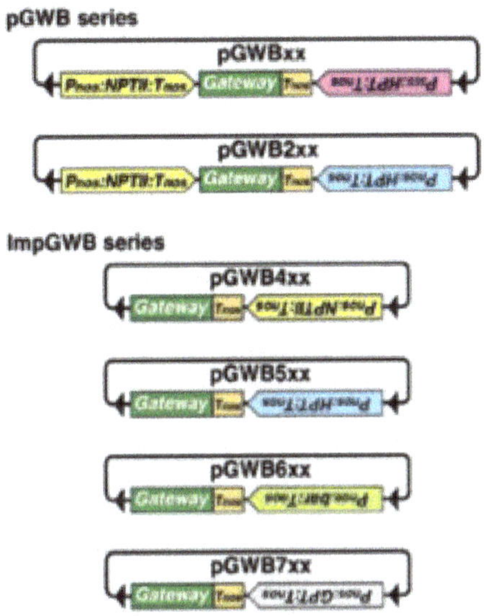

backbone	copy number in *E. coli*	resistance in bacteria	resistance in plants
pBI	low	Km^r^, Hyg^r^	Km^r^, Hyg^r^
pBI	low	Km^r^, Hyg^r^	Km^r^, Hyg^r^
pPZP	high	Spc^r^	Km^r^
pPZP	high	Spc^r^	Hyg^r^
pPZP	high	Spc^r^	BASTA^®r^
pPZP	high	Spc^r^	Tunicamycin^r^

Fig. 6. Characters of pGWBs and ImpGWBs. The Gateway region in vectors represents a variety of acceptor sites as described in the Figure 5. Pnos, nopaline synthase promoter; Tnos, nopaline synthase terminator; P35S, 35S promoter; NPTII, neomycin phosphotransferase II; HPT, hygromycin phophotransferase; *bar*, bialaphos resistance gene; GPT, UDP-*N*- acetylglucosamine: dolichol phosphate *N*-acetylglucosamine-1-P transferase (Koizumi & Iwata, 2008; Koizumi et al., 1999) gene. Kmr, kanamycin-resistance; Hygr, hygromycin- resistance; Spcr, spectinomycin-resistance; BASTA®r, BASTA®-resistance; Tunicamycinr, tunicamycin-resistance

4.4 R4 Gateway binary vector (R4pGWB) series (R4pGWB4xx, R4pGWB5xx, R4pGWB6xx and R4pGWB7xx) for promoter swapping

To assemble multiple DNA fragments in the desired order, an additional four *att* sites (*att*3, *att*4, *att*5 and *att*6) have been developed and applied to MultiSite Gateway cloning (Karimi et al., 2007a; Sasaki et al., 2004). Utilization of these *att* sites (*att*1-6) expanded the availability of cloning technology for more complex gene construction. The cloning system equipped with these *att* sites is useful for swapping of promoters, ORFs and tags, and is also applicable for cloning of multiple transgenes in one vector (Chen et al., 2006). In a typical MultiSite Gateway system, three entry clones containing specialized *att* sites, *att*L4- promoter-*att*R1, *att*L1-ORF-*att*L2, and *att*R2-tag-*att*L3 are simultaneously connected and incorporated into a destination vector carrying *att*R4-Cmr-*ccd*B-*att*R3 acceptor sites to make an *att*B4-promoter-*att*B1-ORF-*att*B2-tag-*att*B3 construct (Figure 7).

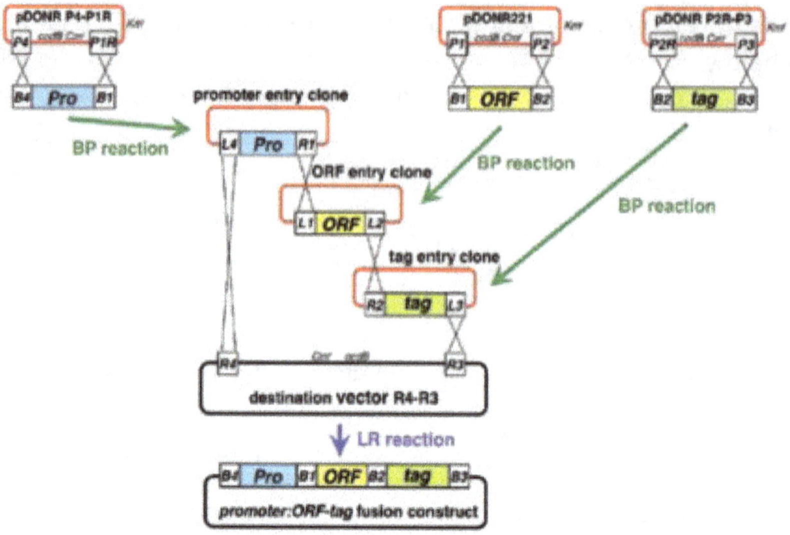

Fig. 7. MultiSite Gateway system. In the MultiSite Gateway system, ***att*1, *att*2, *att*3 and *att*4 sequences are used for cloning of multiple DNA fragments into one vector. A promoter entry clone (L4-Pro-R1), ORF entry clone (L1-ORF-L2), tag entry clone (R2-tag-L3) and destination vector R4-R3 are subjected to an LR reaction. The promoter, ORF and tag sequences are linked and incorporated into the destination vector to form a** *promoter:ORF- tag* **clone.** *B1*, ***att*B1**; *B2*, ***att*B2**; *B3*, ***att*B3**; *B4*, ***att*B4**; *L1*, ***att*L1**; *L2*, ***att*L2**; *L3*, ***att*L3**; *L4*, ***att*L4**; *R1*, ***att*R1**; *R2*, ***att*R2**; *R3*, ***att*R3**; *R4*, ***att*R4**; *P1*, ***att*P1**; *P2*, ***att*P2**; *P3*, ***att*P3**; *P4*, ***att*P4**; *P1R*, ***att*P1R**; *P2R*; ***att*P2R**; *Cm*r, **chloramphenicol-resistance marker;** *ccd*B, **negative selection marker in**

E. coli.; *Pro*, promoter; *Km*r, kanamycin-resistance marker although MultiSite Gateway cloning is an excellent method for building a complicated multigene construct, it is relatively difficult to obtain the desired clone because four recombinations at each *att* site are required for successful cloning. To facilitate multi- fragment cloning, especially for promoter swapping, we developed the R4 Gateway binary vector (R4pGWB) by reducing the number of recombinations needed from four to three (*att*4, *att*1 and *att*2) (Figure 8, left) (Nakagawa et al., 2008). The R4pGWB series was made by replacing the *att*R1 site of ImpGWBs (promoter-less and C-fusion type with four resistance markers) with the *att*R4 site; all tags used in ImpGWB are also available in the R4pGWB system as shown in the Complete List of R4pGWB (http://shimane- u.org/nakagawa/gbv.htm). By an LR reaction with a promoter entry clone (*att*L4-promoter- *att*R1), an ORF entry clone (*att*L1-ORF-*att*L2) and R4pGWB equipped with the appropriate tag, construction of chimeric genes among promoters, ORFs, and tags (*att*B4-promoter-*att*B1- ORF-*att*B2-tag) is achieved very easily. The R4pGWB system is a powerful tool to express an ORF by any desired promoter, *e.g.*, a promoter for strong expression, for tissue or cell specific expression, for developmental stage specific expression, or for induction by biotic or abiotic stimuli.

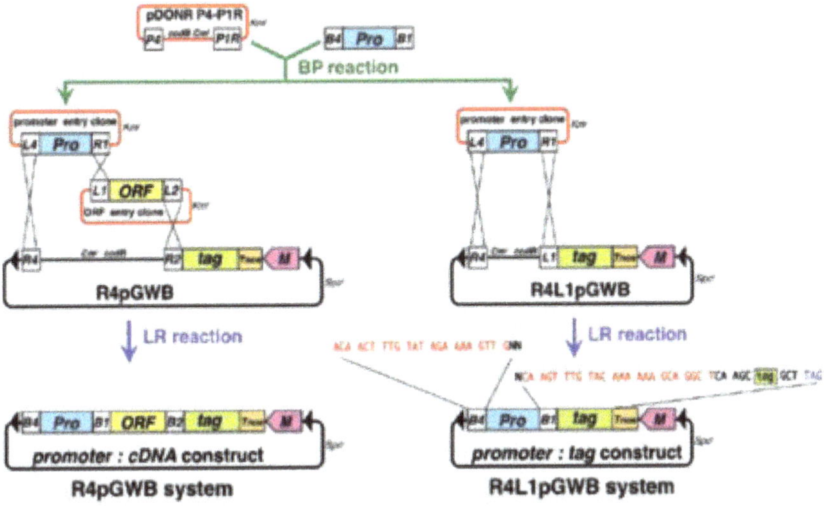

Fig. 8. R4pGWB and R4L1pGWB systems. A promoter entry clone (L4-Pro-R1) is constructed by a BP reaction using pDONR P4-P1R and a B4-Pro-B1 fragment prepared by adapter PCR. Left; in the R4pGWB system, a promoter entry clone (L4-Pro-R1), ORF entry clone (L1-ORF- L2) and R4pGWB are subjected to an LR reaction. The promoter and ORF are linked and incorporated into R4pGWB to form a promoter:ORF-tag clone. Right; in the R4L1pGWB system, only a promoter entry clone (L4-Pro-R1) is used for an LR reaction with an R4L1pGWB. The promoter sequence is incorporated into R4L1pGWB and fused with the tag on the vector. With the R4L1pGWB system using a single LR reaction, a promoter:tag construct is obtained at high efficiency. Nucleotides in red indicate B4 and B1 sequences.

Pro,Promoter; *B1*, *att*B1; *B2*, *att*B2; *B4*, *att*B4; *L1*, *att*L1; *L2*, *att*L2; *L4*, *att*L4; *R1*, *att*R1; *R2*, *att*R2; *R4*, *att*R4; *P4*, *att*P4; *P1R*, *att*P1R; *M*, selection marker for plant; *Cmr*, chloramphenicol- resistance marker; *ccd*B, negative selection marker in *E. coli.*; *Pro*, promoter; *Kmr*, kanamycin- resistance marker

4.5 R4L1 Gateway binary vector (R4L1pGWB) series (R4L1pGWB4xx and R4L1pGWB5xx) for promoter analysis

Due to establishment of the R4pGWB system, many kinds of *att*L4-promoter-*att*R1 entry clones were constructed and have been used as a resource for expression of ORFs in plants. We plan to also utilize these resources of *att*L4-promoter-*att*R1 entry clones for efficient promoter:tag experiments, and developed an R4L1 Gateway binary vector (R4L1pGWB) (Nakamura et al., 2009) containing *att*R4-Cmr-*ccd*B-*att*L1-tag-Tnos. By the simple bipartite LR reaction with *att*L4-promoter-*att*R1 and R4L1pGWB, an *att*B4-promoter-*att*B1-tag-Tnos construct used for promoter assays can be easily obtained in this system (Figure 8, right). The tags in R4L1pGWBs are G3GFP-GUS, GUS, LUC, EYFP, ECFP, G3GFP and TagRFP as shown in the Complete List of R4L1pGWB (http://shimane-u.org/nakagawa/gbv.htm).

5. APPLICATION OF THE PGWB SYSTEM

Because Gateway cloning is efficient, precise, flexible and simple to use, its application will continue to grow in plant research. In this section, we briefly describe two recent advances in our pGWB system, a split reporter for interaction analysis and recycling cloning for multigene constructs.

5.1 Gateway vectors for bimolecular fluorescence complementation (BiFC) assay

BiFC is based on the reconstitution of a fluorescent signal when two interacting proteins or peptides, which are fused to either an N- or C-fragment of a split fluorescent protein, interact. Due to its relative technical simplicity and the ability to use fluorescence microscopes for observation, a growing number of publications describe the use of BiFC to analyze protein-protein interactions. In addition to monitoring protein-protein interactions, this method has expanded to wider application, such as multicolor BiFC to investigate protein complexes (Hu & Kerppola, 2003; Kodama & Wada, 2009; Lee et al., 2008; Waadt et al., 2008), detection *in vivo* (Bracha-Drori et al., 2004; Walter et al., 2004) and combined with bioluminescence resonance energy transfer (BRET; Chen et al., 2008; Gandia et al., 2008; Xu et al., 2007). To date, several BiFC vectors dedicated to plant research have been constructed. Among our efforts in development of Gateway technology, we have generated various destination vectors for BiFC assays. In this section, we introduce our Gateway technology- based BiFC vectors, and describe their application.

5.1.1 Detection of protein-protein interactions in plant cells by BiFC assay

The investigation of protein-protein interactions provides valuable information in cell biology. In addition to BiFC, several other techniques detect protein-protein interactions, such as co-immunoprecipitation assays (Co-IP), *in vitro* binding assays, the yeast two-hybrid system (Y2H; James et al., 1996), the mating-based split-ubiquitin system (mbSUS; Ludewig et al., 2003; Obrdlik et al., 2004), BRET(Chen et al., 2008; Xu et al., 2007), fluorescence resonance energy transfer (FRET; Day et al., 2001), fluorescence lifetime imaging microscopy (FLIM; Bastiaens & Squire, 1999) and fluorescence correlation spectroscopy (FCS; Hink et al., 2002). The imaging-based approaches such as BiFC and FRET have been utilized in plant research because they enable detection in plant cells, in contrast to Y2H and mbSUS, which are functional only in yeast cells, and because they do not require specific antibodies or purification of proteins, unlike Co-IP and *in vitro* binding assays.

The BiFC assay is one of the most convenient techniques among the image-based approaches. Although FRET and FLIM are useful and powerful techniques for detection of protein-protein interactions, FRET requires complicated analysis such as of acceptor bleaching and an exclusive device is necessary for FLIM. Although several considerations are required even for BiFC assays, special devices are not required for detection, and complicated analysis is not necessary after obtaining image data. In addition, the BiFC assay provides information on subcellular location of the interacting proteins.

We used our Gateway vector construction system (Hino et al., 2011; Nakagawa et al., 2008; Nakagawa et al., 2007b) to make destination vectors for BiFC assays. Using these vectors, it is easy to make constructs for detection of protein-protein interactions. These Gateway vectors have worked well in plant cells (Goto et al., 2011; Hino et al., 2011; Singh et al., 2009).

5.1.2 Principles of the BiFC assay

In BiFC assays, a fluorescent reporter, such as CFP, GFP, YFP and RFP, is split into two non- fluorescent fragments, N- and C- fragments (Figure 9A,B). Two proteins or peptides, which are to be tested for interaction, are fused at the N- or C-terminus of each fragment. After expression of both fusion genes simultaneously, if an interaction occurs between the two proteins, the non-fluorescent fragments are reconstituted and behave as an unsplit fluorescent protein. Therefore, the detection of fluorescence means the target proteins interact (Figure 9A).

Once the interaction occurs, the reconstituted molecule does not dissociate into non- fluorescent fragments, leading to enhancement of fluorescence due to accumulation of reconstituted fluorescent proteins.

There are eight potential combinations to be tested for protein-protein interactions in a BiFC assay, taking into account which protein of the two partners tested is fused to the N- or C- terminal end of which N- or C- fragment (Figure 9C). However, improper fusion of a split fragment sometimes abolishes protein function and masks information on subcellular targeting. For example, the peroxisome targeting signal 2 (PTS2) must be fused to the N- terminus of the split fluorescent protein (Singh et al., 2009; Figure 10B). In contrast, PTS1 must be fused to the C-terminus of a split fluorescent protein, because its location at the C- terminus is necessary for its function. In these cases, the number of combinations tested is fewer. However, if there is no information on protein function, all combinations should be tested. Viewed in this light, our destination vectors are useful for construction of several fusion genes at the same time.

5.1.3 Destination vectors for the multicolor and in vivo BiFC assays

Various BiFC vectors have been developed and used in plant research (Bracha-Drori et al., 2004; Diaz et al., 2005; Ding et al., 2006; Goto et al., 2011; Hino et al., 2011; Loyter et al., 2005; Maple et al., 2005; Marrocco et al., 2006; Ohad et al., 2007;

Singh et al., 2009; Waadt et al., 2008; Walter et al., 2004; Zamyatnin et al., 2006). All the vectors, including ours, use P35S to express a fusion gene. There are two ways to insert a target gene into the 5' or 3' end of a split fragment of fluorescent protein gene: (1) cloning into a multicloning site using digestion and ligation, and (2) Gateway technology (Hino et al., 2011; Walter et al., 2004). Our BiFC vectors were developed to be compatible with Gateway technology. We generated four kinds of destination vectors for BiFC assays (Figure 10A), enabling the transfer of a gene of interest from the entry clone to the 5' or 3' end of each split fragment. Therefore, researchers are able to easily fuse a gene of interest to the 5' or 3' end of the split fragment, leading to various convenient constructs.

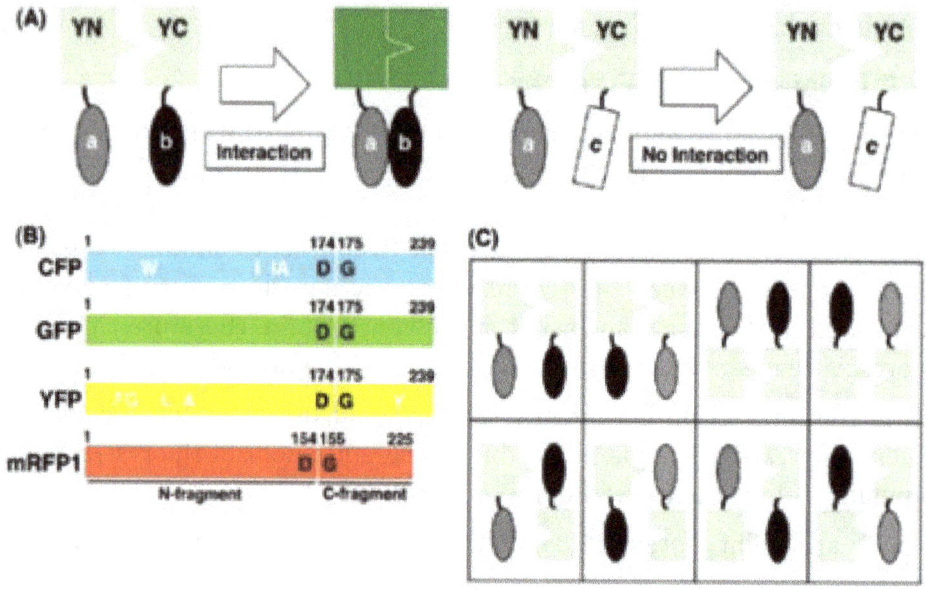

Fig. 9. Principles of the BiFC assay. (A) Nonfluorescent fragments (YN and YC) of a fluorescent protein are brought together through interaction of the tested proteins or peptides (a, b and c) to which they are fused. The interaction of the two proteins causes reconstitution of a fluorescent signal. (B) Diagram of amino acid substitutions among CFP, GFP, YFP and mRFP1, and the positions where they were fragmented. Although there are alternative positions to split a fluorescent protein into two fragments (Hu & Kerppola, 2003; Waadt et al., 2008), the CFP, GFP and YFP in our system were split between residues 174 and 175, and mRFP1, which contains an amino acid substitution of the 66th glutamine to threonine, was split between residues 154 and 155. Amino acids in CFP and YFP that were converted from GFP are depicted in white. In the case of RFP, amino acids that are different from GFP are not represented, since there are many substitutions. (C) Potential combination of two fragments. There are eight possible configurations in the BiFC assay. Each target protein (gray and black) can be fused at its N- or C- terminus to the N- or C-terminal fragment of the fluorescent protein (light green)

The BiFC vectors were initially generated using YFP (Hu et al., 2002). However, other fluorescent proteins, BFP (Hu & Kerppola, 2003), CFP (Kodama & Wada, 2009; Lee et al., 2008), GFP (Hu et al., 2002; Kodama & Wada, 2009), Venus, (Lee et al., 2008), Cerulean (Lee et al., 2008), DsRed-monomer (Kodama & Wada, 2009), mRFP1 (Jach et al., 2006), mCherry (Fan et al., 2008), and a far-red fluorescent protein, mLumin (Chu et al., 2009), have reportedly been useful for BiFC assay. We adopted CFP, GFP, YFP and mRFP1 to generate vectors (Figure 9B), and verified their usefulness for detection of protein-protein interactions (Figure 10B-E).

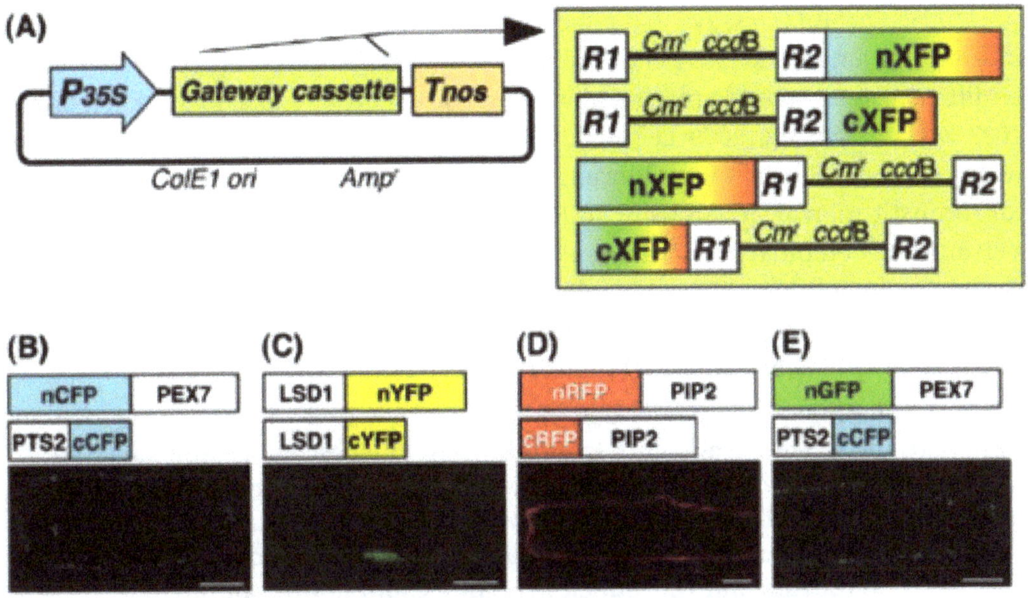

Fig. 10. Schematic representation of the multicolor BiFC vectors and examples of transient expression. (A) Four kinds of destination vectors for transient expression were generated to be compatible with Gateway technology. nXFP and cXFP, the N- or C-fragment, respectively, of CFP, GFP, YFP or RFP; *ColE1 ori*, ColE1 replication origin; *Amp*r, ampicillin- resistance marker used for selection in bacteria; *Cm*r, chloramphenicol-resistance marker; *ccd*B, negative selection marker used in bacteria; P35S, 35S promoter; *Tnos*, nopaline synthase terminator; *R1*, *att*R1; *R2*, *att*R2. (B-E) Fluorescence images of onion epidermal cells expressing the fusion genes indicated above each panel were acquired 18-24 hr after particle bombardment. Bars = 50 µm

PTS2-containing proteins are directed to peroxisomes after binding to a receptor, PEX7, in the cytosol (Hayashi & Nishimura, 2006; Mano & Nishimura, 2005). We were able to observe reconstituted CFP fluorescence as punctate structures, when allowing interaction of *nCFP-PEX7* and *PTS2-cCFP* (Figure 10B), which agrees with a previous report (Singh et al., 2009). Lesion simulating disease 1 (LSD1), a negative regulator of programmed cell death, is a zinc finger protein that forms homodimers. We also tried to detect LSD1 homooligomerization using the combination of LSD1-nYFP and LSD1-cYFP. Reconstituted YFP signals were observed in the cytosol and nucleus (Figure 10C), a result that coincided with previous data (Walter et al., 2004). The localization and interaction of one of the plasma membrane intrinsic proteins, PIP2, which belongs to the aquaporin family, with other PIP members were demonstrated by FRET and FLIM assays in maize cells (Zelazny et al., 2007). An Arabidopsis PIP2 gene, *PIP2;1*, was also fused to split fragments of mRFP1 and used for investigation of homooligomerization (Figure 10D). We were able to detect reconstituted RFP signals at the plasma membrane.

Multicolor BiFC assays have been developed to examine protein-protein interactions among various factors, since some combinations of N- and C-fragments of different fluorescent proteins allow reconstitution of signals (Hu & Kerppola, 2003; Kodama & Wada, 2009; Lee et al., 2008; Waadt et al., 2008). We also investigated which combinations among different fragments in our BiFC vectors are practical for reconstitution of signals using nXFP-PEX7 and PTS2-cXFP (XFP means CFP, GFP, YFP or RFP). Combinations among split fragments of CFP, GFP and YFP enabled the reconstitution of fluorescence (Table 1, Figure 10E), although some combinations did not give reproducible results. In contrast, a reconstituted RFP signal was observed only between split fragments from RFP (Table 1).

Table 1. Summary of the detection of reconstituted signals using various combinations of split fragments from different fluorescent proteins. cC, cG, cY and cR represent the C-fragment of CFP, GFP, YFP and RFP, respectively. nC, nG, nY and nR indicate the N-fragment of CFP, GFP, YFP and RFP, respectively. '+' and '-' denote detection of interaction and inability for interaction, respectively. "±" indicates that reproducible results could not be obtained

	nC-PEX7	nG-PEX7	nY-PEX7	nR-PEX7
PTS2-cC	+	+	+	-
PTS2-cG	+	+	+	-
PTS2-cY	±	±	+	-
PTS2-cR	-	-	-	+

We adapted our BiFC vectors for transient expression (Figure 10A) to binary vectors for *in vivo* BiFC assays (Figure 11A). Using these binary vectors, researchers are able to easily generate transgenic plants expressing a fusion gene of the N- or C-fragment with a gene of interest. We prepared two kinds of binary vectors, containing either Kmr or Hygr markers. Therefore, after crossing transgenic plants expressing either the N- or C-fragment, it will be easier to obtain transgenic plants expressing both N- or C-fragments from screening on medium with both kanamycin and hygromycin.

Agroinfiltration is a powerful technique to express an alien gene *in planta*, and it has been reported that this technique is functional in BiFC assays (Bracha-

Drori et al., 2004; Waadt et al., 2008; Walter et al., 2004). We examined whether our binary vectors could also work well in agroinfiltration using *nYFP-Peroxin 6 (PEX6)* and *cYFP-ABERRANT PEROXISOME MORPHOLOGY 9 (APEM9)* (Figure 11B). We already reported the interaction of PEX6 and APEM9 using transient expression of these fusion genes in onion epidermal cells (Goto et al., 2011). We mixed three cultures of *A. tumefaciens* (strain C58C1RifR) haboring *nYFP-PEX6*, *cYFP-APEM9* or *CFP-PTS1* as peroxisomal markers, and then co-infiltarted into the leaf cells of *Nicotiana tobaccum*. Reconstituted YFP signal was observed as punctate structures (Figure 11C), and these signals surrounded the CFP-labeled peroxisome matrix (Figure 11C-E), showing that BiFC occurs at the peroxisomal membrane, as reported previously (Goto et al., 2011). These results demonstrated that our binary vectors for BiFC assays work well.

5.1.4 Special considerations for BiFC assays using our vectors

In BiFC assays, fluorescence is derived from reconstituted fluorescence or artificial noise. The same is true for our BiFC vectors. Fluorescence is sometimes observed even in combination with an untagged vector as a negative control. Therefore, it is necessary to test expression using a negative control vector. Conversely, when fluorescence is not observed after expression of two fusion genes, there are two views about the result. One is that the interaction does not occur, although the two fusion genes are expressed properly. The other is that gene expression is inefficient or that the genes were inefficiently introduced into the cells. We always express an additional gene, such as *CFP-PTS1* in Figure 11, to investigate the efficiency of gene expression in transient assays. At the same time, this helps visualize cells and organelles so that it is easier to observe introduced cells that are bombarded or agro-infiltrated. The alternative method is the detection of fusion protein by immunoblotting. Some vectors are developed to add the epitope tag to split fragments so that the detection of accumulation of fusion proteins is carried out by immunoblotting (Bracha-Drori et al., 2004; Waadt et al., 2008; Walter et al., 2004). Of course, if specific antibodies against target protein are possible to obtain, they are useful for verification of protein accumulation.

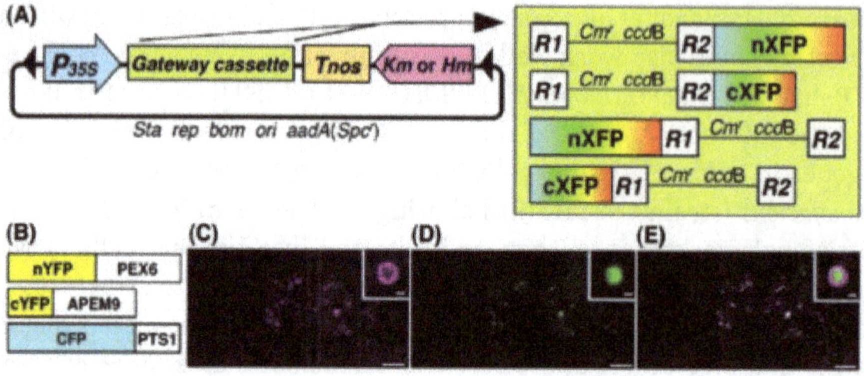

Fig. 11. Schematic representation of the binary vectors for the BiFC assay and examples of an *in vivo* BiFC experiment using an *Agrobacterium*-infiltration technique. (A) Four kinds of destination vectors for an *in vivo* BiFC assay were generated to be compatible with Gateway technology. nXFP and cXFP, the N- or C-fragment, respectively, of CFP, GFP, YFP or RFP; cXFP; *sta*, region conferring stability in *Agrobacterium; rep,* broad host-range replication origin; *bom, cis*-acting element for conjugational transfer; *ori,* ColE1 replication origin; *Cmr*, chloramphenicol-resistance marker; *ccdB,* negative selection marker used in bacteria; *P35S*, 35S promoter; *Tnos,* nopaline synthase terminator; *R1*, *att*R1; *R2*, *att*R2; Black arrowheads indicate right border and left border. (B-E) An example of an *in vivo* BiFC experiment. (B) Three fusion genes, *nYFP-PEX6, cYFP-APEM9* and *CFP-PTS1* as peroxisome markers were expressed in *Nicotiana tobaccum.* (C-E) Fluorescence images of leaf epidermal cells were acquired 3 days after infiltration. (C) Reconstituted YFP signals. (D) Peroxisomes visualized with CFP. (E) A merged image of (C) with (D). Insets represent magnified images of a peroxisome. Bars = 20 μm and 1 μm for each inset

5.1.5 Perspectives

Our BiFC vectors have wide application to analysis of protein-protein interactions. Future introduction of the R4pGWB system (Nakagawa et al., 2008) to these BiFC vectors will allow regulation of each fusion gene under a specific promoter, leading to examination of the interaction with tissue or developmental stage specificity. Additionally, inducible promoters will be used for transient expression in transgenic plants harboring R4pGWB-based BiFC fragments. Since a great variety of fluorescent proteins with different properties, such as large Stokes' shift, are available, more various combinations for the multicolour BiFC assay will be generated by adopting our Gateway technology system to new fluorescent proteins, revealing the relationship among several factors in a complex.

5.2 Recycling cloning system for multigene constructs

Multigene transformation of plants is a powerful technology for molecular breeding because it can simultaneously improve multiple enzymes and factors constituting biological pathways (Ha et al., 2010; Nakayama et al., 2000; Naqvi et al., 2009; Ye et al., 2000). For multigene transformation, methods such as re-transformation, co-transformation, and cross- fertilization are available (Dafny-Yelin & Tzfira, 2007), but the most practical method is the utilization of a multigene construct, a vector carrying multiple expression units (Chen et al., 2006). In this section, we introduce a recycling cloning system for cloning multiple expression units by simple repetitive LR reactions.

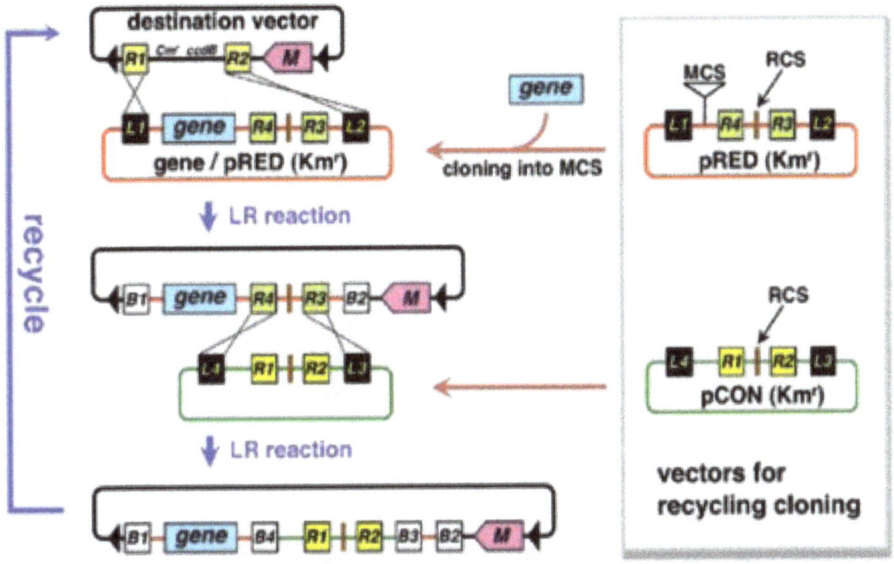

Fig. 12. Schematic illustration of recycling cloning. The pRED vector has the structure L1- MCS-R4-RCS-R3-L2. The gene of interest is cloned into the MCS of pRED419 and subsequently subjected to an LR reaction with a destination vector. In this step, the DNA fragment of gene-R4-RCS-R3 is incorporated into the destination vector and a binary clone carrying B1-gene-R4-RCS-R3-B2 is obtained. Next, the resulting binary clone is subjected to an LR reaction with pCON to introduce the R1-RCS-R2 sequence into the binary clone. The binary clone carrying B1-gene-B4-R1-RCS-R2-B3-B2 is recycled for introduction of another gene by LR reaction with another gene/pRED in a second cycle. The marker gene (M) is transcribed in the opposite orientation to the cloned gene. *B1, att*B1; *B2, att*B2; ***B3, att*B3;** *B4*, ***att*B4; *L1, attL1; L2, attL2; L3, att*L3; *L4, att*L4; *R1, att*R1; *R2, attR2; R3, att*R3; *R4, att*R4; *M,* selection marker for plant; *Cmr,* chloramphenicol-resistance marker; *ccdB,* negative selection marker in *E. coli.;* MCS, multiple cloning site; RCS, rare cutter site**

As shown in the right panel of Figure 12, two vectors are used for each cloning cycle in this system. The recycle donor vector pRED has four *att* sites, a multiple cloning site (MCS) and a rare cutter site (RCS) in the following order: *att*L1-MCS-*att*R4-RCS-*att*R3-*att*L2. The RCS has *Asi*I, *Swa*I, *Fse*I, *Pac*I, *Asc*I and *Pme*I sites. The gene of interest is cloned into the MCS of pRED (gene/pRED) and subjected to an LR reaction with the destination vector containing R1-R2 acceptor sites. In this step, a binary construct carrying gene-*att*R4-RCS-*att*R3 is obtained. Next, conversion vector pCON, containing *att*L4-*att*R1-RCS-*att*R2-*att*L3, is subjected to an LR reaction to introduce the *att*R1-RCS-*att*R2 acceptor site into the resulting binary construct, and the binary construct obtained, which carries *att*R1-RCS-*att*R2, is recycled for the next round of the cloning cycle, together with another gene/pRED clone (Figure 12). Before the LR reactions, binary constructs are digested by a rare cutter to suppress colonies derived from non-recombinants. With these simple repetitive reactions, genes are introduced sequentially into one vector. Using this recycling cloning system, we made a multigene construct containing four expression units of reporter genes and confirmed expression of all four reporters in transformed tobacco BY-2 cells (Kimura, unpublished results).

6. CONCISE SUMMARY

Gateway cloning is a highly efficient, dependable, user-friendly, and adaptable method, which has led to the development and global use of several types of vectors. The pGWBs series comprises many vectors with diverse tags and four resistance markers. These constructs are built using the same vector backbone and offer standardized experimental settings in transgenic research. Due to the simplicity of introducing a tag sequence into pUGW (as shown in Figure 4), the number of vectors available for fusion with new tags is increasing in our Gateway vector system. Split fluorescent protein vectors are crucial tools for BiFC experiments. Our BiFC vectors, which utilize Gateway technology, are valuable for generating multiple fusion genes to detect protein-protein interactions in transient or in vivo studies involving multiple variables. By incorporating the R4pGWB system (Nakagawa et al., 2008) into these BiFC vectors, the range of potential uses will be expanded. Cloning by recycling has the capacity to introduce numerous expression units with great efficiency, so paving the path for a novel approach to genetic engineering in plants.

7. REFERENCES

Bastiaens P.I. & Squire A. (1999). Fluorescence lifetime imaging microscopy: spatial resolution of biochemical processes in the cell. *Trends Cell Biology*, Vol.9, No.2, (Feb 1999), pp. 48-52, ISSN 0962-8924 (Print), 0962-8924 (Linking)

Bevan M. (1984). Binary Agrobacterium vectors for plant transformation. *Nucleic Acids Research*, Vol.12, No.22, (Nov 26 1984), pp. 8711-8721, ISSN 0305-1048 (Print), 0305-1048 (Linking)

Bracha-Drori K., Shichrur K., Katz A., Oliva M., Angelovici R., Yalovsky S., & Ohad N. (2004). Detection of protein-protein interactions in plants using bimolecular fluorescence complementation. *The Plant Journal*, Vol.40, No.3, (Nov 2004), pp. 419- 427, ISSN 0960-7412 (Print), 0960-7412 (Linking)

Brand L., Horler M., Nuesch E., Vassalli S., Barrell P., Yang W., Jefferson R.A., Grossniklaus U., & Curtis M.D. (2006). A versatile and reliable two-component system for tissue- specific gene induction in Arabidopsis. *Plant Physiology*, Vol.141, No.4, (Aug 2006), pp. 1194-1204

Chen H., Zou Y., Shang Y., Lin H., Wang Y., Cai R., Tang X., & Zhou J.M. (2008). Firefly luciferase complementation imaging assay for protein-protein interactions in plants. *Plant Physiology*, Vol.146, No.2, (Feb 2008), pp. 368-376, ISSN 0032-0889 (Print), 0032-0889 (Linking)

Chen Q.J., Zhou H.M., Chen J., & Wang X.C. (2006). A Gateway-based platform for multigene plant transformation. *Plant Molecular Biology*, Vol.62, No.6, (Dec 2006), pp. 927-936

Chu J., Zhang Z., Zheng Y., Yang J., Qin L., Lu J., Huang Z.L., Zeng S., & Luo Q. (2009). A novel far-red bimolecular fluorescence complementation system that allows for efficient visualization of protein interactions under physiological conditions. *Biosensors and Bioelectronics*, Vol.25, No.1, (Sep 15 2009), pp. 234-239, ISSN 1873-4235 (Electronic), 0956-5663 (Linking)

Curtis M.D. & Grossniklaus U. (2003). A gateway cloning vector set for high-throughput functional analysis of genes in planta. *Plant Physiology*, Vol.133, No.2, (Oct 2003), pp. 462-469

Dafny-Yelin M. & Tzfira T. (2007). Delivery of multiple transgenes to plant cells. *Plant Physiology*, Vol.145, No.4, (Dec 2007), pp. 1118-1128, ISSN 0032-0889 (Print), 0032-0889 (Linking)

Day R.N., Periasamy A., & Schaufele F. (2001). Fluorescence resonance energy transfer microscopy of localized protein interactions in the living cell nucleus. *Methods*, Vol.25, No.1, (Sep 2001), pp. 4-18, ISSN 1046-2023 (Print), 1046-2023 (Linking)

Diaz I., Martinez M., Isabel-LaMoneda I., Rubio-Somoza I., & Carbonero P. (2005). The DOF protein, SAD, interacts with GAMYB in plant nuclei and activates transcription of endosperm-specific genes during barley seed development. *The Plant Journal*, Vol.42, No.5, (Jun 2005), pp. 652-662, ISSN 0960-7412 (Print), 0960-7412 (Linking)

Ding Y.H., Liu N.Y., Tang Z.S., Liu J., & Yang W.C. (2006). Arabidopsis GLUTAMINE-RICH PROTEIN23 is essential for early embryogenesis and encodes a novel nuclear PPR motif protein that interacts with RNA polymerase II subunit III. *The Plant Cell*, Vol.18, No.4, (Apr 2006), pp. 815-830, ISSN 1040-4651 (Print), 1040-4651 (Linking)

Earley K.W., Haag J.R., Pontes O., Opper K., Juehne T., Song K., & Pikaard C.S. (2006). Gateway-compatible vectors for plant functional genomics and proteomics. *The Plant Journal*, Vol.45, No.4, (Feb 2006), pp. 616-629

Fan J.Y., Cui Z.Q., Wei H.P., Zhang Z.P., Zhou Y.F., Wang Y.P., & Zhang X.E. (2008). Split mCherry as a new red bimolecular fluorescence complementation system for visualizing protein-protein interactions in living cells. *Biochemical and Biophysical Research Communications*, Vol.367, No.1, (Feb 29 2008), pp. 47-53, ISSN 1090-2104 (Electronic), 0006-291X (Linking)

Gandia J., Galino J., Amaral O.B., Soriano A., Lluis C., Franco R., & Ciruela F. (2008). Detection of higher-order G protein-coupled receptor oligomers by a combined BRET-BiFC technique. *FEBS Letters*, Vol.582, No.20, (Sep 3 2008), pp. 2979-2984, ISSN 0014-5793 (Print), 0014-5793 (Linking)

Goto S., Mano S., Nakamori C., & Nishimura M. (2011). Arabidopsis ABERRANT PEROXISOME MORPHOLOGY9 is a peroxin that recruits the PEX1-PEX6 complex to peroxisomes. *The Plant Cell*, Vol.23, No.4, (Apr 2011), pp. 1573-1587, ISSN 1532- 298X (Electronic), 1040-4651 (Linking)

Ha S.H., Liang Y.S., Jung H., Ahn M.J., Suh S.C., Kweon S.J., Kim D.H., Kim Y.M., & Kim J.K. (2010). Application of two bicistronic systems involving 2A and IRES sequences to the biosynthesis of carotenoids in rice endosperm. *Plant Biotechnology Journal*, Vol.8, No.8, (Oct 2010), pp. 928-938, ISSN 1467-7652 (Electronic), 1467-7644 (Linking)

Hajdukiewicz P., Svab Z., & Maliga P. (1994). The small, versatile *pPZP* family of *Agrobacterium* binary vectors for plant transformation. *Plant Molecular Biology*, Vol.25, No.6, (Sep 1994), pp. 989-994

Hartley J.L., Temple G.F., & Brasch M.A. (2000). DNA cloning using in vitro site-specific recombination. *Genome Research*, Vol.10, No.11, (Nov 2000), pp. 1788-1795

Hayashi M. & Nishimura M. (2006). Arabidopsis thaliana--a model organism to study plant peroxisomes. *Biochimica et Biophysica Acta*, Vol.1763, No.12, (Dec 2006), pp. 1382- 1391, ISSN 0006-3002 (Print), 0006-3002 (Linking)

Helliwell C. & Waterhouse P. (2003). Constructs and methods for high-throughput gene silencing in plants. *Methods*, Vol.30, No.4, (Aug 2003), pp. 289-295

Hilson P., Allemeersch J., Altmann T., Aubourg S., Avon A., Beynon J., Bhalerao R.P., Bitton F., Caboche M., Cannoot B., Chardakov V., Cognet-Holliger C., Colot V., Crowe M., Darimont C., Durinck S., Eickhoff H., de Longevialle A.F., Farmer

E.E., Grant M., Kuiper M.T., Lehrach H., Leon C., Leyva A., Lundeberg J., Lurin C., Moreau Y., Nietfeld W., Paz-Ares J., Reymond P., Rouze P., Sandberg G., Segura M.D., Serizet C., Tabrett A., Taconnat L., Thareau V., Van Hummelen P., Vercruysse S., Vuylsteke M., Weingartner M., Weisbeek P.J., Wirta V., Wittink F.R., Zabeau M., & Small I. (2004). Versatile gene-specific sequence tags for *Arabidopsis* functional genomics: transcript profiling and reverse genetics applications. *Genome Research*, Vol.14, No.10B, (Oct 2004), pp. 2176-2189

Hink M.A., Bisseling T., & Visser A.J.W.G. (2002). Imaging protein-protein interactions in living cells. *Plant Molecular Biology*, Vol.50, No.6, (Dec 2002), pp. 871-883, ISSN 0167-4412

Hino T., Tanaka Y., Kawamukai M., Nishimura K., Mano S., & Nakagawa T. (2011). Two Sec13p homologs, AtSec13A and AtSec13B, redundantly contribute to formation of COPII transport vesicles in Arabidopsis thaliana. *Bioscience, Biotechnology, and Biochemistry*. Vol. 75, No. 9, (Jun 2011), pp.1848-1852

Hu C.D., Chinenov Y., & Kerppola T.K. (2002). Visualization of interactions among bZIP and Rel family proteins in living cells using bimolecular fluorescence complementation. *Molecular Cell*, Vol.9, No.4, (Apr 2002), pp. 789-798, ISSN 1097-2765 (Print), 1097-2765 (Linking)

Hu C.D. & Kerppola T.K. (2003). Simultaneous visualization of multiple protein interactions in living cells using multicolor fluorescence complementation analysis. *Nature Biotechnology*, Vol.21, No.5, (May 2003), pp. 539-545, ISSN 1087-0156 (Print), 1087-0156 (Linking)

Jach G., Pesch M., Richter K., Frings S., & Uhrig J.F. (2006). An improved mRFP1 adds red to bimolecular fluorescence complementation. *Nature Methods*, Vol.3, No.8, (Aug 2006), pp. 597-600, ISSN 1548-7091 (Print), 1548-7091 (Linking)

James P., Halladay J., & Craig E.A. (1996). Genomic libraries and a host strain designed for highly efficient two-hybrid selection in yeast. *Genetics*, Vol.144, No.4, (Dec 1996), pp. 1425-1436, ISSN 0016-6731 (Print), 0016-6731 (Linking)

Karimi M., Inze D., & Depicker A. (2002). GATEWAY vectors for *Agrobacterium*-mediated plant transformation. *Trends in Plant Science*, Vol.7, No.5, (May 2002), pp. 193-195

Karimi M., Bleys A., Vanderhaeghen R., & Hilson P. (2007a). Building blocks for plant gene assembly. *Plant Physiology*, Vol.145, No.4, (Dec 2007a), pp. 1183-1191

Karimi M., Depicker A., & Hilson P. (2007b). Recombinational cloning with plant gateway vectors. *Plant Physiology*, Vol.145, No.4, (Dec 2007b), pp. 1144-1154

Kodama Y. & Wada M. (2009). Simultaneous visualization of two protein complexes in a single plant cell using multicolor fluorescence complementation analysis. *Plant Molecular Biology*, Vol.70, No.1-2, (May 2009), pp. 211-217, ISSN 1573-5028 (Electronic), 0167-4412 (Linking)

Koizumi N., Ujino T., Sano H., & Chrispeels M.J. (1999). Overexpression of a gene that encodes the first enzyme in the biosynthesis of asparagine-linked glycans makes plants resistant to tunicamycin and obviates the tunicamycin-induced unfolded protein response. *Plant Physiology*, Vol.121, No.2, (Oct 1999), pp. 353-361, ISSN 0032-0889 (Print), 0032-0889 (Linking)

Koizumi N. & Iwata Y. (2008). Construction of a binary vector for transformation of *Arabidopsis thaliana* with a new selection marker. *Bioscience, Biotechnology, and Biochemistry*, Vol.72, No.11, (Nov 2008), pp. 3041-3043

Lee L.Y., Fang M.J., Kuang L.Y., & Gelvin S.B. (2008). Vectors for multi-color bimolecular fluorescence complementation to investigate protein-protein interactions in living plant cells. *Plant Methods*, Vol.4, 2008), pp. 24, ISSN 1746-4811 (Electronic), 1746-4811 (Linking)

Loyter A., Rosenbluh J., Zakai N., Li J., Kozlovsky S.V., Tzfira T., & Citovsky V. (2005). The plant VirE2 interacting protein 1. a molecular link between the Agrobacterium T- complex and the host cell chromatin? *Plant Physiology*, Vol.138, No.3, (Jul 2005), pp. 1318-1321, ISSN 0032-0889 (Print), 0032-0889 (Linking)

Ludewig U., Wilken S., Wu B., Jost W., Obrdlik P., El Bakkoury M., Marini A.M., Andre B., Hamacher T., Boles E., von Wiren N., & Frommer W.B. (2003). Homo- and hetero- oligomerization of ammonium transporter-1 NH4 uniporters. *The Journal of Biological Chemistry*, Vol.278, No.46, (Nov 14 2003), pp. 45603-45610, ISSN 0021-9258 (Print), 0021-9258 (Linking)

Mano S. & Nishimura M. (2005). Plant peroxisomes. *Vitamins & Hormones*, Vol.72, 2005), pp.111-154, ISSN 0083-6729 (Print), 0083-6729 (Linking)

Maple J., Aldridge C., & Moller S.G. (2005). Plastid division is mediated by combinatorial assembly of plastid division proteins. *The Plant Journal*, Vol.43, No.6, (Sep 2005), pp. 811-823, ISSN 0960-7412 (Print), 0960-7412 (Linking)

Marrocco K., Zhou Y., Bury E., Dieterle M., Funk M., Genschik P., Krenz M., Stolpe T., & Kretsch T. (2006). Functional analysis of EID1, an F-box protein involved in phytochrome A-dependent light signal transduction. *The Plant Journal*, Vol.45, No.3, (Feb 2006), pp. 423-438, ISSN 0960-7412 (Print), 0960-7412 (Linking)

Miki D. & Shimamoto K. (2004). Simple RNAi vectors for stable and transient suppression of gene function in rice. *Plant and Cell Physiology*, Vol.45, No.4, (Apr 2004), pp. 490-495

Mita S., Suzuki-Fujii K., & Nakamura K. (1995). Sugar-inducible expression of a gene for beta-amylase in Arabidopsis thaliana. *Plant Physiology*, Vol.107, No.3, (Mar 1995), pp. 895-904, ISSN 0032-0889 (Print), 0032-0889 (Linking)

Nakagawa T., Kurose T., Hino T., Tanaka K., Kawamukai M., Niwa Y., Toyooka K., Matsuoka K., Jinbo T., & Kimura T. (2007a). Development of series of Gateway binary vectors, pGWBs, for realizing efficient construction of fusion genes for plant transformation. *Journal of Bioscience and Bioengineering*, Vol.104, 2007a), pp. 31-41

Nakagawa T., Suzuki T., Murata S., Nakamura S., Hino T., Maeo K., Tabata R., Kawai T., Tanaka K., Niwa Y., Watanabe Y., Nakamura K., Kimura T., & Ishiguro S. (2007b). Improved Gateway binary vectors: high-performance vectors for creation of fusion constructs in transgenic analysis of plants. *Bioscience, Biotechnology, and Biochemistry*, Vol.71, No.8, (Aug 2007b), pp. 2095-2100

Nakagawa T., Nakamura S., Tanaka K., Kawamukai M., Suzuki T., Nakamura K., Kimura T., & Ishiguro S. (2008). Development of R4 gateway binary vectors (R4pGWB) enabling high-throughput promoter swapping for plant research. *Bioscience, Biotechnology, and Biochemistry*, Vol.72, No.2, (Feb 2008), pp. 624-629

Nakamura S., Nakao A., Kawamukai M., Kimura T., Ishiguro S., & Nakagawa T. (2009). Development of Gateway binary vectors, R4L1pGWBs, for promoter analysis in higher plants. *Bioscience, Biotechnology, and Biochemistry*, Vol.73, No.11, (Nov 2009), pp. 2556-2559, ISSN 1347-6947 (Electronic), 0916-8451 (Linking)

Nakamura S., Mano S., Tanaka Y., Ohnishi M., Nakamori C., Araki M., Niwa T., Nishimura M., Kaminaka H., Nakagawa T., Sato Y., & Ishiguro S. (2010). Gateway binary vectors with the bialaphos resistance gene, bar, as a selection marker for plant transformation. *Bioscience, Biotechnology, and Biochemistry*, Vol.74, No.6, 2010), pp. 1315-1319, ISSN 1347-6947 (Electronic), 0916-8451 (Linking)

Nakayama H., Yoshida K., Ono H., Murooka Y., & Shinmyo A. (2000). Ectoine, the compatible solute of Halomonas elongata, confers hyperosmotic tolerance in cultured tobacco cells. *Plant Physiology*, Vol.122, No.4, (Apr 2000), pp. 1239-1247, ISSN 0032-0889 (Print), 0032-0889 (Linking)

Naqvi S., Zhu C., Farre G., Ramessar K., Bassie L., Breitenbach J., Perez Conesa D., Ros G., Sandmann G., Capell T., & Christou P. (2009). Transgenic multivitamin corn through biofortification of endosperm with three vitamins representing three distinct metabolic pathways. *P Natl Acad Sci USA*, Vol.106, No.19, (May 12 2009), pp. 7762-7767, ISSN 1091-6490 (Electronic), 0027-8424 (Linking)

Obrdlik P., El-Bakkoury M., Hamacher T., Cappellaro C., Vilarino C., Fleischer C., Ellerbrok H., Kamuzinzi R., Ledent V., Blaudez D., Sanders D., Revuelta J.L., Boles E., Andre B., & Frommer W.B. (2004). K+ channel interactions detected by a genetic system optimized for systematic studies of membrane protein interactions. *P Natl Acad Sci USA*, Vol.101, No.33, (Aug 17 2004), pp. 12242-12247, ISSN 0027-8424 (Print), 0027-8424 (Linking)

Ohad N., Shichrur K., & Yalovsky S. (2007). The analysis of protein-protein interactions in plants by bimolecular fluorescence complementation. *Plant Physiology*, Vol.145, No.4, (Dec 2007), pp. 1090-1099, ISSN 0032-0889 (Print), 0032-0889 (Linking)

Sasaki Y., Sone T., Yoshida S., Yahata K., Hotta J., Chesnut J.D., Honda T., & Imamoto F. (2004). Evidence for high specificity and efficiency of multiple recombination signals in mixed DNA cloning by the Multisite Gateway system. *Journal of Biotechnology*, Vol.107, No.3, (Feb 5 2004), pp. 233-243

Singh T., Hayashi M., Mano S., Arai Y., Goto S., & Nishimura M. (2009). Molecular components required for the targeting of PEX7 to peroxisomes in Arabidopsis thaliana. *The Plant Journal*, Vol.60, No.3, (Nov 2009), pp. 488-498, ISSN 1365-313X (Electronic), 0960-7412 (Linking)

Tanaka Y., Nakamura S., Kawamukai M., Koizumi N., & Nakagawa T. (2011). Development of a series of gateway binary vectors possessing a tunicamycin resistance gene as a marker for the transformation of Arabidopsis thaliana. *Bioscience, Biotechnology, and Biochemistry*, Vol.75, No.4, (May 2011), pp. 804-807, ISSN 1347-6947 (Electronic), 0916-8451 (Linking)

Waadt R., Schmidt L.K., Lohse M., Hashimoto K., Bock R., & Kudla J. (2008). Multicolor bimolecular fluorescence complementation reveals simultaneous formation of alternative CBL/CIPK complexes in planta. *The Plant Journal*, Vol.56, No.3, (Nov 2008), pp. 505-516, ISSN 1365-313X (Electronic), 0960-7412 (Linking)

Walhout A.J., Temple G.F., Brasch M.A., Hartley J.L., Lorson M.A., van den Heuvel S., & Vidal M. (2000). GATEWAY recombinational cloning: application to the cloning of large numbers of open reading frames or ORFeomes. *Methods in Enzymology*, Vol.328, 2000), pp. 575-592

Walter M., Chaban C., Schutze K., Batistic O., Weckermann K., Nake C., Blazevic D., Grefen C., Schumacher K., Oecking C., Harter K., & Kudla J. (2004). Visualization of protein interactions in living plant cells using bimolecular fluorescence complementation. *The Plant Journal*, Vol.40, No.3, (Nov 2004), pp. 428-438, ISSN 0960-7412 (Print), 0960-7412 (Linking)

Xu X., Soutto M., Xie Q., Servick S., Subramanian C., von Arnim A.G., & Johnson C.H. (2007). Imaging protein interactions with bioluminescence resonance energy transfer (BRET) in plant and mammalian cells and tissues. *P Natl Acad Sci USA*, Vol.104, No.24, (Jun 12 2007), pp. 10264-10269, ISSN 0027-8424 (Print), 0027-8424 (Linking)

Ye X., Al-Babili S., Kloti A., Zhang J., Lucca P., Beyer P., & Potrykus I. (2000). Engineering the provitamin A (beta-carotene) biosynthetic pathway into (carotenoid-free) rice endosperm. *Science*, Vol.287, No.5451, (Jan 14 2000), pp. 303-305, ISSN 0036-8075 (Print), 0036-8075 (Linking)

Zamyatnin A.A., Jr., Solovyev A.G., Bozhkov P.V., Valkonen J.P., Morozov S.Y., & Savenkov E.I. (2006). Assessment of the integral membrane protein topology in living cells. *The Plant Journal*, Vol.46, No.1, (Apr 2006), pp. 145-154, ISSN 0960-7412 (Print), 0960-7412 (Linking)

Zelazny E., Borst J.W., Muylaert M., Batoko H., Hemminga M.A., & Chaumont F. (2007). FRET imaging in living maize cells reveals that plasma membrane aquaporins interact to regulate their subcellular localization. *P Natl Acad Sci USA*, Vol.104, No.30, (Jul 24 2007), pp. 12359-12364, ISSN 0027-8424 (Print), 0027-8424 (Linking)

Zheng X., Deng W., Luo K., Duan H., Chen Y., McAvoy R., Song S., Pei Y., & Li Y. (2007). The cauliflower mosaic virus (CaMV) 35S promoter sequence alters the level and patterns of activity of adjacent tissue- and organ-specific gene promoters. *Plant Cell Reports*, Vol.26, No.8, (Aug 2007), pp. 1195-1203

03 ENHANCING THE THERMAL STABILITY OF FIREFLY LUCIFERASES BY THE APPLICATION OF GENETIC ENGINEERING TECHNIQUES

1. INTRODUCTION: AN OVERVIEW

Bioluminescence is the emission of light by living creatures, which occurs through a chemiluminescent reaction facilitated by a special class of enzymes called luciferases. Bioluminescence occurs when luciferin, an organic substrate, undergoes oxidation in the presence of a luciferase enzyme (Wilson & Hastings, 1998). Luciferin and luciferase are words used to describe substrates and enzymes that interact with one other to produce light, without being specific to their structure or function. The emission of light energy in chemical and biological processes is frequently seen, particularly in reactions involving free radicals. However, in the majority of instances, the quantum yield, which refers to the quantity of light quanta emitted when one molecule of a substrate undergoes oxidation, is typically less than 0.1% or 0.01%. The key characteristic and significant benefit of bioluminescent systems is their high quantum yield, ranging from 1 to 60%. This is accomplished by the involvement of luciferase in the reaction. Luciferases function as a substrate that is either covalently or non-covalently attached to the emitting chromophore. However, they serve as a biocatalyst, enabling the creation of electronically excited product. Luciferases typically exhibit a high degree of specificity towards their substrates. The protein architectures of bioluminescent enzymes necessitate certain structural characteristics in luciferins. It is likely essential for achieving the highest level of energy concentration in the enzymatic reaction on the chromophore and for the creation of the electrically excited reaction product, which subsequently reverts back to its original state by emitting a quantum of visible light.

A wide range of species, including insects, fish, bacteria, jellyfish, and protozoa, have been identified as sources of luciferases and thoroughly studied (Wilson & Hastings, 1998). All of them are enzymes that catalyze oxidation-reduction reactions. Molecular oxygen in the air or hydrogen peroxide functions as an oxidizing agent. The molecular compositions of oxidizing substrates, known as luciferins, exhibit significant variations. Luciferins are typically heterocyclic chemicals. The oxidation products have a relatively low excitation energy, which is the main reason why the energy from enzymatic reactions is sufficient to create the product in an electronically excited state. Thus, it can be inferred that luciferin-luciferase systems are exceptional and efficient mechanisms for converting the energy of biochemical reactions into light. The interest in studying the structures and functions of luciferases, as well as understanding their biological significance in nature, remains undiminished (Fraga, 2008; Seliger & McElroy, 1960).

The practical significance of bioluminescent systems also contributes to the curiosity they generate. In order for the enzymatic process to occur, the presence of key metabolites such as flavin mononucleotide (FMN) or adenosine triphosphate (ATP) is required as co-substrates. The high quantum yield of bioluminescence in luciferase-catalyzed processes streamlines the process of light detection. Luciferases are extensively utilized as extremely effective analytical reagents because to their nearly perfect specificity to the substrates. The bioluminescence intensity is directly proportional to the rate at which the electronically excited reaction product is formed. Therefore, it is also directly proportional to the concentration of the substrate, as long as the substrate concentration is below the value of Km. Bioluminescent methods of substrate detection exhibit a high sensitivity, with the ability to detect 10–13–10–18 moles of substrates in a test solution. This sensitivity is significantly greater than that of other detection methods, as each molecule of the substrate entering the reaction emits between 0.3 to 1 quantum of light. Furthermore, the fabrication of luminometers, used for measuring light, is considerably less complex compared to spectrophotometers or spectrofluorimeters. This simplicity also facilitates the practical implementation of bioluminescent systems (Ugarova, 2005).

2. OVERVIEW OF BIOLUMINESCENT SYSTEM OF FIREFLIES

2.1 Scheme of luciferase-catalyzed reaction

Beetle luciferase [luciferin 4-monooxygenase (ATP hydrolysis); luciferin: oxygen 4- oxidoreductase (decarboxylation, ATP hydrolysis), EC 1.13.12.7] catalyzes firefly luciferin oxidation by air oxygen in the presence of MgATP (Fraga, 2008; Ugarova, 1989). The reaction is accompanied by the emission of visible light with the quantum yield of 40-60% (Niwa et al., 2010). As shown in Fig. 1, at the first step the enzyme binds its substrates luciferin (1) and ATP. After the formation of the triple complex luciferin reacts with ATP to form mixed anhydride of carboxylic and phosphoric acids, luciferyl adenylate (2) and pyrophosphate (PPi). The luciferyl adenylate is oxidized with air oxygen to form a cyclic peroxide, dioxytanone (3), through a series of intermediate steps. The dioxytanone molecule has a remarkable feature: one portion of the molecule is a readily oxidizable heterocyclic structure (luciferyl) with low ionizing potential and another portion (peroxide) has a high affinity to electron. Due to the intramolecular electron transfer from the phenolate group to the peroxide, a resonance charge-transfer structure is formed. The break of the O-O bond causes decarboxylation of dioxytanone and formation of biradical (ketone anion-radical and phenolate cation-radical). As a result of intramolecular recombination of the radicals, the reaction product, oxyluciferin (4), is formed in singlet electronically excited state. Depending on the properties of microenvironment, oxyluciferin may exist in the form of ketone (4a), enol (4b), or enolate-anion. The electronically excited oxyluciferin reactivates with the emission of quantum of light with □max from 536 to 623 nm depending on the firefly species and pH (Viviani, 2002).

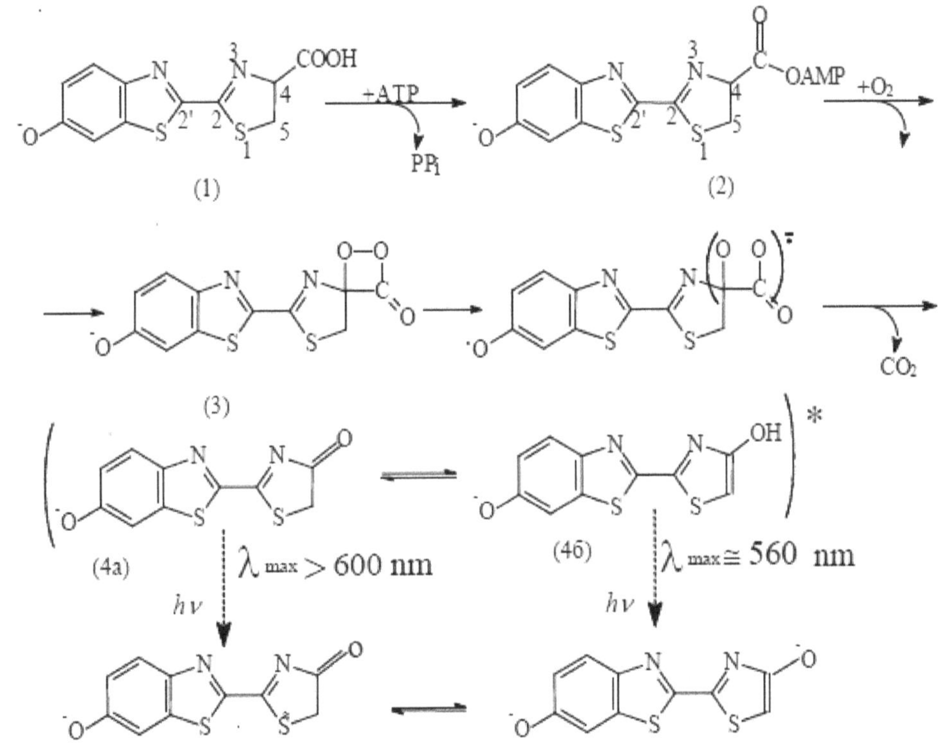

Fig. 1. Scheme of the reaction catalyzed by firefly luciferase (Ugarova et al., 1993)

2.2 Primary structure of firefly luciferases

Up to the middle of 1980s, the studies were carried out with native firefly luciferases, which were isolated from desiccated firefly lanterns. As a rule, composition of enzyme preparations was not homogeneous due to the presence of different modified forms of enzyme, which, probably, appeared during the functioning of the enzyme. For this reason, even primary amino acid sequence of luciferases remained unknown for a long time. A new phase began in 1985 when cDNA of luciferase for the North American *Photinus pyralis* fireflies was isolated (De Wet et al., 1985). In 1987, the amino acid sequence of this luciferase was determined (De Wet et al., 1987), and in 1989 four luciferases from Jamaican click beetles were cloned (Wood et al., 1989). At the beginning of 1990th, the luciferase from the East-European (north-Caucasian) *Luciola mingrelica* fireflies was cloned and the homogeneous preparation of the recombinant *L. mingrelica* luciferase was obtained (*Devine* et al., 1993). At the present time, primary structures of more than 40 luciferase genes are known that were isolated from different species of fireflies, click-beetles, and railroad worms inhabiting in the USA, Russia, Japan, South-American countries, *etc* (Arnoldi et al., 2010; Oba et al., 2010a; Viviani, 2002). The scheme of the chemical reaction catalyzed and the structure of the emitter are identical for all firefly luciferases. Luciferase molecules consist of one polypeptide chain (542-552 residues), do not contain cofactors and have similar amino acid composition. More than half of the amino acids are non-polar or ambivalent. The number of charged residues is also practically identical for all firefly luciferases, and the main difference between them is in the number of Trp and Cys residues. The amino acid sequence identity for different luciferases usually correlates with phylogenetic relationships of the corresponding fireflies (Li et al, 2006). For luciferases from Luciolini tribe of fireflies (*Luciola* and *Hotaria* genus's) the sequence identity is near 80%, and it is more than 90%for the part of the protein after a 200th residue. The structure of *L. mingrelica* luciferase is very close to the structure of Japanese *Hotaria* (*Luciola*) *parvula* luciferase (98% identity) despite the huge geographic separation of the original firefly species. The *Luciola* luciferases show less identity with luciferases from other firefly subfamilies (67% identity), for example, *P. pyralis* or *L. noctiluca*, and with luciferases from click beetles (43% identity) (Ugarova & Brovko, 2002). At the same time, the identity between the luciferases isolated from Jamaican and Brazil beetles reaches 80% (Oba et al, 2010b; Viviani et al., 1999).

Beetle luciferases belong to the superfamily of enzymes catalyzing formation of acyladenylates from ATP and compounds with a carboxyl group. This superfamily includes the families of non-ribosomal peptide synthetases, acyl-CoA-ligases, *etc* (Gulick, 2009). In 1990s the search of conserved motifs in the amino acid sequences of the proteins within this superfamily has been carried out in our laboratory using computer analysis. In addition to the previously known motif 1 (residues 197-210 in the *L. mingrelica* luciferase), another conserved motif 2 (residues 410-460) has been found. It was proposed that the conserved amino acid residues in motifs 1 and 2 perform key functions in the catalysis and that both motifs belong to a single conserved structural element. It was also proposed that the 197-220 region forms the ATP binding center and the 410-460 region is important for the interaction of the luciferase with its activator coenzyme A (Morozov & Ugarova, 1996). Later structural findings have confirmed these conclusions.

2.3 Spatial structure of firefly luciferase and its complexes with substrates

In 1996 the spatial structure of the luciferase from the American *P. pyralis* fireflies without substrates was obtained using X-Ray analysis (Conti et al., 1996). This opened a new stage in the studies of firefly luciferases. The X-Ray data confirmed the supposition given above about the important role of motifs 1 and 2. Both motifs are loops with undefined electron density indicating high mobility of these fragments, and the motif 2 is partly involved in a polypeptide connecting the two domains of the protein globule. Mutual arrangement of the domains plays an important role in the catalysis as it will be shown below. Unfortunately, one may only assume the possible binding centers for luciferin and ATP. In 1997 the data about the crystal structure of another enzyme belonging to the superfamily of adenylating proteins (adenylase) were published (Conti et al., 1997). In this case the crystals of the complex of the enzyme with its substrates AMP and phenylalanine (analog of the complex of the enzyme with the adenylation product) were analysed. As one may expect, spatial structures of adenylase and luciferase, which both have the same ability for adenylation of substrate carboxyl group with ATP, appeared to be rather similar although the amino acid sequence homology of these enzymes is low. Both enzymes are composed of two domains, large N-domain, which, in turn, may be divided into three sub-domains, and small C-domain. The two domains are connected with very flexible and disordered loop. Each domain and even sub-domain of the adenylase has similar topology with the corresponding domain or sub-domain of the luciferase. The only significant difference is mutual rearrangement of the N- and C-domains. The domains of adenylase are drawn together and rotated by 90o with respect to their orientation in the firefly luciferase (Conti et al., 1997). It was proposed that the observed rotation of domains relative to each other is a consequence of change in the globule conformation upon binding of substrates. In this case the three- dimensional structure of the luciferase-luciferin-ATP complex should be similar to that of the adenylase-phenyl alanine-AMP complex. The computer modeling on the basis of this hypothesis allowed to construct a model of the enzyme-substrate complex for the *P. pyralis* luciferase (Sandalova & Ugarova, 1999). The structures of the luciferase before and after binding of substrates are shown in Fig. 2.

Neither ATP, nor luciferin can bind with the enzyme without the rotation of the domains. The catalytically important residues of the C-domain (Lys529 and Thr527) approach the substrate molecules only after the rotation of the domains. Similar model structure was obtained by Branchini and coworkers (Branchini et al.

1998). All the residues that are in direct contact with the substrates are absolutely conservative. These models were further confirmed by the independent studies: the residual activity of the luciferase mutant, in which Lys529 was changed to Ala, was less than 0.1% (Branchini et al., 2000). The role of others residues of the luciferase active site was supported by mutagenesis methods (Branchini et al., 1999, 2000, 2001, 2003).

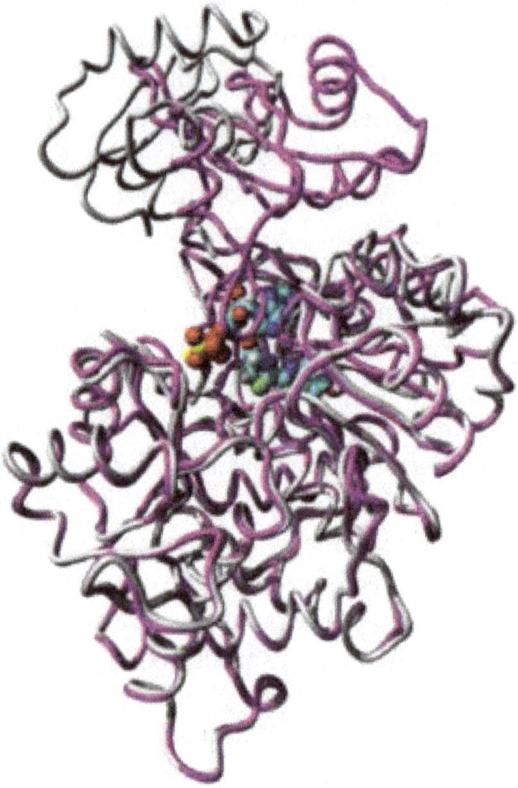

Fig. 2. Superimposition of the spatial structures of firefly luciferase without substrates (grey tubes) and the luciferase-luciferin-ATP complex (magenta tubes). The substrates are shown as CPK models (Sandalova & Ugarova, 1999). Molecular graphics were created with YASARA (www.yasara.org) and PovRay (www.povray.org) software

In 2006 three structures of *Luciola cruciata* luciferase were solved (Nakatsu et al., 2006): in complex with Mg-ATP, in complex with the analog of intermediate product of the adenylation step – DLSA, and in complex with the reaction products – LO and AMP. This allowed to correct the data obtained previously by homology modeling. Firefly luciferase consists of two domains, large N-terminal domain (1-436 amino acid residues) and small C- terminal domain (443-544 amino acid residues), which are connected by the flexible loop (337-442 amino acid residues). By-turn N-terminal domain is composed of two distinct subdomains (Wang et al.,

2001): A (1-190) and B (191-436) forming a strong hydrophobic interface (Fig. 3).

When luciferase is in complex with DLSA, the C-domain is rotated by 90° relative to the N- domain compared to the free enzyme. This leads to the transformation from open to closed conformation of luciferase so that both domains are in close contact. The structures obtained allowed to determine the structure of the firefly luciferase active site and the environment of the substrates at the adenylation step of the reaction. Later findings revealed that yet another rotation of the C-domain by 140° is required for the oxidation step (Gulick, 2009). In this conformation the residue K445 assumes the role of the group K531 (Branchini et al., 2005; Branchini et al., 2011).

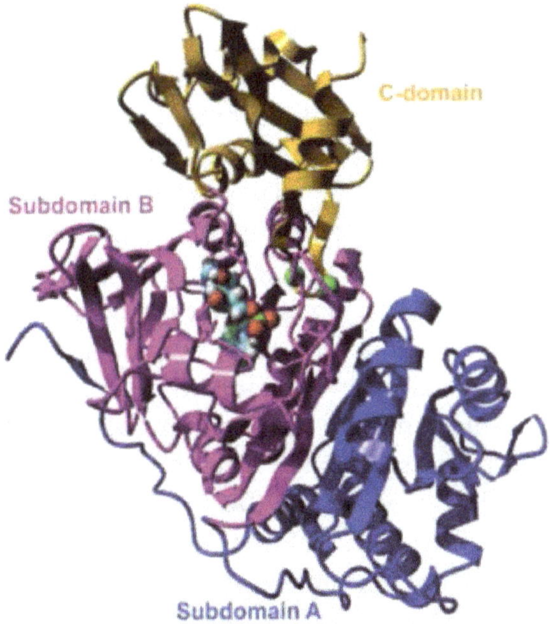

Fig. 3. 3D-structure of *L. cruciata* firefly luciferase in complex with DLSA (Nakatsu et al., 2006). Subdomains A, B and C are depicted in blue, magenta and orange, respectively

3. STABILITY OF FIREFLY LUCIFERASES IN SOLUTION

Firefly luciferases are currently used for microbial contamination testing, as very sensitive reporter genes and in pyrosequencing. They are also promising labels in immunoassays and biosensors for the study of protein-protein interactions (Binkowski et al., 2009; Lundin, 2000; Viviani & Ohmiya, 2006). The stability and activity of enzymes are basic parameters defining the efficiency of their applications. Nevertheless, applications of wild-type firefly luciferases are often limited by the insufficient stability of the enzyme. Firefly luciferases demonstrate low stability in water solutions (Dementieva et al., 1989; White et al., 1996) and can aggregate as well as can be easily adsorbed on the surfaces (Hall & Leach, 1988; Herbst et al., 1998; Suelter & DeLuca, 1983).

The *P. pyralis* luciferase loses up to 99% of activity within 24 hours at 4-8°C as a result of protein adsorption on the container surface when low concentrations of luciferases are used. The addition of bovine serum albumin (0.1 mg/ml) reduces activity loss associated with adsorbtion to 10% even at very low concentrations of luciferase (10 ng/ml). An increase of ionic strength increased the stability of luciferase and an addition of Triton X-100 (0.2 mM) and glycerol (10-50%) prevented the adsorption of luciferase on the surface (Suelter & DeLuca, 1983). The stabilizing effect of salt addition was also observed for *L. mingrelica* firefly luciferase. In presence of 0.4 M MgSO4 and 0.5 M Na2SO4 the thermal stability of *L. mingrelica* luciferase was 2 and 10 times higher respectively compared with a buffer solution without the salt added (Brovko et al., 1982). However, even in the presence of stabilizing salts at 10°C *L. mingrelica* luciferase loses about 90% of activity after 10-15 days storage at the concentration of 0.1 mg/ml. But after the addition of 10% ethylene glycol and the antimicrobial agent (0,02% NaN3) the luciferase can be stored more than 100 days without loss of activity (Dementieva et al., 1989). The presence of osmolytes in the solution such as glycine betaine and polyols can lead to a significant stabilization of luciferase at the elevated temperatures (Eriksson et al, 2003; Mehrabi et al., 2008; Moroz et al., 2008). Luciferases from the *Luciola* genus readily inactivate at the low ionic strength though *P. pyralis* luciferase is crystallized at the same conditions (Dementieva et al., 1989; Simomura et al., 1977). Native firefly luciferases as well as recombinant enzyme during eukaryotic expression are localized in peroxisomes (De Wet et al., 1987; Gould et al., 1987; Keller et al., 1987) due to the presence of C-terminal signal peptide (AKM, SKL) that is recognized by the cellular peroxisomal transport system (Rainer, 1992).

Firefly luciferase is sensitive to proteolysis, which leads to a short *in vivo* half-life of the enzyme. This effect is especially pronounced at high temperatures when partial enzyme unfolding leads to the higher accessibility of proteolytic sites (Frydman, 1999). Two protease sensitive regions (206-220 and 329-341 residues) were identified in *P. pyralis* luciferase amino acid sequence (Thompson et al., 1997). Half-life of luciferase activity *in vivo* in mammalian cells is only 3-4 hours at 37°C (Leclerc et al., 2000; Thompson et al., 1991). Substrate and competitive inhibitors change the conformation of luciferase, when introduced to prokaryotic and eukaryotic cells, leading to a several-fold decrease of enzyme degradation (Thompson et al., 1991). *P. pyralis* luciferase quickly inactivates in eukaryotic cells at 40-45°C with a half-life of 4-20 min (Forreitor et al., 1997; Souren et al., 1999b). Nevertheless, the stability can be relatively high at moderate temperatures. For example, *L. mingrelica* luciferase produced in frog oocytes had a half-life of about 2 days (Kutuzova et al., 1989).

At the same time even in the solutions with high ionic strength and in the presence of glycerol (10%) and bovine serum albumin (10 mg/ml) *P. pyralis* luciferase loses >90% of activity *in vitro* within 6-20 min at 37-42°C (Nguyen et al., 1989; Thulasiraman & Matts, 1996; Tisi et al., 2002; White et al., 1996). Similar results were obtained for *L. mingrelica* luciferase (Lundovskikh et al., 1998; Ugarova et al., 2000). It was demonstrated that inactivated luciferase is virtually unable to restore activity after cooling and usually aggregates (Minami & Minami, 1999; Schumacher et al., 1996; Souren et al., 1999a, 1999b). Appreciable spontaneous reactivation was only observed for diluted solutions of luciferase that was fully inactivated by guanidine chloride. But reactivation rate was very low and equilibrium was reached after 72 hours. In case of partial denaturation by guanidine chloride the degree of reactivation was low apparently due to aggregation of the enzyme (Herbst et al., 1998). Denatured luciferase can still be effectively reactivated in the presence of different chaperone systems. For instance, in rabit reticulocyte lysate a refolding of luciferase occurs with a half-period of 8 minutes (Nimmesgern et al., 1993). A detailed mechanism of luciferase inactivation in solution is still unknown. In several works (Frydman et al., 1999; Herbst et al., 1998; Wang et al., 2001) different unfolding intermediates were studied that are formed during chemical denaturation and refolding of *P. pyralis* luciferase. Thermoinactivation kinetics was studied in detail in case of *L. mingrelica* luciferase (Brovko et al., 1982; Ugarova et al., 2000). It was shown that at elevated temperature *L. mingrelica* luciferase undergoes two-step inactivation: at the first stage a reversible dissociation of luciferase homodimers probably occurs, which is followed by an irreversible inactivation of monomers (Lundovskikh et al., 1998). Luciferases of *Luciola cruciata* and *Luciola lateralis* were isolated and purified in 1992. *L. cruciata* luciferase was found to be similar to *P. pyralis* luciferase in stability, whereas *L. lateralis luciferase* showed significantly higher thermal and pH stability compared with the other two. *L. cruaciata* and *P. pyralis* luciferases were completely inactivated after incubation at 50°C for 30 min. Under the same conditions (10 mM Na- phosphate, 0.2 mM EDTA, 100 mM NaCl, 10% glycerol) the remaining activity of *L. lateralis* luciferase was about 20% (Kajiyama et al., 1992).

Thus, an optimization of the enzyme microenvironment in the solution may increase the stability of luciferases. The more promising approach is thermostabilization of luciferases by genetic engineering methods, which opens wide opportunities of new enzyme forms with enhanced resistance to such environment factors as temperature, organic solvents, and other chemical agents (Ugarova, 2010). It is important to emphasize that comparison of thermostability of mutant and wild-type luciferases must be performed under identical conditions, because the buffer composition and different stabilizing agents (BSA, ammonium sulfate, phosphates, glycerol, etc.) can greatly affect the rate of thermal inactivation.

4. THERMOSTABILIZATION OF FIREFLY LUCIFERASES BY RANDOM AND SITE-DIRECTED MUTAGENESIS

In the absence of structural information on a luciferase, random mutagenesis was the most common and efficient approach to obtain mutant proteins with improved stability to the action of temperature, pH, etc. Firefly luciferase is particularly suitable for activity screens owing to the ease with which its bioluminescence activity can be detected (Wood & DeLuca, 1987). Kajiyama and Nakano were the first to use random mutagenesis in an attempt to isolate thermostable mutants of the luciferase from *L. cruciata* fireflies (Kajiyama & Nakano, 1993). The mutagen-treated plasmid was transformed into *E.coli* JM101. After 12 h at 37°C, colonies on LB/ampicillin plates were transferred on nitrocellulose filters, which were put on new agar plates and incubated at 60°C for 30 min. Remaining luciferase activity was determined on filters by photographic method using X-ray films. Three isolated mutants carried the same an amino acid substitution Thr217Ile and were superior to wild-type in thermal and pH stability. Furthermore, its specific activity was increased to 130% of that of the wild-type. In order to examine the effect of amino acid residue substitution at position 217 on the thermostability of *L. cruciata* luciferase, the authors have replaced the residue Thr217 with all possible amino acid residues by site-directed mutagenesis. The thermostability of these mutants correlated well with hydrophobicity of the substituted amino acid residue. Especially, Thr217Leu and Thr217Val luciferases still retained over 75% of activity after 10 min incubation at 50°C. The fact that hydrophilic and large hydrophobic (Trp and Tyr) substitution resulted in an expression yield >1000-fold less than wild-type is also consistent with residue 217 being in a buried, hydrophobic environment (Kajiyama & Nakano, 1993). The amino acid sequences of *L. cruciata* and *L. lateralis* luciferases are 94% identical. To examine the effect of hydrophobic amino acid at position 217 on the thermostability of *L. lateralis luciferase*, three mutants, Ala217Ile, Leu, Val were constructed, and all of them demonstrated enhanced thermostability (Kajiyama & Nakano, 1994). The *L. lateralis* luciferase mutant Ala217Leu retained over 70% of the initial activity after 60 min incubation at 50°C. Its half-life was about 20 times longer than that of the wild type *L. lateralis* luciferase. Its thermostability was superior to that of the *L. cruciata* luciferase mutant Thr217Leu.

Random mutagenesis was also used to obtain thermostable mutant of *P.pyralis* luciferase. The substitution Glu354Lys increased thermostability of the enzyme 5-fold (White et al., 1996). The substitution of Glu354 with all possible amino acid residues by site-directed mutagenesis showed that the most stable mutants contained Lys or Arg residues. Thus, the substitution of negatively charged residue to positive one in this part of enzyme molecule increased the thermostability of *P.pyralis* luciferase. Thermostable *P.pyralis* luciferase was also obtained by a combination of random and site-directed mutagenesis. The double mutant was constructed that contained the substitutions Glu354Lys and Ala215Leu (similar to Ala217Leu in *L. lateralis* luciferase). In this case the effect of thermostabilization was not as high as for *. lateralis* luciferase. At 37°C the single mutants retained 10-15% of activity after 5 hours, whereas the wild type luciferase was completely inactivated. The double mutant combined the thermostability gains of the single mutants and retained greater than 50% activity for over 5 h. At 42°C the half life of the double mutant was reduced to 20 minutes. At 50°C it was only 4 min (Price et al., 1996). Other point mutations have been identified (largely by random mutagenesis) that significantly increase the thermostability of the *P.pyralis* luciferase: T214A, I232A and F295L. Combining these point mutations with the E354K mutation into the *P.pyralis* gene resulted in mutant luciferase (rLucx4ts) that had an increase in thermostability of about 7°C relative to the wild-type enzyme. Hence, in this case the multiple point mutations led to a cumulative increase in thermostability (Tisi et al., 2002).

After the spatial structure of luciferase was published, it became possible to rationally select specific positions for mutagenesis. For example, in molecule of *P.pyralis* luciferase five bulky hydrophobic solvent-exposed residues, which are all non-conserved and do not participate in secondary-structure formation, were substituted by hydrophilic ones, in particular by charged groups. These substitutions (F16R, L37Q, V183K, I234K and F465R) led to the enzyme with greatly improved pH-tolerance and stability up to 45°C. The mutant showed neither a decrease in specific activity relative to the wild-type luciferase (Law et al., 2006). Introduction of almost all known point mutations (12 residues) enhancing the thermostability of *P. pyralis* luciferase resulted in a highly stable mutant with half-time of inactivation of 15 min at 55°C, whereas wild-type luciferase inactivates within seconds at this conditions (Tisi et al., 2007).

5. RATIONAL PROTEIN DESIGN APPROACH TO PRODUCE THE STABLE AND ACTIVE ENZYME

Mutations that are efficient in one particular luciferase do not always lead to successful results when applied to other homologous luciferases. For example, the mutation E354R increased the thermal stability of *P. pyralis* luciferase, whereas the corresponding E356R substitution did not affect *H. parvula* luciferase. The substitution A217L in *L. lateralis*, *L. cruciata* and in *P. pyralis* (A215L) firefly luciferases produced fully active and thermostable mutants, but in the case of *H. parvula luciferase* this mutation decreased activity to about 0.1% of the wild type in spite of some increase in thermal stability (Kitayama, et al. 2003). These results are of particular interest for the *L. mingrelica* luciferase because it shares 98% homology with *H. parvula*. Hence, both enzymes are considered to be almost identical, and the similar effect of this mutation could be expected for *L. mingrelica* luciferase.

A rational protein design approach was used to increase thermal stability of *L. mingrelica* luciferase and prevent the detrimental effect of the of the A217L mutation on its activity by combining the mutation A217L with additional substitutions in its vicinity. The three-dimensional structure of the firefly luciferase and the multiple sequence alignment of beetle luciferases were analyzed to identify these additional substitutions (Koksharov & Ugarova, 2011a). Comparison of the A217 environment in *L. mingrelica* luciferase with that of *L. cruciata* and *L. lateralis* luciferases showed only 3 significant differences: G216N, I212L, S398M. Another difference was the change I212L, but it is unlikely to be important because the properties of Leu and Ile are very close. On the other hand, the neighboring residue G216 and the more remote S398 are characteristic for the small subgroup of luciferases very close in homology to *L. mingrelica* luciferase (including *H. parvula* luciferase). We surmised that the elimination of these differences between two groups of luciferases would lead to the A217 environment similar to that of *L. cruciata* and *L. lateralis* luciferases, which could possibly prevent the loss of activity accompanying the substitution A217L. First, we assumed that that changing the neighboring residue G216 would be sufficient to retain the enzyme activity/ Therefore, the double mutant G216N/A217L was constructed. Since this double mutant still showed low activity, we introduced the additional substitution S398M of the less close residue. This led to a stable and active mutant of *L. mingrelica* luciferase (Table 1).

Table 1. Thermal stability of luciferases with substitution of the residue 217 in a 0.05 M Na- phosphate buffer, containing 0.4 M (NH4)2SO4, 2 mM EDTA, 0.2 mg/ml BSA, pH 7.8

Enzyme	Mutant	Relative specific activity%	Temperature of inactivation	Half-life, min	Reference
Luciola cruciata luciferase	wild-type	100	50 °C	~ 4	Kajiyama& Nakano, 1993
	T217I	130		~ 28	
Luciola lateralis luciferase	wild-type		50 °C	~ 6	Kajiyama & Nakano, 1994
	A217L			~ 125	
Hotaria parvula luciferase	wild-type	100	45°C	~ 18	Kitayama et al., 2003
	A217L	0.074		~ 60	
Luciola mingrelica luciferase	wild-type	100	45°C	13 ± 1	Koksharov & Ugarova, 2011a
	G216N/A217L	10		280 ± 28	
	S398M	106		16.1 ± 1.6	
	G216N/A217L/S398M	60		276 ± 28	

The residues 216, 217, 398 are located near one of the walls of the luciferin-binding channel (Fig. 4). In the majority of beetle luciferases position 216 is normally occupied with a residue having a side group but in *L. mingrelica* and *H. parvula* luciferases it is occupied with Gly. Glycine is known to be a very destabilizing residue when in internal position of α-helices because of the absence of side group and excessive conformational freedom (Fersht & Serrano, 1993).

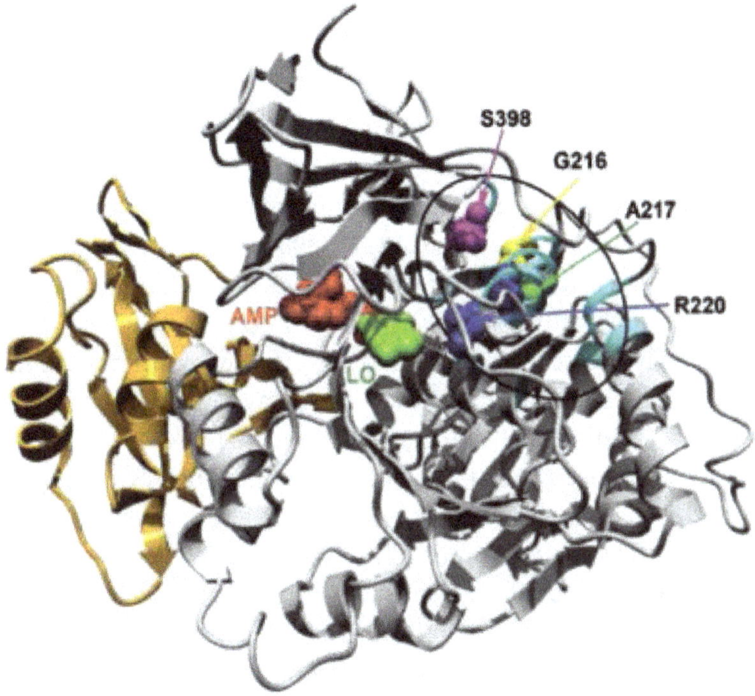

Fig. 4. Structure of *L. mingrelica* **luciferase in complex with oxyluciferin (LO) and AMP. The residues G216, A217, R220 and S398 are indicated by arrows. 7 Å microenvironment of A217 is indicated by ellipse (Koksharov & Ugarova, 2011a). The large N-terminal and the smaller C-terminal domains are depicted in grey and orange, respectively**

Since the G216 is located in the α-helix (Fig. 4) it can be suggested that it makes the surrounding structure less stable and more sensitive to the substitutions of the neighboringresidues. This can explain the unusual decrease in activity in case of the A217L mutation in *Hotaria parvula* luciferase (Kitayama, et al. 2003). The double mutation G216N/A217L resulted in the significant increase of the thermal stability of *L. mingrelica* luciferase, but this mutant retained only 10% of the wild-type activity. The comparison of the environment of residue 217 in the crystal structure of *L. cruciata luciferase* (Nakatsu, et al., 2006) with the homology model of *L. mingrelica* luciferase (Koksharov & Ugarova, 2008) (Fig. 4) shows that internal cavities probably exist in *L. mingrelica* luciferase near the 216 and 398 positions because of the smaller size side groups of the residues in this positions compared to *L. cruciata luciferase*. Additional cavity in the vicinity of S398 could potentially decrease the local conformational stability, make it more flexible and sensitive to the mutations and the changes in the environment. This hypothesis is supported by the higher resistance of the bioluminescence spectrum of the S398M mutant to pH and temperature, which indicates more rigid and stable microenvironment (Ugarova & Brovko, 2002).

The lowered local conformational stability in the vicinity of G216 and S398 residues can explain why the A217L mutation in *H. parvula* and *L. mingrelica* luciferaess leads to the decline in activity and red shift of λ_{max} that were not observed in the cases of *L. cruciata*, *L. lateralis*, *P. pyralis* luciferases containing Asn or Thr at the position 216 and Met at the position 398. In the former case the enzymes are much more likely to loose the conformation optimal for the activity as a result of residue substitutions. As can be seen the G216, A217, S398 residues are located in one plane with the neighboring residue R220 (Fig. 5). The residue R220 (the residue R218 in *P.pyralis* luciferase) is highly conservative and necessary for the green emission of firefly luciferases. Its substitutions led to the red bioluminescence, 3-15-fold decrease in activity, extended luminescence decay times and dramatic increase in *Km* values (Branchini et al., 2001). The G216N/A217L double substitution in *L. mingrelica* luciferase caused the similar type of effects but of less extent. Thus, in *L. mingrelica* and *H. parvula* luciferases the proper alignment of the R220 residue can be affected by the substitution of A217L and lead to the observed detrimental effects. Placing Asn and Met at positions 216 and 398 respectively (as in the triple mutant G216N/A217L/S398M of *L. mingrelica* luciferase and in native *L. cruciata*, *L. lateralis* luciferases) makes local microenvironment of A217 sufficiently rigid to retain active conformation in the case of the A217L mutation.

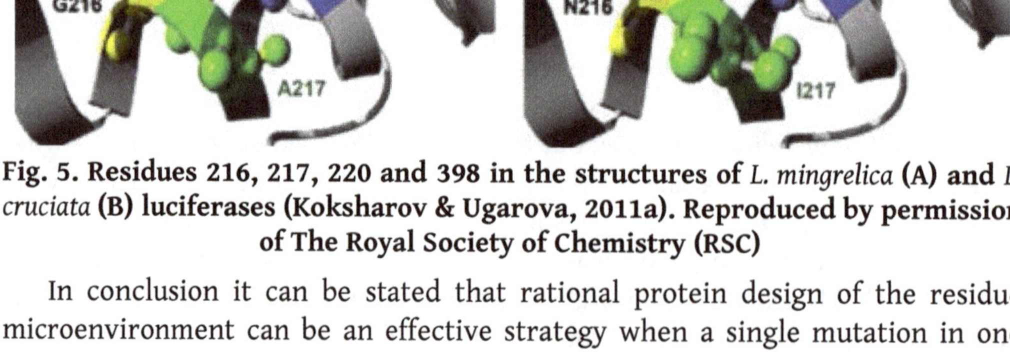

Fig. 5. Residues 216, 217, 220 and 398 in the structures of *L. mingrelica* **(A) and** *L. cruciata* **(B) luciferases (Koksharov & Ugarova, 2011a). Reproduced by permission of The Royal Society of Chemistry (RSC)**

In conclusion it can be stated that rational protein design of the residue microenvironment can be an effective strategy when a single mutation in one firefly luciferase does not lead to the desirable effect reported for the mutation of the homologous residue in the another firefly luciferase. The constructed triple mutant G216N/A217L/S398M showed significantly improved thermal stability, high activity and bioluminescence spectrum close to that of the wild-type enzyme. The improved characteristics of this mutant make it a promising tool for *in vitro* and *in vivo* applications.

6. SITE-DIRECTED MUTAGENESIS OF CYSTEINE RESIDUES OF *LUCIOLA MINGRELICA* FIREFLY LUCIFERASE

The number of Cys residues of luciferases is highly varied (from 4 to 13 residues) depending on the firefly species. Luciferases contain three absolutely conservative SH groups that do not belong to the active site. However their mutagenesis was shown to affect activity and stability of luciferases (Dement'eva et al., 1996; Kumita et al., 2000). For example, the mutant *Photinus pyralis* luciferase in which all the four Cys residues were substituted with Ser, retained only 6.5 % of activity, whereas mutants with single substitutions lost 20-60% of activity (Kumita et al., 2000; Ohmiya & Tsuji, 1997).

The *Luciola mingrelica* firefly luciferase contains eight cysteine residues, three of which correspond to the conservative cysteine residues of *P. pyralis* firefly luciferase - 82, 260, and 393. Mutant forms of *L. mingrelica* C than that of the wild-type enzyme. The substitutions C82A, C260A did not affect the thermal stability of luciferase. The pLR plasmid, encoding firefly luciferase with the structure identical to that of the native enzyme, was previously used for the preparation of the mutant forms of the enzyme with single substitutions of the non-conserved cysteine residues C62S, C146S (Lomakina et al., 2008) and C164S (Modestova et al., 2010). These substitutions also had no significant effect on the catalytic and spectral properties of the luciferase, but they resulted in an increase of the enzyme thermal stability and in a decrease of the dependence of inactivation rate constant on the enzyme concentration (unlike the wild-type enzyme). Moreover, the DTT influence on luciferase stability was diminished. These effects were most pronounced for the enzyme with the substitution C146S.□luciferase containing single substitutions of these cysteine residues to alanine were obtained previously (Dement'eva et al., 1996). These substitutions had no effect on bioluminescent and fluorescent spectra of the enzyme and on enzyme activity. The stability of the C393A mutant was 2-fold higher at 5-35 The purification of recombinant luciferase obtained using the plasmid pLR is a complicated multistage process. Therefore, the recombinant *L. mingrelica* luciferase with C-terminal His6- tag was used for mutagenesis of cysteine residues (Modestova et al., 2011). The wild-type enzyme and its mutant forms were expressed in *E. coli BL21(DE3) cells* transformed with the pETL7 plasmid (Koksharov & Ugarova, 2011a). This approach led to the simpler scheme of the luciferase purification and to the increase of the enzyme yield due to the use of the highly efficient pET expression system. The influence of polyhistidine tag on luciferase properties was not previously analyzed in detail according to the literature. A number of publications indicate that while his-tags often don't affect enzyme function, in many cases the biological or physicochemical properties of the histidine tagged proteins are altered compared to their native counterparts (Amor-Mahjoub et al., 2006; Carson et al., 2007; Efremenko et al., 2008; Freydank et al., 2008; Klose et al., 2004; Kuo & Chase, 2011). The goal of this study was to elucidate the role of non-conserved cysteine residues in the *L. mingrelica* firefly luciferase, to study the mutual influence of these residues and the effect of His6-tag on the activity and thermal stability of luciferase (Modestova et al., 2011).

6.1 Analysis of the fragments of luciferase amino acid sequences containing cysteine residues

Among the firefly luciferases those amino acid sequences are known, firefly luciferases from *Luciola* and *Hotaria* genera, and the *Lampyroidea maculata* firefly luciferase form a separate group with more than 80% amino acid identity (Fig. 6). The second group includes luciferases from firelies of various genera: *Nyctophila, Lampyris, Photinus, Pyrocoelia,* etc. The sequence identity of luciferases from the first and the second group does not exceed 70%.

Amino acid sequences of the firefly luciferases belonging to these groups vary significantly. One of the most evident distinctions is the amount and location of cysteine residues. The residue C82 is absolutely conserved in all beetle luciferases, and the residue C260 is absolutely conserved in all firefly luciferases. The residue C393 is conserved in all beetle luciferases except the *Cratomorphus distinctus* (Genbank AAV32457) and one (Genbank U31240) of the *P. pennsylvanica* luciferases. The C62, 86, and 284 residues are also absolutely conserved in all luciferases from the first group. The residue C146 is conserved in all luciferases of the first group, except for the *L. lateralis* and *L. cruciata* luciferases, in which alanine and tyrosine are located at the position 146. The residue C164 is conserved in luciferases of the first group except for the *L. lateralis* luciferase, which contains S146. The C86 residue is located in a highly conserved region of luciferases of the first group, near the C82 residue, which in its turn is located not far from the active site of the enzyme. Besides, the C86 residue is located near the surface of the protein, and the surface area of its side chain, that is accessible to the solvent, is about 11 Å2. The residue C146 is of particular interest because of its surface location. Its side chain is exposed to the solvent with the accessible surface area as high as 48 Å2. As a whole the *Luciola* luciferases possess high amino acid sequence identity. However, there are several small areas in their amino acid sequences the composition of which varies significantly. It is in these areas that the residues C62 and C164 are located.

These residues are positioned in two α-helixes and are in close proximity with each other (Fig. 7).

Origin	C62	C82, C86	C146	C164	C260	C284	C393
First group of luciferases							
Luciola mingrelica	FDITCRLAEAM	IALCSENCEEFF	VQKTVTCIKKIVI	NFGGHDCMETFI	LGYFACGYRVVML	TLQDYKCTSVILV	RRGEICVKGPS
Luciola cruciata	LEKSCCLGKAL	IALCSENCEEFF	VQKTVTTIKTIVI	DYRGYQCLDTFI	LGYLICGFRVVML	TLQDYKCTSVILV	RRGEVCVKGPM
Hotaria parvula	FDITCRLAEAM	IALCSENCEEFF	VQKTVTCIKTIVI	NFGGHDCMETFI	LGYFACGYRVVML	TLQDYKCTSVILV	RRGEICVKGPS
Hotaria unmunsana	FDITCRLAEAM	IALCSENCEEFF	VQKTVTCIKTIVI	NFGGYDCMETFI	LGYFACGYRVVML	TMQDYKCTSVILV	RRGEICVKGPS
Hotaria tsushimana	FDITCHLAEAM	IALCSENCEEFF	VQKTVTCIKTIVI	NFGGYDCMETFI	LGYFACGYRVVML	TMQDYKCTSVILV	RRGEICVKGPS
Luciola italica	FDITCRLAEAM	IALCSENCEEFF	VQKTVTCIKTIVI	NFGGYDCVETFI	LGYFACGYRIVML	TLQDYKCTSVILV	RRGEICVKGPS
Lampyroidea maculata	FDISCRLAEAM	IALCSENCEEFF	VQKTVTCIKTIVI	NFGGYDCVETFI	LGYFACGYRIVML	TMQDYKCTSVILV	RRGEICVKGPS
Luciola lateralis	LEKSCCLGEAL	IALCSENCEEFF	VQKTVTAIKTIVI	DYRGYQSMDNFI	LGYLTCGFRIVML	TLQDYKCSSVILV	RRGEVCVKGPM
Luciola terminalis	LDVSCRLAQAM	IALCSENCEEFF	VQKTVTCIKTIVI	DYQGYDCLETFI	LGYLICGFRIVML	TLADYKCNSAILV	RRGEICVKGPM
Second group of luciferases (illustrated by *Photinus pyralis* luciferase)							
Photinus pyralis	FEMSVRLAEAM	IVVCSENSLQFF	VQKKLPIIQKIII	DYQGFQSMYTFV	LGYLICGFRVVLM	SLQDYKIQSALLV	QRGELCVRGPM

Fig. 6. Fragments of amino acid sequence alignment of various firefly luciferases (the regions containing Cys residues). The numbering corresponds to that of *Luciola mingrelica* luciferase

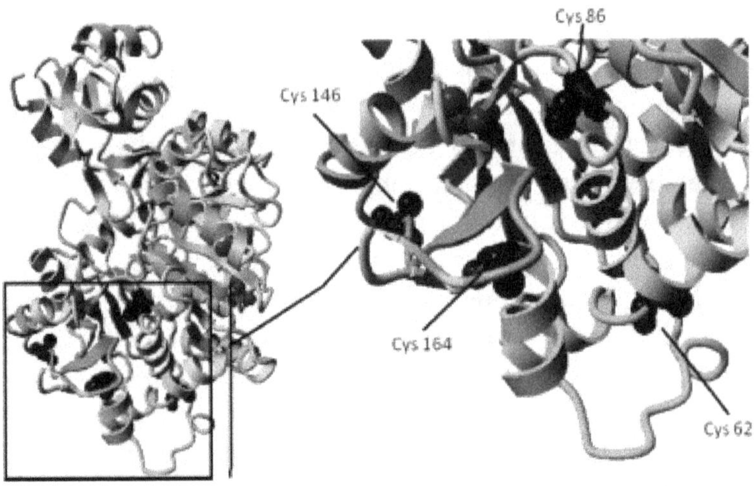

Fig. 7. Fragment of the 3D structure of *Luciola mingrelica* firefly luciferase containing the residues C62 and C164

The cysteine residues 62, 86, 146, and 164 of *L. mingrelica* luciferase were chosen for the site- specific mutagenesis. In terms of the molecule topology the most suitable substitutions of the Cys are Ser (hydrophilic amino acid) and Val (hydrophobic amino acid). The side chain sizes of these residues are similar to that of Cys. We considered Ser as the most suitable substitution for C86 and C146 residues because the side chains of these residues are in contact with aqueous solution. The residue C164 was also substituted by Ser because its microenvironment is weakly hydrophilic. Moreover, our previously results (Modestova et al., 2010) suggest that in certain conditions this residue becomes available to the solvent. In case of the residue Cys62 two mutants were obtained: C62S and C62V.

6.2 Preparation and physicochemical properties of mutant luciferases

The recombinant *L. mingrelica* firefly luciferase encoded by the plasmid pETL7 (GenBank No. HQ007050) (Koksharov & Ugarova, 2011a) served as the parent enzyme (wild-type). This form contains 4 additional amino acid residues (MASK) on N-terminus as compared to the native sequence of *L. mingrelica* firefly luciferase (GeneBank No. S61961). The sequence AKM at its C-terminus is replaced by the sequence SGPVEHHHHHH. A number of mutants were obtained by site-directed mutagenesis of the plasmid pETL7: the mutant luciferases with the single substitutions C62S, C62V, C86S, C146S, C164S, double substitutions C62/146S, C62/164S, C86/146S, and C146/164S; the triple substitution C62/146/164S. The wild-type luciferase and its mutant forms were purified using metal chelate chromatography. The expression level and the specific activity of wild-type and its mutants C62S, C62V, C164S, C62/146S, and C146S/C164S were the same within an experimental error. Specific activity of the mutant C146S was ~15% higher than that of the wild-type, while its expression level was unaltered. Meanwhile, the substitution C86S resulted in the decrease of the enzyme expression level (62% compared to wild-type) and its specific activity (30% compared to wild-type). The properties of the firefly luciferase with the double substitution C86S/146S were similar to those of the mutant C86S. Drastic decrease of the expression level and of the enzyme specific activity was observed at the introduction of the double mutation C62S/C164S and the triple mutation C62S/C146S/C164S. Bioluminescence and intrinsic fluorescence spectra of the wild-type luciferase and its mutant forms were identical. Single mutations had almost no effect on the *K*m values for both substrates (*K*m and *K*m 2) with the exception of the mutant C86S, for which, as well as for the mutant LH C86S/C146S, 1.5-fold increase of both parameters was observed. The simultaneous substitution of the residues C62S and C164S in both double and triple mutants led to 30% increase of K_m^{ATP}, but didn't affect K_m^{LH2}.

The irreversible inactivation of the wild-type luciferase and its mutant forms was measured in 0.05 M Tris-acetate buffer (2 mM EDTA, 10 mM MgSO4, pH 7.8) at 37° and 42°C at concentration range of 0.01-1.0 μM. The inactivation of the wild-type luciferase and its mutant forms followed the monoexponential first-order kinetics at all enzyme concentrations assayed. The *k*in values of the wild-type luciferase and its mutant forms did not depend on the initial luciferase concentration. The enzyme stabilization was only observed for the mutant C146S: the *k*in C and by 30% - at 42°C (Table 2). At 37°C the k_{in} value decreased 2-fold at 37 *k*in values of the mutants C62V, C164S and C146S/C164S were similar to the *k*in of the wild-type luciferase, but at 42°C the *k*in values of these mutants were higher than that of the wild-type enzyme. All other mutants were less stable than the wild-type enzyme. The substitution C86S caused a significant destabilizing effect on the enzyme: the *k*in value increased twofold both at 37° and 42°C. The double mutant C62S/C164S and the triple mutant C62S/C146S/C164S were the least stable among the mutants obtained.

Table 2. Rate constants of irreversible inactivation of wild-type luciferase and its mutant forms with single and multiple substitutions of the 62, 86, 146, 164 cysteine residues at 37 and 42°C

Enzyme	*k*in, min-1	
	37°	42°
wild-type	0,022 ± 0,004	0,074 ± 0,006
C62V	0,024 ± 0,004	0,135 ± 0,004
C62S	0,036 ± 0,004	0,127 ± 0,004
C86S	0,040 ± 0,002	0,160± 0,006
C146S	0,011 ± 0,002	0,058 ± 0,003
C164S	0,018 ± 0,003	0,108 ± 0,005
C62S/C146S	0,042 ± 0,005	0,108 ± 0,005
C62S/C164S	0,052 ± 0,003	0,153 ± 0,005
C86S/C146S	0,047 ± 0,004	0,120 ± 0,006
C146S/C164S	0,023 ± 0,006	0,086 ± 0,005
C62S/C146S/C164S	0,055 ± 0,005	0,142 ± 0,006

6.3 The effect of polyhistidine tag on the properties of firefly luciferase

Comparison of the physicochemical properties of luciferases with single substitutions of the residues C62S, C146S and C164S that were obtained for *L. mingrelica* luciferase without His6- tag (Lomakina et al., 2008) with that of the mutant enzymes containing C-terminal His6-tag (Modestova et al., 2011) led to a conclusion that the His6-tag shows significant influence on the luciferase properties. Introduction of the His6-tag into the luciferase structure leads to the increase of the Km^{ATP} and Km^{LH2} values. The interaction of the enzyme with the substrates is known to involve the rotation of a big N-domain and a small C-domain of the luciferase against each other at almost 90° (Sandalova & Ugarova, 1999). This movement is necessary for the participation of the residue K531 from C-domain in the formation of enzyme-ATP- luciferin active complex. The presence of the flexible His6-tag on the C-terminus of the protein molecule might somewhat impede the process of domains rotation, that may result in a slight increase of Km values for the both substrates.

Thermal inactivation of the firefly luciferase without His6-tag is a two-step process, which includes a fast and a slow inactivation stages. The *k*in values of both stages are dependent on the enzyme concentration, which is known to be a characteristic feature of oligomeric enzymes. The single mutations C62S, C146S, C164S result in stabilization of the enzyme at the slow stage of inactivation and in a decrease of *k*in dependence on the enzyme concentration (Lomakina et al., 2008). The thermal inactivation of the His6-tag containing wild-type luciferase and its mutants is a one-step process. The *k*in values of these enzymes do not depend on luciferase concentration and coincide with the *k*in values of the respective mutants without His6-tag that were measured at the increased enzyme concentration (1 μM). This influence of the His6-tag on the inactivation kinetics of the wild-type luciferase and its mutants may be due to the fact that the presence of the His6-tag considerably alters the process of luciferase oligomerization.

6.4 Effect of the cysteine substitutions on luciferase structure and thermal stability

The substitution C146S results in a 2-fold stabilization of the enzyme at 37°C and in a 30% increase of the enzyme stability at 42°C. This effect is associated with the surface location of the side chain of this residue, its large solvent accessible area and the lack of interactions with other amino acid residues of the enzyme. The C164S substitution doesn't alter the enzyme stability at 37°C, but leads to some destabilization at 42°C, though this destabilization is less than that caused by the substitutions C62V, C62S and C86S. This effect is, on the one hand, due to the fact, that the C164 residue is located in an area, which is distant from the enzyme active site. On the other hand, the raise of temperature causes the increase of solvent accessibility and the replacement of cysteine residue by the hydrophilic serine improves interactions with the solvent.

Analysis of the luciferase 3D-model shows that it is hard to unambiguously estimate the properties of the C62 residue microenvironment. This residue contacts with both hydrophilic and hydrophobic amino acids. Therefore, two enzymes were obtained that carry a hydrophilic and a hydrophobic side chain in the position 62. The specific activity, the expression level and the kinetic parameters of the mutants C62S and C62V were similar to those of the wild-type enzyme. The *k*in values at 42°C were also similar, but the mutant C62V turned out to be 2-fold more stable than the mutant C62S at 37°C. Therefore, the hydrophobic valine residue is more advantageous at 37°C in terms of the enzyme stability. However, at temperature of 42°C the role of the amino acid residue microenvironment in the enzyme stabilization becomes less pronounced and both modifications – serine or valine – result in destabilization of the protein globule.

The substitution C86S shows the most significant influence on the luciferase properties. It results in a decrease of the luciferase expression level and the specific activity, a deterioration of the Km values for both substrates, and a decrease of the enzyme thermal stability. The C86 residue is located within an unstructured area of the amino acid chain of the enzyme (Fig. 8). The amino acid sequence forms a loop in this area due to the formation of a hydrogen bond between the SH-group of the residue C86 and the oxygen atom OE1 belonging to the residue E88. The SH-group of cysteine residue is known to have a tendency to form non-linear hydrogen bonds due to fact that the deformation of the valence angle has a relatively small energy cost (Raso et al., 2001). The OH-group of serine residues has no such tendency. Thereby it may be possible that the hydrogen bond between S86 and E88 residues can't be formed in the mutant C86S. This may lead to an increase in mobility of the chain fragment containing the abovementioned residues.

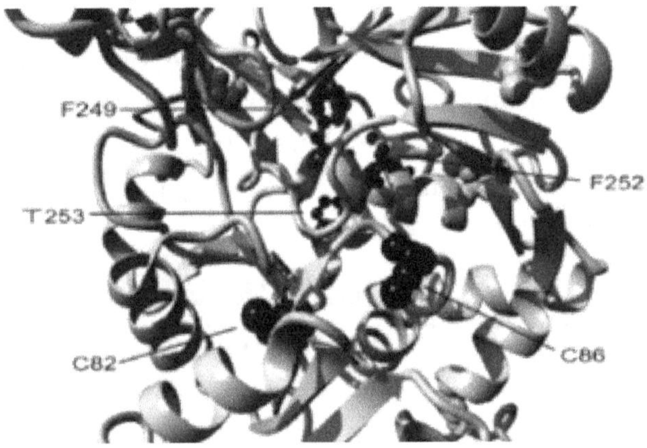

Fig. 8. Fragment of the 3D structure of *Luciola mingrelica* firefly luciferase containing C82 and C86 residues (Modestova et al., 2011)

It is important to underline that the C86 residue is located in an absolutely conserved area of luciferases *Luciola* genus, not far from the enzyme active site and at the distance of ~15 Å from T253, F249, F252 residues. These residues participate in the process of luciferase substrates binding, and it is known that their mutations lead to a drastic alteration of the enzyme catalytic properties and, in certain cases, to the disturbance of the enzyme expression process (Freydank et al., 2008). On the basis of the experimental data one can conclude that disturbance stripping-down of the protein structure (the "untwisting" of the helix) in the area of the localization of the residue C86 disrupts the native structure of the firefly luciferase active site area and leads to the deterioration of the luciferase activity and stability.

Analysis of the properties of the mutants with multiple amino acid substitutions indicates that in most of the cases the effect of such substitutions is additive. For instance, the C86S/C146S mutant possesses the properties of the luciferase with single C86S substitution, because it is the C86S substitution that affects the enzyme properties most significantly. The mutants C62S/C146S and C146S/C164S also possess the characteristic properties of the respective mutants with single replacements. However, the combination C62S/C164S leads to the drastic decrease of the enzyme expression level, to the lowering of its specific activity and stability and to the increase of the Km^{ATP} in comparison with the enzymes with the single substitutions C62S and C164S. These facts indicate that the effect of these substitutions is nonadditive. The analysis of luciferase 3D structure shows that C62 and C164 residues belong to two closely located α-helixes (Fig. 8). The single mutations of these residues have no significant effect on the enzyme properties, which is probably due to the enzyme ability to compensate the effects of these substitutions. Meanwhile, the double substitutions affect the mutual disposition of two α-helixes, in which these residues are located.

Thus, the role of each cysteine residue in luciferase molecule is different and is determined by its location relative to the active site, its microenvironment and even the oligomerization state of luciferase. For example, in some cases the introduction of Cys residues into internal protein core can increase the luciferase stability after replacement of hydrophilic residue by more hydrophobic Cys. Such examples will be shown below.

7. INCREASE OF *P. PYRALIS* LUCIFERASE THERMOSTABILITY BY INTRODUCTION OF DISULFIDE BRIDGES

It was mentioned above that luciferases are peroxisomal enzymes. They do not form structural disulfide bonds despite of containing SH-groups (Ohmiya & Tsuji, 1997). When expressed in *E. coli*, firefly luciferases cannot form any disulfide bonds due to the reducing environment of the cytoplasm. On the other hand, introduction of disulfide bridges was found to be one of the most efficient strategies for increasing protein stability (Eijsink et al., 2004). Recently, disulfide bridges were introduced into *P. pyralis* firefly luciferase (Imani et al., 2010) by site-directed mutagenesis. Two different mutant proteins were made with a single bridge. *P.pyralis* firefly luciferase contains four cysteine residues at the positions 81, 216, 258 and 391. To find the residues capable to form disulfide bridges after their mutation to cysteine, the crystal structure of *P. pyralis* luciferase was uploaded to the NCBS integrated Web Server. The results from server showed that there are 150 pairs that could potentially be selected for disulfide bridge formation. But only two pairs of residues were chosen due to their similar size to the Cys residues: A103 and S121, located distant from active site region of the enzyme, and A296 and A326, situated in the vicinity of the active site region. The ability of mutated sites to form disulfide bridges was analyzed in Swiss-PDB Viewer.

Two mutant luciferases, each containing one S-S bridge, were obtained: A103C/S121C and A296C/A326C. Relative specific activity showed a 7.25-fold increase for the mutant A296C/A326C whereas the mutant A103C/S121C showed only 80% of wild-type specific activity. Both mutants were more stable then the wild-type enzyme. For example, after incubation at 40°C for 5 min the mutants A296C/A326C and A103C/S121C retained ~88% and 22% of activity respectively, whereas the wild-type enzyme lost nearly all of its activity. Using circular dichroism spectropolarimetric and fluorescence spectroscopic analysis, the conformational changes of the enzyme structure were revealed, showing the more fixed structure of aromatic residues, more compactness of tertiary structure, and a remarkable increase in α-helix content.

It can be concluded that disulfide bridge formation in mutant A296C/A326C did not have a destabilizing effect on the enzyme and caused a remarkable change in both secondary and tertiary structure that is reflected in active site structure. These changes endow the enzyme with properties that show an increased resistance to pH and temperature without any stabilizer. On the other hand, the thermal stability of the mutant A103C/S121C arises from the change of tertiary structure. Finally, these results showed that the engineered disulfide bridge not only did not destabilize the enzyme but also in one mutant it improved the specific activity and led to pH-insensitivity of the enzyme (Imani et al., 2010).

8. THERMOSTABILIZATION OF THE *LUCIOLA MINGRELICA* FIREFLY LUCIFERASE BY *IN VIVO* DIRECTED EVOLUTION

Firefly luciferase can be simply screened for its *in vivo* bioluminescence activity (Wood & city, and activity in non- natural environments (Jäckel et al., 2008; Turner, 2009). The critical part of a directed DeLuca, 1987). This makes a directed evolution approach the most promising for optimization of different luciferase properties including thermostability. This strategy was shown to successful improve of a wide range of properties for different enzymes, for example, thermal stability, enantioselectivity, substrate specievolution experiment is the availability of a sensitive and efficient screening procedure. Otherwise identifying the desired mutants within large libraries can become very laborious and costly. However, there is only one example known when directed evolution was used for enhancing the thermostability of firefly luciferase. Wood & Hall obtained the exceptionally stable mutant of *Photuris pennsylvanica* luciferase by this approach. This mutant still remains the most stable firefly luciferase to date. In this case a sophisticated automatic robotic system was implemented to screen mutant libraries. It limits the possibility of wide application of this technique. However, that system was able to screen more than 10000 mutants per cycle with a precise measurement of *in vitro* properties of the mutants generated such as activity and Km. The developed ultra-stable mutant contained 28 substitutions and demonstrated a half-life of about 27 h at 65°C (Wood & Hall, 1999). The more simple, but efficient screening strategy was successfully used here to evolve a thermostable form of *L. mingrelica* luciferase (Koksharov & Ugarova, 2011b).

8.1 Directed evolution of luciferase

Wild-type *L. mingrelica* luciferase displays rather low thermostability with a half-life of 50 minutes at 37°C. So, the consecutive rounds of random mutagenesis and screening were used to considerably improve thermostability of *L. mingrelica* luciferase without compromising its activity. The fact that E. *coli* cells withstand temperatures up to about 55°C (Jiang *et al.*, 2003) and the availability of *in vivo* bioluminescence assay, allowed to identify thermostable mutants by a simple non-lethal *in vivo* screening of *E. coli* colonies that contained mutant luciferases. The incubation of *E. coli* colonies at elevated temperatures resulted in the inactivation of less stable luciferase mutants. Therefore, thermostable mutants displayed higher residual bioluminescence activity and could be efficiently detected by a simple photographic registration of *in vivo* bioluminescence of colonies. *E. coli* cells remained viable after the subjection to elevated temperatures and the subsequent detection of *in vivo* bioluminescence. Therefore, there was no need in using replica plates, which simplified the procedure. Each round of screening could be carried out in a simple and rapid manner (Koksharov & Ugarova, 2010, 2011b).

The plasmid pLR3 (GenBank No. HQ007051) (Koksharov & Ugarova, 2008), which contains *L. mingrelica* luciferase gene, was used in random mutagenesis performed by error-prone PCR. A mutation rate of about 1 amino acid change (2-3 base changes) per the region mutated is reported to be most desirable for an efficient selection of improved mutant (Cirino et al., 2003). It generally gives 30-40% of active clones in the library (Cirino et al., 2003), so this frequency was targeted in our work. Mutagenesis was applied to a 785 bp region of the luciferase gene, which corresponds to amino acid residues 130-390 out of 548 residues of *L. mingrelica* luciferase. This region was chosen because of the convenient restriction sites available (XhoI and BglII) and because most reported mutants, that increase the thermostability of firefly luciferases, are located in this region. The results indicate that the screening of 1000 colonies typically gives a couple of different thermostable mutants. Up to 2000-3000 mutant colonies could be conveniently screened on a single 90 mm Petri dish. The mutant S118C was used as a parent enzyme for directed evolution because it demonstrated slightly higher thermostability compared with the wild-type enzyme (Koksharov & Ugarova, 2008). The most thermostable mutant identified in each cycle of mutagenesis was used as a starting point in the following cycle (Table 3).

Table 3. Mutants of *Luciola mingrelica* firefly luciferase obtained during four cycles of directed evolution

Cycle	Parent enzyme	Number of clones screened	Active clones ratio, %	Incubation temperature before screening	Mutant enzyme*)	Substitutions compared with the parent enzyme
1.	S118C	800	53%	37°C	**1T1**	T213S S364C
					1T2	S364A
					1T3	
2.	1T1	900	53%	50°C	**2T1**	K156R A217V
					2T2	E356V
3.	2T1	600	65%	50°C	**3T1**	C146S E356K
					3T2	
					3T3	E356V
4.	3T1	1400	65%	55°C	4TS	R211L

*) For each cycle, the mutant showing the highest stability is shown in bold and underlined. It was used as a parent for the following cycle.

At the first cycle of mutagenesis the screening of the mutant colonies was performed directly after their growth at 37°C. The wild-type *L. mingrelica* luciferase is insufficiently stable at these conditions, so the *in vivo* bioluminescence of its colonies is rather dim. Three clones were identified during screening that produced distinctly brighter colonies because of the increased thermostability (Table 3). During the second and third cycles of mutagenesis an additional incubation at 50°C for 40 min was required to detect mutants showing higher stability. Three mutants obtained at the third cycle displayed similar brightness after incubation at 50°C but increasing the incubation temperature to 55°C showed that the mutants 3T1, 3T2 are more stable than 3T3. After the fourth round of directed evolution the mutant 4TS was identified, which showed the highest *in vivo* thermostability among the mutants described in this study. It retained noticeable brightness of bioluminescence after incubation of its colonies at 55°C for 40 min while all the other mutants were completely inactivated. Moreover, the mutant 4TS displayed decreased but noticeable *in vivo* bioluminescence when its colonies were heated for 20 min at 60°C. *E. coli* cells completely lost their viability after 2 min at 60°C. Therefore, further selection of mutants with even higher stability will require the of replica plates.

8.2 Expression and purification of mutant and wild-type luciferases

The wild-type *L. mingrelica* luciferase and the mutant 4TS were expressed using the plasmid pETL7, which was described earlier. Average yields of the purified proteins (mg per 1 L of culture) were 160 mg for wild-type and 300 mg for te mutant 4TS. As a result of purification the enzymes were obtained in 20 mM Na-phosphate buffer containing 0.5 M NaCl, pH 7.5 containing 300 mM imidazole, 2 mM EDTA, 1 mM DTT. Generally the luciferases proteins remained fully active for at least 1 month in this buffer. For the long-term storage the proteins were transferred to 50 mM Tris-acetate buffer (pH 7.3) containing 100 mM Na2SO4, 2 mM EDTA and frozen at -80°C. This way they retained full activity for at least 2 years and tolerated several freeze-thaw cycles without inactivation. Despite the fact that the catalytic efficiency of the intermediate mutants was not monitored, the resultant mutant 4TS demonstrated the significant improvement of specific activity as well as *Km* for ATP.

8.3 Thermostability

Comparison of 4TS and wild-type *L. mingrelica* luciferase thermal stability at 42°C in Tris- acetate buffer TsB1 (50 mM Tris-acetate buffer containing 20 mM MgSO4, 2 mM EDTA, 0.2 mg/ml BSA, pH 7.8) showed a 65-fold the increase in the half-life of *L. mingrelica* luciferase at 42°C (from 9.1 to 592 min). Thermal inactivation of the wild-type enzyme and 4TS was also studied in Na-phosphate buffer TsB2 (50 mM Na-phosphate buffer containing 410 mM (NH4)2SO4, 2 mM EDTA, 0.2 mg/ml BSA, pH 7.8) to compare these results with other literature data (Kajiyama & Nakano, 1994; Kitayama, et al., 2003; White, et al., 1996). At all the temperatures studied the mutant 4TS was significantly more stable than the wild-type. As can be seen from the Arrhenius plot, TsB2 buffer causes significant stabilization of both the wild-type enzyme and 4TS compared with TsB1 buffer (Fig. 9)

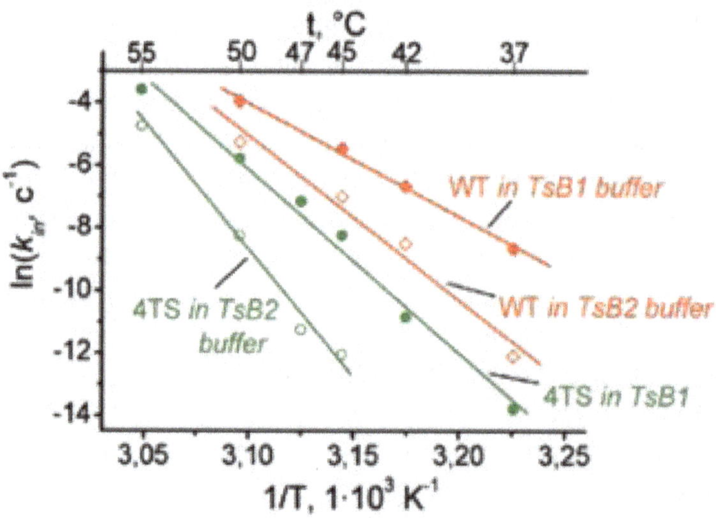

Fig. 9. Arrhenius plot showing the dependence of rates of inactivation on temperature for the wild-type luciferase (diamonds) and the mutant 4TS (circles) in buffer TsB1 (closed symbols) and TsB2 (open symbols) (Koksharov & Ugarova, 2011b). C(enzyme)=13 μg/ml

8.4 Structural analysis

The mutant 4TS contains 7 new substitutions compared with its parent form S118C: T213S, K156R, R211L, A217V, C146S, E356K, and S364C. All the substitutions are non-conservative among firefly luciferases. Judging from the order of appearance of these substitutions in the course of directed evolution (Table 3), literature data and their location in the 3D structure of the enzyme (Fig. 10), four of these substitutions were suggested to be the key mutations that cause the high stability of the mutant 4TS: R211L, A217V, E356K, and S364C. The mutations of the residues A217 (Kajiyama & Nakano, 1993) and E356 (White, et al., 1996) are known to significantly increase the thermostability of firefly luciferases according to the previous studies. The effect of the residues R211 and S364 on thermostability is identified for the first time. The increase in stability by the substitutions R211L, A217V, S364C, and S364A, can be attributed to the improvement of the internal hydrophobic packing (Fersht & Serrano, 1993).

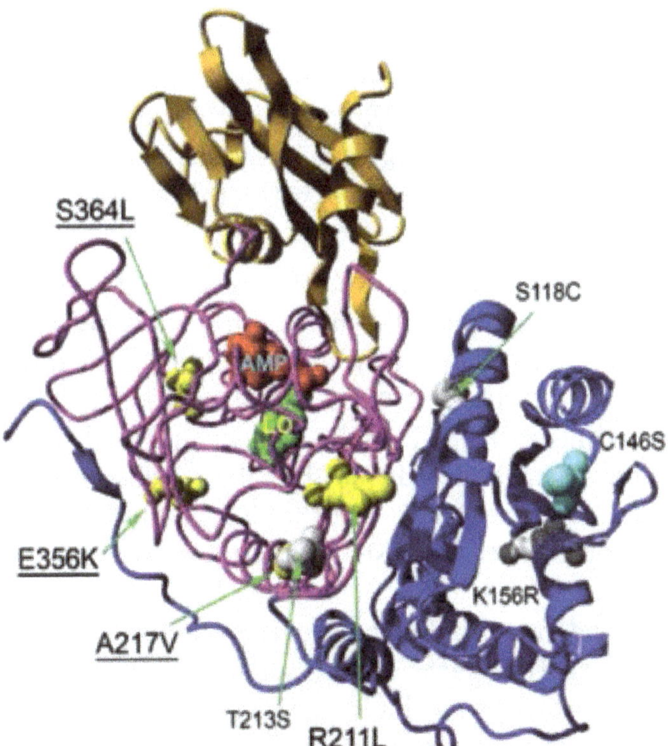

Fig. 10. Homology model of *L. mingrelica* luciferase showing the location of substitutions in the mutant 4TS. Four key thermostabilizing mutations are

underlined. LO *u* AMP – luciferyl and adenylate groups of DLSA (5'-*O*-[N-(dehydroluciferyl)-sulfamoyl] adenosine).

In the case of R211L, S364C, and S364A, the increase of hydrophobicity of the protein core is achieved by the substitution of the non-conservative buried polar residues by the hydrophobic ones. As a result of the substitution A217V the larger side group of Val fills the internal cavity, which is otherwise occupied by a water molecule (Conti et al., 1996). The surface mutation C146S is known to increase the resistance to oxidative inactivation (Lomakina *et al.*, 2008). This mutation can explain the increased storage stability of 4TS in the absence of DTT compared with wild-type. The WT luciferase loses 70% of its activity within two weeks, whereas the mutant 4TS was remained fully active within one month at the same conditions (Koksharov & Ugarova, 2011b). The mutants T213S/S364C and S364A displayed similar *in vivo* properties. There, it the substitution T213S is unlikely to affect thermostability. The substitution of the surface residue 156 from positively charged Lys to similar in properties Arg is also unlikely cause a significant effect on luciferase. The starting mutant S118C showed only small 1.5-fold increase in stability at 42°C. The mutant 4TS and its variant without the mutation S118C showed indistinguishable *in vivo* thermostability at 60°C. Thus, the contribution of S118C seems insignificant. Interestingly, Ser118 is highly Subdomains A, B and C are depicted in blue, magenta and orange, respectively conservative in firefly luciferases. The only exceptions are the similar substitution S118C in the recently cloned juvenile luciferase from *L. cruciata* (Oba et al, 2010a) and the substitution S118T in the luciferase from *Lampyroidea maculata* (Emamzadeh et al., 2006). However, in luciferases from non-firefly beetles this position is usually occupied by His or Val.

All four key thermostabilizing substitutions (R211L, A217V, E356K, and S364C) are located in the second subdomain of firefly luciferase. According to the results of Frydman and coworkers (Frydman *et al.*, 1999), the fragments of firefly luciferase comprising residues 1- 190 and 422-544 possess high intrinsic stability. These fragments mainly correspond to the subdomains A and C of firefly luciferase (Fig. 10). That study demonstrated that the middle subdomain B (192-435) was significantly less stable and that it was the first to unfold under denaturating conditions. Hence, it likely that the stability of the second subdomain is the less stable "bottleneck" that determines the stability of the firefly luciferase protein. Therefore, most of the thermostabilizing mutations would tend to be located in the second subdomain or at the interface of this subdomain and the remaining parts of the protein. It is noteworthy that almost all thermostable mutants reported in the literature are located in this part of the luciferase structure, which is consistent with this hypothesis.

9. CONCISE SUMMARY

We have proven that the in vivo directed evolution technique is a straightforward and effective approach to enhance the thermal stability of firefly luciferase. This method enables the production of extremely stable mutants without compromising catalytic effectiveness. The ultimate mutant developed in this study exhibited enhanced catalytic characteristics, including higher specific activity, decreased Km for ATP, and an increased temperature optimum. Beetle luciferases are commonly employed in usual applications such as ATP-related assays or reporter genes, where they are utilized at either room temperature or 37°C. The mutant 4TS maintains 70% of its activity after being incubated at 37°C for two days. Hence, its stability is adequate for the majority of typical in vivo and in vitro uses. The mutant 4TS is an excellent tool for determining ATP due to its high specific activity, catalytic efficiency, and enhanced protein production (Ugarova et al., 2010). The enhanced temperature tolerance of this mutant can be advantageous for in vivo imaging and in high temperature applications. The newly found locations in this work can effectively be utilized for stabilizing other firefly luciferases, particularly those belonging to the Luciola and Hotaria genus. The described in vivo screening approach, which is non-lethal, has the potential to be used to different luciferases, both beetle and non-beetle, where there is a need for the production of thermostable variants of the enzyme.

10. REFERENCES

Amor-Mahjoub, M.; Suppini, J.; Gomez-Vrielyunck, N. & Ladjimi, M. (2006). The effect of the hexahistidine-tag in the oligomerization of HSC70 constructs. *Journal of Chromatography. B. Analyt. Technol. Biomed. Life Sci.*, Vol.844, No.2, (December 2006), pp. 328–334, ISSN 1570-0232

Arnoldi, F.; Neto, A. & Viviani, V. (2010). Molecular insights on the evolution of the lateral and head lantern luciferases and bioluminescence colors in *Mastinocerini* railroad- worms (Coleoptera: Phengodidae). *Photochemical and Photobiological Sciences*, Vol.9, No.1, (December 2009), pp. 87-92, ISSN 1474-905X

Binkowski, B.; Fan, F. & Wood, K. (2009). Engineered luciferases for molecular sensing in living cells. *Curr. Opin. Biotechnol.*, Vol.20, No.1, (February 2009), pp. 14-18, ISSN 0958-1669

Branchini, B.; Magyar, R.; Murtiashow, M.; Anderson, S. & Zimmer, M. (1998). Site-directed mutagenesis of histidine 245 in firefly luciferase: a proposed model of the active site. *Biochemistry,* Vol.37, No.44, (November 1998), pp. 15311-15319, ISSN 0006-2960

Branchini, B.; Magyar, R.; Murtiashaw, M.; Anderson, S.; Helgerson, L. & Zimmer, M. (1999). Site-directed mutagenesis of firefly luciferase active site amino acids: a proposed model for bioluminescence color. *Biochemistry,* Vol.38, No.40, (October 1999), pp. 13223-13230, ISSN 0006-2960

Branchini, B.; Murtiashaw, M.; Magyar, R. & Anderson, S. (2000). The role of lysine 529, a conserved residue of the acyl-adenylate-forming enzyme superfamily, in firefly luciferase. *Biochemistry,* Vol.39, No.18, (May 2000), pp. 5433-5440, ISSN 0006-2960

Branchini, B.; Magyar,R.; Murtiashaw, M. & Portier N. (2001). The role of active site residue arginine 218 in firefly luciferase bioluminescence. *Biochemistry,* Vol.40, No.8, (February 2001), pp. 2410-2418, ISSN 0006-2960

Branchini, B.; Southworth. T.; Murtiashaw, M.; Boije, H. & Fleet S. (2003). A mutagenesis study of the putative luciferin binding site residues of firefly luciferase. *Biochemistry,* Vol.42, No.35, (September 2003), pp. 10429-10436, ISSN 0006-2960

Branchini, B.; Southworth, T.; Murtiashaw, M.; Wilkinson, S.; Khattak, N.; Rosenberg, J. & Zimmer, M. (2005). Mutagenesis evidence that the partial reactions of firefly bioluminescence are catalyzed by different conformations of the luciferase C- terminal domain. *Biochemistry*, Vol.44, No.5, (January 2005), pp. 1385-1393, ISSN 0006-2960

Branchini, B.; Rosenberg, J.; Fontaine, D.; Southworth, T.; Behney, C. & Uzasci, L. (2011). Bioluminescence Is Produced from a Trapped Firefly Luciferase Conformation Predicted by the Domain Alternation Mechanism. *Journal of the American Chemical Society*, Vol.133, No.29, (June 2011), pp. 11088-11091, ISSN 0002-7863

Brovko, L.; Belyaeva, E. & Ugarova, N. (1982). Subunit interactions in luciferase from the firefly *Luicola mingrelica*. Their role in the manifestation of enzyme activity and the process of thermal inactivation. *Biochemistry (translation from Biokhimiya, USSR),* Vol.47, No.5, (May 1982), pp. 760-766, ISSN 0320-9725

Carson, M.; Johnson, D.; McDonald, H.; Brouillette, C. & Delucas, L. (2007). His-tag impact on structure. *Acta Cryst. D*, Vol.63, No.3 (March 2007), pp. 295–230, ISSN 0907-4449

Cirino, P.; Mayer, K. & Umeno, D. (2003). Generating mutant libraries using error-prone PCR. *Methods in Molecular Biology,* Vol.231, (April 2003), pp. 3-9, ISSN 1064- 3745

Conti, E.; Franks, N. & Brick P. (1996). Crystal structure of firefly luciferase throws light on a superfamily of adenylate-forming enzymes. *Structure,* Vol.4, No.3, (March 1996), pp. 287-298, ISSN 0969-2126

Conti, E.; Stachelhaus, T.; Marahiel, M. & Brick P. (1997). Structural basis for the activation of phenylalanine in the non-ribosomal biosynthesis of gramicidin S. *EMBO J.*, Vol.16, (July 1997), pp. 4174-4183, ISSN 0261-4189

Dement'eva, E.; Kutuzova, G. & Ugarova, N. (1989). Biochemical properties and stability of homogeneous luciferase of *Luciola mingrelica* fireflies. *Moscow Universitet Chemistry Bulletin (translation)*, Vol.44, No.6, (November 1989), pp. 601-606. ISSN 0027-1314

Dement'eva, E.; Zheleznova, E.; Kutuzova, G.; Lundovskikh, I. & Ugarova, N. (1996). Physicochemical properties of recombinant *Luciola mingrelica* luciferase and its mutant forms, *Biochemistry (Moscow)*, Vol.61, No.1, (January 1996), pp. 115-119, ISSN 0320-9725

Devine J.; Kutuzova, G.; Green V.; Ugarova N. & Baldwin, T. (1993). Luciferase from the East European firefly *Luciola mingrelica*: cloning and nucleotide sequence of the cDNA, overexpression in *Escherichia coli* and purification of the enzyme. *Biochimica Biophysica Acta, Gene Structure and Expression,* Vol.1173, No.2, (May 1993), pp. 121- 132, ISSN 0006-3002

De Wet, J.; Wood, K.; DeLuca, M.; Helinskii, D. & Subramani, S. (1987). Firefly luciferase gene: structure and expression in mammalian cells. *Mol. Cell. Biol.*. Vol.7, No.2, (February 1987), pp. 725-737, ISSN 0270-7306

De Wet, J.; Wood, K.; Helinskii, D. & De Luca, M. (1985). Cloning of firefly luciferase cDNA and the expression of active luciferase in *Escherichia coli*. *Proc. Natl. Acad. Sci. USA,* Vol.82, No.23, (December 1985), pp. 7870-7873. ISSN 0027-8424

Efremenko, E.; Lyagin, I.; Votchitseva, Y.; Gudkov, D.; Peregudov, A.; Aliev, T. & Varfolomeev, S. (2008). The influence of length and localization of polyhistidine tag in the molecule of organophosphorus hydrolase on the biosythesis and behavior of fusion protein. In: *Biotechnology: state of the art and prospects for development,* G.E. Zaikov, (Ed.), 87-101, Nova Science Publishers Inc., ISBN 978-1-60456-015-2, New- York, USA

Eijsink, V.; Bjork, A.; Gaseidnes, S.; Sirevag, R.; Synstad, B.; van den Burg, B. & Vriend, G. (2004). Rational engineering of enzyme stability. *Journal of Biotechnology,* Vol.113, No.1-3, (September 2004), pp. 105-120, ISSN 0168-1656

Emamzadeh, A.; Hosseinkhani, S.; Sadeghizadeh, M.; Nikkhah, M.; Chaichi, M. & Mortazavi, M. (2006). cDNA cloning, expression and homology modeling of a luciferase from the firefly *Lampyroidea maculata*. *Journal of Biochemistry and Molecular Biology*. Vol.39, No.5, (September 2006), pp. 578-585, ISSN 1225-8687

Eriksson, J.; Nordstrom, T. & Nyren,P. (2003). Method enabling firefly luciferase-based bioluminometric assays at elevated temperatures. *Anal. Biochem.* Vol.314, No.1, (March 2003), pp. 158-161, ISSN 0003-2697

Fersht, A. & Serrano, L. (1993). Principles of protein stability derived from protein engineering experiments. *Curr. Op. Struct. Biol.*, Vol.3, No.1, (February 1993), pp. 75-83, ISSN 0959-440X

Forreiter, C.; Kirschner, M. & Nover, L. (1997). Stable transformation of an arabidopsis cell suspension culture with firefly luciferase providing a cellular system for analysis of chaperone activity *in vivo*. *The Plant Cell*, Vol.9, No.12, (December 1997), pp. 2171- 2181, ISSN 1040-4651

Fraga, H. (2008). Firefly luminescence: A historical perspective and recent developments. *Photochemical and Photobiological Sciences*, Vol.7, No.2, (February 2008), pp. 146-158, ISSN 1474-905X

Freydank, A.; Brandt, W. & Dräger, B. (2008). Protein structure modeling indicates hexahitidine-tag interference with enzyme activity, *Proteins*, Vol.72, No.1, (July 2008), pp. 173–183, ISSN 0887-3585

Frydman, J.; Erdjument-Bromage, H.; Tempst P. & Hartl, F. (1999). Co-translational domain folding as the structural basis for the rapid *de novo* folding of firefly luciferase. *Nat. Struct. Mol. Biol.*, Vol.6, No.7, (July 1999), pp. 697-705, ISSN 1072-836

Gould, S.; Keller, G. & Subramani, S. (1987). Identification of a peroxisomal targeting signal at the carboxy terminus of firefly luciferase. *J. Cell Biol.*, Vol.105, No.6, (December 1987), pp. 2923-2931, ISSN 0021-9525

Gulick, A. (2009). Conformational Dynamics in the Acyl-CoA Synthetases, Adenylation Domains of Non-ribosomal Peptide Synthetases, and Firefly Luciferase. *ACS Chemical Biology*, Vol.4, No.10, (July 19, 2009), pp. 811-827, ISSN 1554-8929

Jäckel, C.; Kast, P. & Hilvert, D. (2008). Protein Design by Directed Evolution. *Annual Review of Biophysics*, Vol.37, No.1, (June 2008), pp. 153-173, ISSN 1936-122X

Jiang, X.; Morgan, J. & Doyle, M. (2003). Thermal inactivation of *Escherichia coli O157:H7* in cow manure compost. *J. Food. Prot.*, Vol.66, No.10, (October 2003), pp. 1771-1777, ISSN 0362-028X

Hall, M. & Leach, F. (1988). Stability of firefly luciferase in Tricine buffer and in a commercial enzyme stabilizer. *J. Biolum. Chemilum.*, Vol.2, No.1, (January 1988), pp. 41-44, ISSN 0884-3996

Herbst, R.; Gast, K. & Seckler, R. (1998). Folding of Firefly *Photinus pyralis* Luciferase: Aggregation and Reactivation of Unfolding Intermediates. *Biochemistry*, Vol.37, No.18, (May 1998), pp. 6586-6597, ISSN 0006-2960

Imani, M.; Hosseinkhani, S.; Ahmadian, S. & Nazari, M. (2010). Design and introduction of a disulfide bridge in firefly luciferase: increase of thermostability and decrease of pH sensitivity. *Photochemical and Photobiological Sciences*, Vol.9, No.8, (August 2010), pp. 167-1177, ISSN 1474-905X

Kajiyama, N.; Masuda, T.; Tatsumi, H. & Nakano, E. (1992). Purification and characterization of luciferases from fireflies, *Luciola cruciata* and *Luciola lateralis*. *Biochim.Biophys. Acta*, V. 1120, No.2, (April 1992), pp. 228-232, ISSN 0006-3002

Kajiyama, N. & Nakano, E. (1993). Thermostabilization of firefly luciferase by a single amino acid substitution at position 217. *Biochemistry*, Vol.32, No.50, (December 1993), pp. 13795-13799, ISSN 0006-2960

Kajiyama, N. & Nakano, E. (1994). Enhancement of thermostability of firefly luciferase from *Luciola lateralis* by a single amino acid substitution. *Biosci. Biotechnol. Biochem*. Vol.58, No,6, (June 1994), pp. 1170-1171, ISSN 0916-8451

Keller, G.; Gould S.; Deluca M. & Subramani, S. (1987). Firefly Luciferase is Targeted to Peroxisomes in Mammalian Cell, *PNAS*, Vol.84, No.10, (May 1987), pp. 3264-3268, ISSN 0027-8424

Kitayama, A.; Yoshizaki, H.; Ohmiya, Y.; Ueda, H. & Nagamune, T. (2003). Creation of a thermostable firefly luciferase with pH-insensitive luminescent color. *Photochemistry and Photobiology*, Vol.77, No.3, (March 2003), pp. 333-338, ISSN 0031- 8655

Klose, J.; Wendt, N.; Kubald, S.; Krause, E.; Fechner, K.; Beyermann, M.; Bienert, M.; Rudolph, R. & Rothemund, S. (2004). Hexahistidin tag position influences disulfide structure but not binding behavior of *in vitro* folded N-terminal domain of rat corticotropin-releasing factor receptor type 2a. *Protein Sciences*, Vol.13, No.9, (August 2004), pp. 2470-2475, ISSN 0961-8368

Koksharov, M. & Ugarova, N. (2008). Random mutagenesis of *Luciola mingrelica* firefly luciferase. Mutant enzymes whose bioluminescence spectra show low pH- sensitivity. *Biochemistry (Moscow)*, Vol.73, No.8, (August 2008), pp. 862-869, ISSN 0006-2979

Koksharov, M. & Ugarova, N. (2010). Thermostability enhancement of *Luciola mingrelica* firefly by *in vivo* directed evolution. *Luminescense*, Vol.25, No.1, (January 2010), pp. 135-136, ISSN 1522-7243

Koksharov, M. & Ugarova, N. (2011a). Triple substitution G216N/A217L/S398M leads to the active and thermostable *Luciola mingrelica* firefly luciferase. *Photochemical and Photobiologocal Sciences*, Vol.10, No.6, (July 2011), pp. 931-938, ISSN 1474-905X

Koksharov, M. & Ugarova, N. (2011b). Thermostabilization of firefly luciferase by *in vivo* directed evolution. *Protein Eng. Des. Sel.*, Vol.24, No.11, (November 2011), pp. 835- 844, ISSN 1741-0126

Kumita, J.; Jain, L.; Safroneeva, E. & Woolley, G. (2000). A cysteine-free firefly luciferase retains luminescence activity. *Biochem. Biophys. Res. Commun.*, Vol.267, No.4, (January 2000), pp. 394-397, ISSN 0006-291X

Kuo, W & Chase, H. (2011). Exploiting the interactions between poly-histidine fusion tags and immobilized metal ions. *Biotechnology Letters*, Vol.33, No.6, pp. 1075-1084, ISSN 0141-5492

Kutuzova, G.; Skripkin, E.; Tarasova, N.; Ugarova, N. & Bogdanov, A. (1989). Synthesis and pathway of *Luciola mingrelica* firefly luciferase in *Xenopus laevis* frog oocytes and in cell-free systems. *Biochimie,* Vol.71, No.4, (April 1989), pp. 579-583, ISSN 0300-9084

Law, G.; Gandelman, O.; Tisi, L.; Lowe, C. & Murray, J. (2006). Mutagenesis of solvent-exposed amino acids in *Photinus pyralis* luciferase improves thermostability and pH tolerance. *Biochem. J.*, Vol.397, No.2, (July 2006), pp. 305-312, ISSN 0264-6021

Leclerc, G.; Boockfor, F.; Faught, W. & Frawley, L. (2000). Development of a destabilized firefly luciferase enzyme for measurement of gene expression. *Biotechniques,* Vol.29, No.3, (September 2000), pp. 590-601, ISSN 0736-6205

Li, X.; Yang, S. & Liang, X. (2006). Phylogenetic relationship of the firefly, *Diaphanes pectinealis* (Insecta, Coleoptera, Lampyridae), based on DNA sequence and gene structure of luciferase. *Zoological Research*, Vol.27, No.4, (August 2006), pp. 367-374, ISSN 0254-5853

Lomakina, G.; Modestova, Y. & Ugarova N. (2008). Enhancement of thermostability of the east european firefly (*Luciola mingrelica*) luciferase by site-directed mutagenesis of nonconservative cysteine residues Cys62 and Cys146. *Moscow University Chemistry Bulletin*, Vol.63, No.2, (May 2008), pp. 63-66, ISSN 0027-1314

Lundin, A. (2000). Use of firefly luciferase in ATP-related assays of biomass, enzymes, and metabolites. *Methods Enzymol.*, Vol.305, (December 2000), pp. 346-370, ISSN 0076- 6879

Lundovskikh, I.; Leontieva, O.; Dementieva, E. & Ugarova, N. (1998). Recombinant *Luciola mingrelica* firefly luciferase. Folding *in vivo*, purification and properties. *Proceedings of the 10th International Symposium on Bioluminescence and Chemiluminescence,* pp. 420-424, ISBN 0-471-98733-6, Bologna, Italy, September 4-8,1998

Mehrabi, M.; Hosseinkhani, S. & Ghobadi S. (2008). Stabilization of firefly luciferase against thermal stress by osmolytes. *International Journal of Bioogical Macromolecules*, Vol.43, No.2, (August 2008), pp. 187-191, ISSN 0141-8130

Minami, Y. & Minami, M. (1999). Hsc70/Hsp40 chaperone system mediates the Hsp90- dependent refolding of firefly luciferase. *Genes to Cells,* Vol.4, No.4, (December 1999), pp. 721-729, ISSN 1356-9597

Modestova, Y., Lomakina, G. & Ugarova N. (2010). Temperature dependence of thermal inactivation of *L.mingrelica* firefly luciferase and its mutant forms with C62S, C146S and C164S single point mutations. *Luminescence*, Vol.25, No 2, (February 2010), pp. 184-185, ISSN 1522-7243

Modestova, Y.; Lomakina G. & Ugarova N. (2011). Site-direcred mutagenesis of cysteine residues of *Luciola mingrelica* firefly luciferase. *Biochemistry (Moscow),* Vol.76, No.10, (September 2011), pp. 1407-1415, ISSN 0320-9725

Moroz, N.; Gurskii, D. & Ugarova, N. (2008). Stabilization of ATP reagents containing firefly *L. mingrelica* luciferase by polyols, *Moscow Univ. Chem. Bull.*,Vol.63, No.2, (March 2008), pp. 67-70, ISSN 0027-1314

Morozov, V. & Ugarova, N. (1996). Conserved motifs in a superfamily of enzymes catalyzing acyl adenylate formation from ATP and compounds with a carboxyl group. *Biochemistry (Moscow),* Vol.61, No.8, (August 1996), pp. 1068-1072, ISSN 0027-1314

Nakatsu, T.; Ichiyama, S.; Hiratake, J.; Saldanha, A.; Kobashi, N.; Sakata, K. & Kato, H. (2006). Structural basis for the spectral difference in luciferase bioluminescence. *Nature,* Vol.440, No.7082, (March 2006), pp. 372-376, ISSN 0028-0836

Nguyen, V.; Morange, M. & Bensaude, O. (1989). Protein denaturation during heat shock and related stress. *Escherichia coli* beta-galactosidase and *Photinus pyralis* luciferase inactivation in mouse cells. *Journal of Biological Chemistry,* Vol.264, No.18, (June 1989), pp. 10487-10492, ISSN 0021-9258

Nimmesgern, E. & Hartl, F. (1993). ATP-dependent protein refolding activity in reticulocyte lysate : Evidence for the participation of different chaperone components. *FEBS Letters,* Vol.331, No.1-2, (September 1993), pp. 25-30, ISSN 0014-5793

Niwa, K.; Ichino, Y.; Kumata, S.; Nakajima, Y.; Hiraishi, Y.; Kato, D.; Viviani, V. & Ohmiya, Y. (2010). Quantum yields and kinetics of the firefly bioluminescence reaction of beetle luciferases. *Photochemistry and Photobiology,* Vol.86, No.5, (November 2010), pp. 1046-1049, ISSN 0031-8655

Oba, Y.; Mori, N.; Yoshida, M. & Inouye, S. (2010a). Identification and Characterization of a Luciferase Isotype in the Japanese Firefly, *Luciola cruciata,* Involving in the Dim Glow of Firefly Eggs. *Biochemistry,* Vol.49, No.51, (November 2010), pp. 10788- 10795, ISSN 0006-2960

Oba, Y.; Kumazaki, M. & Inouye, S. (2010b). Characterization of luciferases and its paralogue in the Panamanian luminous click beetle *Pyrophorus angustus*: A click beetle luciferase lacks the fatty acyl-CoA synthetic activity. *Gene*, Vol.452, No.1, (February 2010), pp. 1-6, ISSN 0378-1119

Ohmiya, Y. & Tsuji, F. (1997). Mutagenesis of firefly luciferase shows that cysteine residues are not required for bioluminescence activity. *FEBS Letters*, Vol.404, No.2, (March 1997), pp. 115-117, ISSN 0014-5793

Price, R.; Squirrell. D.; Murphy, M. & White, P. (1996). Genetic engineering of firefly luciferase towards its use as a label in gene probe assays and immunoassays. *Proceedings of the 9th International Symposium on Bioluminescence and Chemiluminescence*, pp. 220-223, ISBN 0 471 97502 8, Woods Hole, Massachusetts, October, 1996

Rainer, R. (1992). Targeting signals for protein import into peroxisomes. *Cell Biochemical Functions*, Vol.10, No.3, (February 1992), pp. 193-199, ISSN 0263-6484

Raso, S.; Clark, P.; Haase-Pettingell, C.; King, J. & Thomas, G. (2001). Distinct cysteine sulfhydryl environments detected by analysis of Raman S-hh markers of Cys→Ser mutant proteins. *Journal of Molecular Biology*, Vol.307, No.3, (March 2001), pp. 899- 911, ISSN 0022-2836

Sandalova, T. & Ugarova, N. (1999). Model of the active site of firefly luciferase. *Biochemistry (Moscow)*, Vol.64, No.8, (August 1999), pp. 962-967, ISSN 0320-9725

Schumacher, R.; Hansen, W.; Freeman, B.: Alnemri E.; Litwack G. & Toft D. (1996). Cooperative Action of Hsp70, Hsp90, and DnaJ Proteins in Protein Renaturation. *Biochemistry*, Vol.35, No.47, (November 1996), pp. 14889-14898, ISSN 0006-2960

Seliger, H. & McElroy, W. (1960). Spectral emission and quantum yield of firefly bioluminescence. *Arch. Biochem. Biophys*. Vol.88, No.1, (May 1960), pp. 136-141, ISSN 0003-9861

Shimomura, O.; Goto, T. & Johnson, F. (1977). Source of Oxygen in the CO2 Produced in the Bioluminescent Oxidation of Firefly Luciferin. *Proceeding of National Academy of Sciences, USA*, Vol.74, No.7, (July 1977), pp. 2799-2802, ISSN 0027-8424

Souren, J.; Wiegant, F. & Van Wijk, R. (1999a). The role of hsp70 in protection and repair of luciferase activity *in vivo*; experimental data and mathematical modelling. *Cell. Mol. Life Sci.* Vol.55, No.5, (May 1999), pp. 799-811, ISSN 1420-682X

Souren, J.; Wiegant, F.; Van Hof, P.; Van Aken, J. & Van Wijk, R. (1999b). The effect of temperature and protein synthesis on the renaturation of firefly luciferase in intact H9c2 cells. *Cell. Mol. Life Sci.,* Vol.V55, No.11, (August 1999), pp. 1473-1481, ISSN 1420-682X

Suelter, C. & DeLuca, M. (1983). How to prevent losses of protein by adsorption to glass and plastic. *Analytical Biochemistry,* Vol.135, No.1, (November 1983), pp. 112-119, ISSN 0003-2697

Thompson, J.; Hayes, L. & Lloyd. D. (1991). Modulation of firefly luciferase stability and impact on studies of gene regulation. *Gene,* Vol.103, No.2, (July 1991), pp. 171-177, ISSN 0378-1119

Thompson, J.; Geoghegan, K.; Lloyd, D.; Lanzetti, A.; Magyar, R.; Anderson, S. & Branchini, B. (1997). Mutation of a protease-sensitive region in firefly luciferase alters light emission properties. *Journal of Biological Chemistry,* Vol.272, No.30, (July 1997), pp. 18766-18771, ISSN 0021- 9193

Thulasiraman, V. & Matts, R. (1996). Effect of geldanamycin on the kinetics of chaperone- mediated renaturation of firefly luciferase in rabbit reticulocyte lysate. *Biochemistry,* Vol.35, No.41, (October 1996), pp. 13443-13450, ISSN 0006-2960

Tisi, L.; White, P.; Squirrell, D.; Murphy, M.; Lowe, C. & Murray, J. (2002). Development of a thermostable firefly luciferase. *Analytica Chimica Acta,* Vol.457, No.1, (April 2002), pp. 115-123, ISSN 0003-2670

Tisi, L.; Lam, G.; Gandelman, O. & Murray, J. (2007). pH tolerant luciferase. *PCT Patent No WO 2007/1017684 A2*

Turner, N. (2009), Directed evolution drives the next generation of biocatalysts. *Nature Chemical Biology*, Vol.5, No.8, pp. 567-573, ISSN 1552-4450

Ugarova, N. (1989). Luciferase of *Luciola mingrelica* fireflies. Kinetics and regulation mechanism. *Journal of Bioluminescence and Chemiluminescence,* Vol.4, No.1, (July 1989), pp. 406-418, ISSN 0884-3996

Ugarova, N. (2005). Structure and functions of firefly luciferase. In: *Chemical and Biological Kinetics,* S.D.Varfolomeev & E.B.Burlakova (Eds.), Vol.2, pp. 205-233, Koninklijke, Leiden, The Netherlands, ISBN: 90 6764 4315

Ugarova, N. (2010). Stabilization of firefly luciferase *Luciola mingrelica* by genetic engineering methods. *Moscow Univ. Chem. Bull.,* Vol.65, No.3, (May, 2010), pp. 139- 143, ISSN 0027-1314

Ugarova, N.; Brovko, L. & Kutuzova, G. (1993). Bioluminescence and bioluminescent analysis: recent development in the field. *Biochemistry (Moscow),* Vol.58, No.9, (September 1993), pp. 976-992, ISSN 0320-9725

Ugarova, N., Dementieva, E. & Lundovskikh, I. (2000). Thermoinactivation and reactivation of firefly luciferase, *Proceedings of 11th International Symposium on Bioluminescence and Chemiluminescence.* ISBN 981-02-4679-X, pp. 193-196, Pacific Grove, California, USA, September 6-10, 2000

Ugarova, N. &. Brovko, L. (2002). Protein structure and bioluminescent spectra for firefly bioluminescence, *Luminescence,* Vol.17, No.5, (October 2002), pp. 321-330, ISSN 1522-7243

Ugarova, N.; Koksharov, M. & Lomakina, G.. (2010). Reagent for determining adenosine-5'- triphosphate. *PCT Patent Appl. WO 2010/134850*

Viviani, V. (2002). The origin, diversity, and structure function relationships of insect luciferases. *Cell Mol Life Sci.,* Vol.59, No.11, (November 2002), pp. 1833-1850, ISSN 1420-682X

Viviani, V.; Silva, A.; Perez, G.; Santelli, R.; Bechara, E. & Reinach, F. (1999). Cloning and molecular characterization of the cDNA for the Brazilian larval click-beetle *Pyrearinus termililluminans* luciferase. *Photochem. Photobiol.,* Vol.70, No,2, (August 1999), pp. 254-260, ISSN 0031-8655

Viviani, V. & Ohmiya,Y. (2006). Beetle Luciferases: Colorful Lights on Biological Processes and Diseases, In: *Photoproteins in Bioanalysis,* S. Daunert & S. K. Deo, (Eds), 49-63, Wiley-VCH, Weinheim, ISBN 978-3-527-31016-6

Wang, W.; Xu, Q.; Shan, Y. & Xu, G. (2001). Probing local conformational changes during equilibrium unfolding of firefly luciferase: fluorescence and circular dichroism studies of single tryptophan mutants. *Biochem. Biophys. Res. Commun.,* Vol.282, No.1, (March 2001), pp. 28-33, ISSN 0006-291X

White, P.; Squirrell, D.; Arnaud, P.; Lowe, C. & Murray, J. (1996). Improved thermostability of the North American firefly luciferase: saturation mutagenesis at position 354. *Biochem. J.,* Vol.319, No.2, (October 2006), pp. 343-350, ISSN 0264- 6021

Wilson, T. & Hastings, J. (1998). Bioluminescence. *Annual Review of Cell and Developmental Biology,* Vol.14, No.1, (November 1998), pp. 197-230, ISSN 1081-0706

Wood, K. & DeLuca, M. (1987). Photographic detection of luminescence in *Escherichia coli* containing the gene for firefly luciferase. *Anal. Biochem.,* Vol.161, No.2, (March 1987), pp. 501-507, ISSN 0003-2697

Wood, K.; Lam, Y.; Seliger, H. & McElroy, W. (1989). Complementary DNA coding click beetle luciferases can elicit bioluminescence of different colors. *Science,* Vol.244, No.4905, pp. 700-702, ISSN 0036-8075

Wood, K. & Hall, M. (1999), Thermostable luciferase and methods of production. *PCT Patent Appl. WO 1999/014336*

04 MANIPULATION OF THE PHENYLPROPANOID PATHWAY USING GENETIC ENGINEERING IN LEUCAENA LEUCOCEPHALA

1. INTRODUCTION: AN OVERVIEW

Trees serve as repositories for numerous commercially and biotechnologically valuable substances. Wood, a natural resource, is composed of lignin, hemicelluloses, and cellulose. The biochemistry of lignin, one of the most prevalent biopolymers on Earth, has been intensively studied due to its significant industrial importance and the interest in understanding its properties (Boerjan et al., 2003). Phenylpropanoid metabolism, which is a significant pathway for lignin formation in plants, has wide-ranging effects on agriculture, industry, and the environment, making it a globally significant aspect of plant chemistry.

In addition to its crucial function in maintaining plant health and growth, excessive amounts of lignin pose challenges in the agro-industrial utilization of different plant species. The elimination of this component in paper making is deemed undesirable due to the high cost, energy consumption, and pollutant generation involved (Baucher et al., 2003; Boerjan, 2005; Chiang, 2002). Therefore, it is crucial to modify the lignin content of designer plant species in order to reduce the reliance on natural forest resources for domestication purposes in the future. Significant scientific attention has been directed towards enhancing the quality of wood in trees by modifying various genes involved in the production of lignin. This could have significant implications for enhancing the usability of wood material (Chiang, 2006; Higuchi, 2006). Leucaena leucocephala, a highly adaptable and rapidly growing tree species, is of great significance to the paper and pulp industry in India. It accounts for around 25% of the overall raw material used. The increased utilization of this particular tree species in the pulp industry can be attributed to its exceptional capacity for biomass production and its remarkable adaptability to diverse soil and climatic conditions. Each advancement in the growth of this particular type of tree, resulting in higher biomass and lower lignin content, would greatly benefit the pulp and paper sector. This would lead to a drop in costs and the emission of harmful chemicals during the production of paper pulp.

2. LIGNIN: OCCURRENCE, STRUCTURE AND FUNCTION

Lignin, derived from the Latin word "lignum" which means wood, is a highly prevalent organic polymer found in nature, second only to cellulose (Boerjan et al., 2003). Cellulose is an essential structural component of the cell walls in all vascular plants and typically makes up around 25% of the total biomass on land. Cellulose is essential for maintaining the structural integrity of the cell wall and providing stiffness and strength to the stem (Chabannes et al., 2001; Jones et al., 2001). Lignin is predominantly produced and stored in the secondary cell wall of specialized cells, including xylem vessels, tracheids, and fibers. Additionally, it is found in small quantities in the periderm, where its connection with suberin serves a protective function against infections. Furthermore, lignin serves the purpose of rendering the cell wall impermeable to water, thereby facilitating the transportation of water and solutes through the vascular system (Sarkanen & Ludwig, 1971). Lignins are intricate, racemic, aromatic heteropolymers formed from the dehydrogenative polymerization of monolignols, specifically coumaryl, coniferyl, and sinapyl alcohol monomers, which vary in their level of methoxylation (Freudenberg & Neish, 1968). The monolignols produce p-hydroxyphenyl (H), guaiacyl (G), and syringyl (S) phenylpropanoid units when they are included in the lignin polymer (Fig. 1).

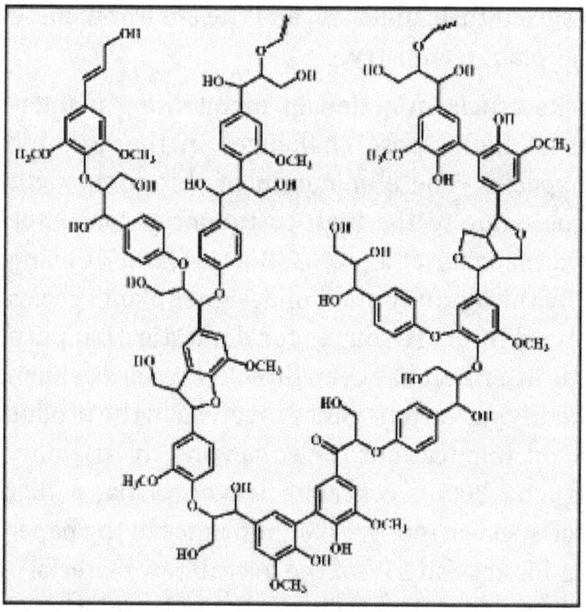

Fig. 1. Structure of a lignin oligomer. These structures consist of phenylpropanoid units H, G and S. A number of such oligomers cross-polymerizes to form a complex structure of lignin

The quantity and makeup of lignins differ among different groups of organisms, types of cells, and specific layers of cell walls. These characteristics are affected by both developmental processes and environmental factors (Campbell & Sederoff, 1996). Softwoods have a greater lignin content, ranging from 27-33%, compared to hardwoods and grasses, which have lignin contents of 18-25% and 17-24%, respectively. Compression wood located in the lowest section of branches and leaning stems contains the greatest quantities of lignin, ranging from 35% to 40% (Sarkanen & Ludwig, 1971).

2.1 Lignin biosynthesis

Despite being studied for more than a century, there are still many unsolved aspects about the production of lignin. The monolignol biosynthetic pathway has undergone multiple revisions, yet it continues to be a subject of controversy. Over the past twenty years, substantial progress has been achieved in the identification and characterization of several genes involved in the production of monolignols in several plant species. There are also other articles discussing the progress made in understanding the routes involved in monolignol biosynthesis (Boerjan et al., 2003; Humphreys & Chapple, 2002; Whetten & Sederoff, 1995; Whetten et al., 1998). Lignins in plants are formed through the polymerization of monolignols, specifically p-coumaryl, coniferyl, and sinapyl alcohol monomers, which vary in their level of methoxylation (Higuchi, 1985; Sederoff & Chang, 1991). This process occurs via the phenylpropanoid biosynthetic route (Gross, 1985). This pathway consists of an intricate sequence of diverging metabolic events that are responsible for the production of various substances such as lignin, flavonoids, and hydroxycinnamic acid conjugates. Several intermediates and end products of this route have significant functions in plants, including phytoalexins, antiherbivory chemicals, antioxidants, ultraviolet (UV) protectants, pigments, and fragrance compounds. The various roles of lignin and its derivatives in protecting plants from both living organisms and environmental factors highlight the crucial importance of this pathway for plant well-being and survival.

The synthesis of lignin represents one of the most energy demanding biosynthetic pathways in plants, requiring large quantities of carbon skeletons. Deposition of lignin in plants proceeds via the following steps:

- The biosynthesis of monolignols
- Transport of monolignols from the site of synthesis to the site of polymerization
- Dehydrogenation & Polymerization of monolignols.

2.1.1 Biosynthesis of monolignols

The biosynthesis of monolignols proceeds through the phenylpropanoid pathway starting with deamination of phenylalanine to produce cinnamic acid and involves successive hydroxylation reactions of the aromatic ring, followed by phenolic *o*-methylation and conversion of the side chain carboxyl to an alcohol group (Fig. 2). Immense amount of work has been done in characterizing the monolignol biosynthesis pathway in past two decades. It is a complex pathway comprising of enzymes with functions like methyltransferase, hydroxylase, reductase and dehydrogenase. Some of the important enzymes involved in monolignol biosynthesis have been discussed below.

2.1.1.1 Palanine ammonia-lyase (PAL)

The enzyme phenylalanine ammonia-lyase (PAL; EC: 4.3.1.5) that catalyzes the conversion of phenylalanine to transcinnamic acid, is the initial step towards monolignol biosynthesis and other phenolic secondary plant metabolites. Genes encoding PAL have been studied in *Populus* species (Kao et al. 2002; Osakabe et al., 1995), loblolly pine and other plant species (Bate et al., 1994; Hatton et al., 1995; Jones, 1984; Kumar & Ellis, 2001; Leyva et al., 1992; Ohl et al., 1990). *PAL* exists as a multiple member gene family and the individual members can be involved in different metabolic pathways as suggested by their expression patterns in association with certain secondary compounds accumulated in specific tissue or developmental stage. The biochemical activity of all known PALs is verified as specific deamination of phenylalanine, but genetic and physiological function may vary among different PAL members. The expression of PAL genetic function is controlled by various genetic circuits and signaling pathways.

2.1.1.2 Cinnamate 4-Hydroxylase (C4H)

C4H (EC: 1.14.13.11) constitutes the CYP73 family of the large group of Cyt P450 monooxygenases. It catalyzes the 4-hydroxylation of *trans*-cinnamate, the central step in the generation of phenylalanine-derived substrates for the many branches of phenylpropanoid metabolism. The first and the last enzymes of this short sequence of closely related reactions, termed the general phenylpropanoid metabolism, are PAL and 4CL, respectively. A second metabolic link couples C4H to the membrane-localized Cyt P450 Reductase (CPR). The expression patterns of all three C4H-linked enzymes, PAL, 4CL, and CPR, and of the corresponding mRNAs have been analyzed in cell-suspension cultures and various intact tissues of parsley (Logemann et al., 1995) and *Arabidopsis* (Mizutani and Ohta, 1997). A reduction in PAL levels leads to an increase in the S/G ratio, whereas reduced C4H activity leads to a decrease in the S/G ratio. These observations support the existence of some sort of metabolic channeling between the enzymes of the central phenylpropanoid pathway and those of monolignol biosynthesis and also provide a basis for the development of new strategies for modified or reduced lignin content.

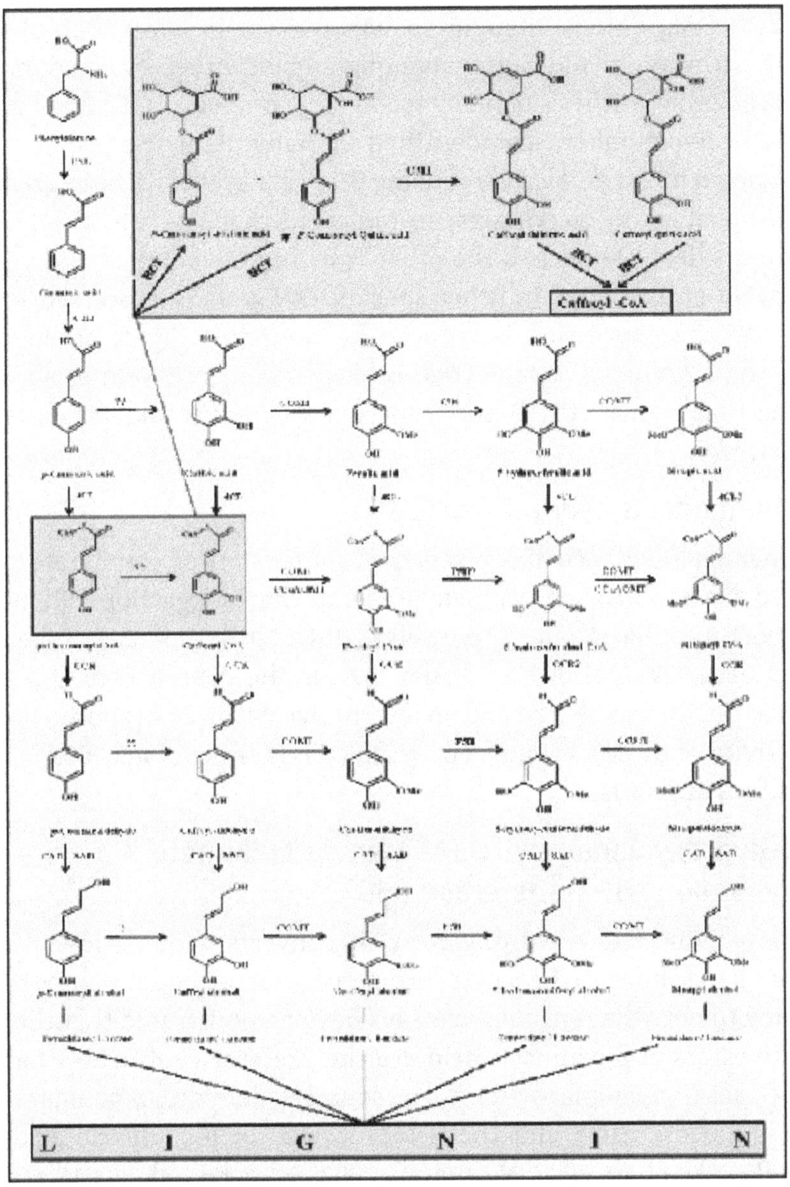

Fig. 2. An overview of monolignol biosynthesis pathway: PAL, Phenylalanine ammonia- lyase; C4H, Cinnamic acid 4-hydroxylase; C3H, p-Coumarate 3-hydroxylase; HCT, Hydroxycinnamoyltransferase; 4CL, 4-Coumarate-CoA ligase; CCoAOMT, Caffeoyl-CoA O- methyltransferase; CCR, Cinnamoyl CoA reductase; CAld5H, Coniferyl aldehyde 5- hydroxylase; AldOMT, 5- hydroxyconiferyl aldehyde O-methyltransferase; CAD, Cinnamyl alcohol dehydrogenase; SAD, Sinapyl alcohol dehydrogenase developmental stage. The biochemical activity of all known PALs is verified as specific deamination of phenylalanine, but genetic and physiological function may vary among different *PAL* members. The expression of *PAL* genetic function is controlled by various genetic circuits and signaling pathways.

Similar to PAL, C4H is thought to be involved in a number of secondary metabolism pathways in addition to monolignol biosynthesis as *p*-coumarate is an intermediate for biosynthesis of many secondary compounds (Croteau et al., 2000). Multiple *C4H* gene members are identified in many plant species, however, only one *C4H* is known in the *Arabidopsis* genome (Raes et al., 2003). The expression study of two *C4H* members in quaking aspen indicated that one is strongly expressed in developing xylem tissues and the other is more active in leaf and young shoot tissues (Shanfa et al., 2006). In other species, *C4H* gene is expressed in a variety of tissues and the expression is induced by wounding, light, pathogen attacks and other biotic & abiotic stimuli (Bell-Lelong et al., 1997; Raes et al., 2003). The mechanisms that regulate the genetic function of *C4H* gene and its family members are yet unknown.

2.1.1.3 Coumarate 3-hydroxylase (C3H)

Early biochemical evidence suggested that conversion of coumarate to caffeate is catalyzed by a nonspecific phenolase, but that suggestion did not receive much support in other studies (Boniwell & Butt, 1986; Kojima & Takeuchi, 1989; Petersen et al., 1999; Stafford & Dresler, 1972). The gene encoding *p*-coumarate 3-hydroxylase (*C3H*) was cloned and an alternative pathway proposed based on the enzyme activity of *CYP98A3* gene from *Arabidopsis* (Franke et al., 2002 a; Nair et al., 2002; Schoch et al., 2001).

2.1.1.4 p-hydroxycinnamoyl-CoA: quinate shikimate p-hydroxycinnamoyl - transferase (HCT)

The enzyme *p*-coumarate 3-hydroxylase (C3H) converts *p*-coumaric acid into caffeic acid and has been shown to be a cytochrome p450-depenedent monooxygenase. It is interesting to note that enzymatic assays have demonstrated that the shikimate and quinate esters of *p*-coumaric acid are the preferred substrates for C3H over *p*-coumaric acid, *p*-coumaroyl-CoA, *p* coumaraldehyde, *p*-coumaryl alcohol, nor the 1–*O*-glucose ester and the 4–*O*-glucoside of *p*-coumaric acid are good substrates (Franke et al., 2002; Nair et al., 2002; Schoch et al., 2001). *p*-Coumarate is first converted to *p*-coumaroyl-CoA by 4CL, with subsequent conversion to *p*-coumaroyl-shikimate and *p*-coumaroyl-quinate, the substrates for C3H, by *p*-hydroxycinnamoyl-CoA:quinate-(CQT) or *p*-hydroxycinnamoyl-CoA:shikimate *p*-hydroxycinnamoyl transferase (CST) (Schoch et al., 2001). These enzymes, described as reversible enzymes, can convert caffeoyl-shikimate or caffeoyl-quinate (chlorogenic acid) into caffeoyl-CoA, the substrate for CCoAOMT. A reversible acyltransferase with both CQT and CST activity, designated HCT, has been purified and the corresponding gene cloned from tobacco (Hoffmann et al., 2002). Silencing of *HCT* through RNA interference (RNAi) lead to reduction in lignin, hyper accumulation of flavonoids and growth inhibition in *Arabidopsis* (Chapple, 2010; Hoffmann et al., 2004, 2007)

2.1.1.5 Coumarate Coenzyme-A ligase (4CL)

Genetic and biochemical functions of 4-Coumarate Coenzyme A ligase (4CL; EC: 6.2.1.12) genes have been clearly demonstrated in association with monolignol biosynthesis (Lewis and Yamamoto, 1990; Lee et al., 1997; Hu et al., 1998, 1999; Harding et al., 2002). *4CL* genes usually exist as a multi-gene family. Four *4CL* genes were detected in the *Arabidopsis* genome and the expression of each member was regulated differentially in tissues and development stages (Raes et al., 2003). In aspen trees, two *4CL* genes were cloned and their expression were clearly distinct, with one in epidermal & leaf tissue and the other specifically in developing xylem tissue (Harding et al., 2002; Hu et al., 1998). Furthermore, the enzymatic activities of *4CL* members from aspen, loblolly pine, tobacco, soybean, *Arabidopsis*, and many other species were found to have distinct substrate specificities (Hu et al., 1998; Lindermayr et al., 2003; Voo et al., 1995). Whether the substrate specificity of the *4CL* members relates to different metabolic pathways is unknown. As the 4CL catalytic kinetics vary among species, it is also likely that the mainstream pathway mediated by 4CL may not be exactly the same in all plant species or tissues. Nevertheless, monolignol biosynthesis is tightly controlled by *4CL*.

2.1.1.6 O-methyltransferases (O-MT)

S-adenosyl-L-methionine methyltransferases are key enzymes in the phenylpropanoid, flavanoid and many other metabolic pathways in plants. The enzymes Caffeate 3-O methyltransferase (COMT; EC: 2.1.1.68) and Caffeoyl CoA 3-O methyltransferase (CCoAOMT; EC: 2.1.1.104) control the degree of methoxylation in lignin precursors *i.e. p*- hydroxyphenyl (H), guaiacyl (G) and syringyl (S) units of lignin (Higuchi, 1990; Ralph et al., 1998; Boerjan et al., 2003). The methylation reactions at the C3 and C5 hydroxyl functions of the lignin precursors were thought to occur mainly at the cinnamic acid level via a bi- functional COMT. However, the association of *CCoAOMT* expression with lignification (Pakusch et al., 1991; Ye et al., 1994; Ye & Varner, 1995; Ye, 1997; Martz et al., 1998; Chen et al., 2000) and the observation that down-regulation of *COMT* preferentially affected the amount of S units suggested the existence of an alternative pathway for the methylation of the lignin precursors at the hydroxycinnamoyl CoA level and specific O-methyltransferase *i.e.* CCoAOMT converts caffeoyl CoA into feruloyl CoA and 5-hydroxyferuloyl CoA into sinapyl CoA (Martz et al., 1998).

2.1.1.7 Cinnamoyl CoA reductase (CCR)

The reduction of cinnamoyl CoA esters to cinnamaldehydes is the first metabolic step committed to monolignol formation (Lacombe & Hawkins, 1997). This first reductive step in lignin biosynthetic pathway is performed by Cinnamoyl CoA reductase (EC: 1.2.1.44) and it controls the over-all carbon flux towards lignin. CCR activity is found to be generally low in plants so it is hypothesized that it may play a crucial role as a rate limiting step in regulation of lignin biosynthesis (Ma & Tian, 2005). CCR is apparently encoded by a single gene per haploid genome in *Eucalyptus* (Lacombe & Hawkins, 1997), poplar (Leple et al., 1998), ryegrass (Larsen, 2004; McInnes et al., 2002), *Triticum* (Ma, 2007) and tobacco (Piquemal & Lapierre, 1998) and by two genes in maize (Pichon, Courbou et al., 1998), and *Arabidopsis* (Lauvergeat & Lacomme, 2001). The *CCR* genes in various species appear as a multiple member family. In the *Populus* genome, there exist 8 *CCR*-homolog or *CCR*-like gene sequences (Li & Cheng, 2005). *Triticum* (Ma & Tian, 2005; Ma, 2007), maize (Pichon & Courbou, 1998), switchgrass (Escamilla-Trevino & Shen, 2010), *Medicago* (Zhou & Jackson, 2010) and *Arabidopsis* (Lauvergeat & Lacomme, 2001) have been shown to possess two or more than two isoforms (*CCR1* and *CCR2*) which are involved in mutually exclusive or redundant functions like, constitutive lignifications and defense. Several other *CCR* gene sequences have been deposited in the GenBank database, but their functions have still not been demonstrated. It is proposed that all CCR enzymes have a similar catalyzing mechanism for converting the CoA esters to aldehydes in monolignol biosynthesis.

2.1.1.8 Coniferaldehyde 5-hydroxylase (CAld5H)

CAld5H enzyme like C4H belongs to cytochrome P450 monoxygenase family. The hydroxylation reaction in the biosynthesis of S-unit (syringyl) was first considered to occur at the ferulate level (Grand, 1984), and hence, the enzyme was called Ferulate 5-hydroxylase (F5H). However, studies later have revealed that F5H can also function at later steps in the pathway, mainly at the coniferyl aldehyde or coniferyl alcohol level (Humphreys et al., 1999; Li et al., 2000). This enzyme was therefore alternatively renamed coniferaldehyde- 5- hydroxylase (CAld5H) (Osakabe et al., 1999). F5H/CAld5H is unusual in that it is a multifunctional plant P450 with three physiologically relevant substrates. The *Km* for the substrates such as coniferaldehyde, coniferyl alcohol and for the ferulic acid are 1 μM, 3 μM and 1000 μM respectively. This study demonstrates that the coniferaldehyde is the most preferred substrate for the enzymes (Humphrey et al., 1999). Considerable evidence is now available that shows that in angiosperm trees, the syringyl monolignol pathway branches out from guaiacyl pathway through coniferaldehyde and is regulated in sequence by three genes encoding coniferaldehyde 5-hydroxylase (*CAld5H*), 5-hydroxyconiferaldehyde *O*-methyltransferase (*COMT*) and sinapyl alcohol dehydrogenase (*SAD*).

2.1.1.9 Cinnamyl/Sinapyl alcohol dehydrogenases (CAD/SAD)

In gymnosperm wood, coniferyl alcohol is the major monolignol units while both coniferyl and sinapyl alcohols are present in angiosperm wood. CAD (E.C: 1.1.1.195), depicts a class of NADPH dependent oxidoreductase, suggested to catalyze multiple cinnamyl alcohol formations from their corresponding cinnamaldehydes (Lewis &

Yamamoto, 1990; Whetten & Sederoff, 1995; Whetten et al., 1998). This reduction of aldehydes to corresponding alcohols has been considered to be an indicator of lignin biosynthesis because of its specific role at the end of the monolignol biosynthesis pathway (Baucher et al., 1996). When the *Populus* tree was studied for monolignol biosynthesis in wood forming tissue, in addition to *CAD*, it was found in aspen that another gene, its sequence similar to but distinct from *CAD*, is also associated with lignin biosynthesis (Li et al., 2001). The biochemical characterization of the recombinant protein encoded by this gene indicated that the enzymatic activity has specific affinity toward sinapaldehyde, therefore it was named SAD. Compared with SAD enzyme kinetics, CAD showed a catalytic specificity towards coniferaldehyde instead. The catalytic specificities of the two enzymes have been further verified in protein structure analysis (Bomati & Noel, 2005). Furthermore, it was demonstrated that the expression of *CAD* is associated with G-lignin accumulation while *SAD* was associated with S-lignin formation during xylem differentiation (Li et al., 2001). The evidence from molecular, biochemical and cellular characterizations strongly suggest that *CAD* is involved in G-monolignol biosynthesis and *SAD* in S-monolignol biosynthesis in aspen wood formation.

2.1.2 Transport of monolignols

After the synthesis, the lignin precursors or monolignols are transported to the cell wall where they are oxidized and polymerized. The monolignols formed are insoluble and toxic to the plant cell and hence are converted to their respective glucosides by the action of UDP- glycosyltransferases (UDP-GT). This conversion renders the monolignols, soluble and less toxic to the plant cells, which can be stored in plant vacuoles, and transported to the cell wall as the need arises. It has been hypothesized that these monolignol glucosides are storage or transport forms of the monolignols (Steeves et al., 2001).

2.1.3 Dehydrogenation and polymerization

After transport of the monolignols to the cell wall, lignin is formed through dehydrogenative polymerization of the monolignols (Christensen et al., 2000). The dehydrogenation of monolignol radicals has been attributed to different class of enzymes, such as peroxidases (POX), laccases (LAC), polyphenol oxidases, and coniferyl alcohol oxidase. Lignin is a hydrophobic and optically inactive polymer, which is highly complex and heterogeneous in nature. Lignin polymerization is a radical coupling reaction, where the monolignols are first activated into phenoxy radicals in an enzyme catalyzed dehydrogenation reaction. These radicals couple to form dimers, oligomers and eventually the lignin polymer (Freudenberg, 1968). Peroxidases are heme-containing oxidoreductases that use H2O2 as the ultimate electron acceptor. The natural electron donor molecules in a peroxidase catalyzed reaction vary and include, monolignols, hydroxycinnamic acids (Zimmerlin et al., 1994), tyrosine residues in extensions (Brownleader et al., 1995) and auxin (Hinman & Lang, 1965). Several reports on peroxidase activity or gene expression in lignin- forming tissues have appeared, but only a few isoenzymes or genes have been specifically associated with lignification (Christensen et al., 2001; Marjamaa et al., 2006; Quiroga et al., 2000; Sato et al., 1993).

2.2 Regulation of monolignol biosynthesis

Developmental program of lignification associated with certain types of plant cells, such as xylem and fibers, require coordinated regulation of different lignin biosynthesis genes, as well as with genes controlling other aspects of plant growth and development. Different transcription factors such as R2R3-MYB, KNOX, MADS, LIM have been found to be regulating lignin biosynthesis in many plants, although the understanding of the molecular mechanism of pathway regulation is still limiting (Campbell & Rogers, 2004; Zhou et al., 2006, 2008). Lignification can be modified in a more efficient and precise way by understanding the regulation of these pathways via altering the expression of relevant transcription factors.

3. LIGNIN AS BARRIER FOR PAPER PRODUCTION

From an agro-economical point of view, lignin is considered to have a negative impact because it affects the paper manufacture and limits digestibility of forage crops. High quantity and low Syringyl (S) to Guaiacyl (G) lignin ratio plays a detrimental role in economy and ecology of paper production. Every unit increase in S/G ratio decreases the cost of paper production by two and half times. Both lignin content and composition are known to have impact on pulp & paper because residual lignin in the wood fibers causes a discoloration and a low brightness level of the pulp (Chaing et al., 1988). Consequently, for the production of high quality paper, lignin has to be removed from cellulose during the pulping process without damaging the polysaccharide component of wood. During chemical (Kraft) pulping, a large amount of Sodium hydroxide (NaOH) and Sodium sulfide (Na2S) are required to extract lignin from the pulp (Axegard et al., 1992). Subsequently, the residual lignin is further removed with bleaching agents, such as Chlorine dioxide (ClO2), Hydrogen peroxide (H2O2), Sodium hypochlorite(NaOCl), Oxygen(O2), or Ozone (O3) (Axegard et al.,1992; Biermann, 1993; Christensen et al., 2000).

These lignin extraction & bleaching procedures can partly degrade cellulose and consequently, reduce pulp quality and paper strength. Lignin extraction consumes large quantities of chemicals and energy leading to poor environmental image for these industries (Biermann, 1996; Higuchi, 1985; Odendahl, 1994). For this reason, engineering of plants with cell wall structures that are more susceptible to the krafting, and thus, more amenable to hydrolysis, or are sufficiently altered so as to shunt the above processes is an attractive approach to improve pulping efficiency and potentially alleviate some of the negative environmental impacts of the paper making industry. Apart from the great deal of work in the lignin field for improving the pulping process, many examples can be found based on research aimed at altering the lignin content for improving bio-fuel production (Chen, 2006; Chen & Dixion, 2007; Davision, 2006; Franke et al., 2002), as well as for improving forage crop digestibility (Table 1).

4. GENETIC ENGINEERING OF PHENYLPROPANOID BIOSYNTHETIC PATHWAY

Despite the extensive literature on genetic modifications of lignin biosynthesis in a variety of plants, only a few studies have reported the impact of modified lignin content and composition on pulping and bleaching processes. Nevertheless, significant progress has been made in this field, as summarized in Table 1.

Emerging genetic engineering strategies *in planta* including manipulation of lignin biosynthesis at regulatory level, controlling monolignol polymerization enzymes, and modifications of lignin polymer structure, together with exploration of lignin degrading enzymes from other organisms provide us the necessary tools for producing designer plant species with reduced/altered lignin traits, so as to meet the needs of paper, livestock industries, etc. However, traditional genetic engineering strategies such as upregulation and downregulation of monolignol biosynthetic genes are still applied and have been successful in facilitating lignin decomposition by altering both lignin content and composition. One such example based on genetic modifications in *Leucaena* is described here, providing insights into the reactions, and regulation of genes involved in lignin biosynthesis, and its impact on determining lignin quality for paper industries.

Table 1. Genetic engineering of different lignin biosynthetic genes, and their effects on lignin contents and composition

Plant	Gene	Lignin Content	Lignin Composition	Phenotype	References
Medicago	CAD	Reduced	S/G Ratio↓	Forage Digestibility↑	Baucher et al., 1999
	HCT	"	-	"	Shadle et al., 2007
	C3H/ HCT	"	H/G↑	"	Ziebell et al., 2010
	CP 450 enzymes	"	-	"	Reddy et al., 2005
	COMT	"	-	Cellulose content↑	Marita et al., 2003

Plant	Gene	Lignin Content	Lignin Composition	Phenotype	References
	CCoAOMT	"	-	Forage Digestibility↑	Guo et al., 2001
Nicotiana	PAL/ CCoAOMT	"	S/G Ratio↑	'	Sewalt et al., 1997
	CCR	"	S/G Ratio↑	Pulping Efficiency↑	O'Connell et al., 2002
	CAD	Unchanged	Aldehyde units↑	"	Halpin et al., 1994
	CAD/ CCoAOMT	"	S/G Ratio↓	Forage Digestibility↑	Vailhe et al., 1998
Arabidopsis Alfalfa	POX CCR C4H	Reduced " "	S/G Ratio↑ - S/G Ratio↓	Saccharification↑ Forage Digestibility↑ "	Kavousi et al., 2010 Goujon et al., 2003 Reddy et al., 2005
	HCT	"	High H	"	Shadle et al., 2007
	C3H	"	"	"	Reddy et al., 2005
	CCoAOMT	"	S/G Ratio↑	"	Guo et al., 2001a, b
	Cald5H	Unchanged	S/G Ratio↓	Unchanged	Reddy et al., 2005
Populus	CAD COMT	" Reduced	" "	Forage Digestibility↑ Pulping Efficiency↓	Baucher et al., 1999 Jouanin et al., 2000
	CCoAOMT	"	S/G Ratio↑	Pulping Efficiency↑	Petit, Conil et al., 1999
	CAD	Unchanged	Aldehyde units↑	"	Lapierre et al., 2004
Picea Festuca Maize *Linum*	4CL CCR CAD COMT	Reduced " " "	free phenolics ↑ S/G Ratio↓ H/G Ratio↓ - S/G Ratio↓	Cellulose content↑ Pulping Efficiency↑ Forage Digestibility↑ "	Voelker et al., 2010 Wadenback et al., 2007 Chen et al., 2003 He et al., 2003
	CCoAOMT	"	"	-	Day et al., 2009

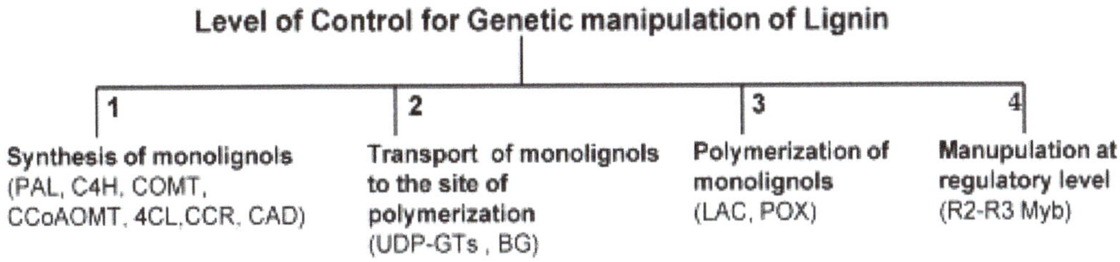

Fig. 3. Levels of control for manipulation of lignin biosynthesis. The control level 1 involves manipulation of lignin biosynthetic genes; level 2 and 3 involves manipulation of genes involved in transport and polymerization of monolignols; while level 4 deals with manipulations of different regulatory factors

4.1 Leucaena leucocephala as a source of pulp

A great deal of knowledge on the molecular biology and regulation of phenylpropanoid biosynthesis has been derived from investigations in plants such as *Arabidopsis*, Alfalfa etc.

While these models will continue to serve as platforms for studying lignifications, a number of other plant species, e.g., *Leucaena leucocephala* have recently been selected for such studies. The paper industry in India mainly uses bamboos, *Eucalyptus* sp., *Casuarina* sp. and *Leuacaena* sp. as a source for paper pulp. Selection of the species depends upon availability, price and acceptability by any given industrial unit. However *Leucaena* sp. is extensively used in India and nearly a quarter of raw materials for paper and pulp industry comes from this plant (Srivastava et al., 2011). *Leucaena* is also valued as an excellent source of nutritious forage.

(a) (b)

Fig. 4. (a) *Leucaena leucocephala* cv. K-636 growing at NCL premises, (b) A transgenic plant cultivated in our green-house. (Photographs courtesy of Shakeel Abbassi, NCL)

It has been estimated that dried leaves of *Leucaena* contain nearly 28-35% of protein content of high nutritional quality. Besides this, *Leuceana* is also an excellent source of firewood, industrial fuel, organic fertilizer, timber and gum (Cottom et al., 1977). A native of Central America, *Leucaena* has been naturalized pan-tropically, with members of its genera being vigorous, drought tolerant, highly palatable, high yielding & rich in protein and grow in wide range of soils (Hughes, 1998; Jones, 1979). However, these attributes are limited by the occurrence of anti-nutritive factors in the fodder, such as tannins and mimosine (Hammond et al., 1989; Hegarty et al., 1964; Jones, 1979).

4.2 Phenylpropanoid biosynthesis genes from Leucaena leucocephala

Different genes involved in phenylpropanoid biosynthesis were studied in detail from *Leucaena*. Table 2 summarizes the details of such genes isolated from *Leucaena* along with their GenBank accession numbers.

4.3 Regeneration system of Leucaena leucocephala

Genetic transformation of plants needs reproducible robust regeneration system. Regeneration of complete plants through tissue culture has made it possible to introduce foreign genes in to plant genome and recover transgenic plants. The limited success rate for regeneration of leguminous trees has been reported in few research works.

Table 2. Phenylpropanoid biosynthesis genes from *Leucaena*: C4H: Cinnamate 4-Hydroxylase; 4CL: 4 Coumarate CoA Ligase; CCoAOMT: Caffeoyl CoA O-Methyl Transferase; COMT: Caffeate *O*-Methyl Transferase; CCR: Cinnamoyl CoA Reductase; CAld5H: Coniferaldehyde 5-Hydroxylase; CAD: Cinnamyl Alcohol Dehydrogenase; POX: Peroxidase; *Partial sequence

Name of Genes	Accession Nos.	cDNA, kb	Amino acid no.	Mm, Da	Annotations
PAL	JN540043*	--	--	--	*Ll-PAL*
C4H	GU183363	1.836	505	57.94	*LlC4H1*
	HQ191221	1.761	505	57.93	*LlC4H2*
	HQ191222	1.760	505	57.99	*LlC4H3*
4CL	FJ205490	1.935	542	58.87	*4CL 1*
	FJ205491*	1.831	519	56.74	*4CL2*
CCoAOMT	DQ431234	0.735	244	*27.55*	*CCoAOMT2*
	DQ431233	0.735	244	*27.57*	*CCoAOMT1*
CCR	EU195224	1.005	334	36.30	*Ll-CCRH2*
	DQ986907	1.011	336	36.52	*Ll-CCRH1*
CAld5H	EU041752	1.826	511	57.42	*Ll-Cald5H*

Name of Genes	Accession Nos.	cDNA, kb	Amino acid no.	Mm, Da	Annotations
CAD	EU870436	1.178	357	38.87	*Ll-CAD*
Lignin POX	GU143879	0.951	316	33.94	*POX1*
	GU143878	0.951	316	34.03	*POX2*
	EU649680	0.951	316	34.20	*Ll-POX*
Glucosylhydrolase	EU328158	1.524	507	57.57	*Ll-GH Family 1*
R2R3 MYB factor	GU901209	0.911	235	26.50	*Ll_R2R3_MYB2_SSM*
Cellulose	FJ871987	3.391	1075	123.30	*Ll-7CesA*
Synthase(CesA)	GQ267555	3.368	1073	119.97	*Ll-8CesA*

The complete *in vitro* plantlet regeneration of *Leucaena* (Cultivar, K-67) from lateral bud explants has been optimized with the maximum shoot multiplication rate of 22 shoots per explants. These regenerated plantlets were transplanted *ex vitro* with 80% survival rates (Goyal et al., 1985). In addition to the regeneration from lateral bud explants, an alternative (both direct and callus mediated/indirect) plantlet regeneration system has been successfully demonstrated with 100% regeneration frequency using cotyledon explants from 3-4 days old plants (Cultivar, K-636). It is interesting to note that the plantlets regenerated from cotyledonary explants rooted without any requirements of growth regulators on basal media (Saafi et al., 2002). Addition of Thidiazuron (TDZ) to the shoot induction medium has substantially improved the number of *in vitro* shoots per explants as compared to the basal shoot induction medium with N6-Benzyladenine (BA). Liquid pulse treatment of the induced shoots with TDZ resulted in the improvement in the subsequent rooting. The plantlets regenerated in this manner showed more than 90% survival rate *ex vitro* when grown in coco-peat mixture (Shaik et al., 2009).

In order to improve overall *in vitro* plantlet regeneration efficiency, attempts to propagate elite (cultivar K-8, K-636) and wild type varieties of *Leucaena* were made by supplementing the basal shoot induction media with puterscin. It has been observed that putriscine (9.3 μM) significantly enhanced the number of regenerated shoots from hypocotyls explants when compared to the induction medium containg only BA (22.2 μM). The incidence of yellowing and leaf abscission was successfully abridged by addition of glutamine (685 μM) or adenine (540 μM) which indirectly added *ex vitro* survival of the plants. All the regenerated plantlets from hypocotyls explants exhibited 100% *in vitro* rooting and were subsequently transplanted *ex vitro* (Sirisha et al., 2008). *In vitro* regeneration system for some other cultivars of *Leucaena* (K-8, K-29, K-68 and K-850) from mature trees derived nodal explants as well as seedlings derived cotyledonary node explants have also been reported, where cultivar K-29 gave the best response *in vitro*. Indirect (through callus phase) somatic embryogenesis of cv. K-29 using 40.28 μM NAA and 12.24 μM were also established. These somatic embryos were further matured in full strength medium (Rastogi et al., 2008).

Considering all above discussed reports, it can be suggested that our reports (Shaik et al., 2009) of improved method also have produced a consistent regeneration system for *Leucaena* which will be beneficial for the mass propagation and genetic transformation of *Leucaena* species. Lignin content in wood pulp adversely influences the quality of paper produced. In *Leucaena* which is an important paper pulp wood crop, it becomes important to identify and multiply elite clones having naturally low lignin content. In addition to this approach there is a need to develop transgenic plants with altered or reduced lignin content for its efficient and eco-friendly removal from pulp. The above mentioned multiple regeneration pathways are an excellent tool to introducing foreign genes. Out of all these methods shoot regeneration from cotyledonary node explants are more responsive to multiple shoot induction (Hussain et al., 2007). Genetic transformation procedures particularly particle bombardment which is considered as most effective means of gene delivery can be applied to the transformation of these shoot meristems. Cotyledonary explant derived multiple shoots form most suitable tissue for genetic transformation due to their higher regeneration frequency. Therefore, our recent study dealt with multiple shoot induction from the cotyledonary nodes of *Leucaena* in response to cytokinins, thidiazuron (TDZ) and N6- benzyladenine (BA) supplemented in half strength MS (½-MS) medium and also their effect on *in vitro* rooting of the regenerated shoots (Fig. 5).

The addition of cytokinins to the medium was found essential for multiple shoot induction. *Leucaena* cotyledonary nodes carried a high potential for rapid multiple shoot regeneration on medium containing lower concentrations of TDZ (0.05 or 0.23 μM) (Shaik et al., 2009). As multiple shoots originated from the mass of closely placed shoot initials of axillary meristems (Fig. 5), this system could be efficiently used for particle bombardment mediated transformation. This efficient and high frequency *in vitro* regeneration system is highly reproducible and can be used for mass propagation and genetic transformation of *Leucaena*.

4.4 Genetic transformation of Leucaena

The genetic transformation protocols based on *Agrobacterium*-mediated and/or direct gene transfers by biolistic bombardment have been successfully applied to numerous woody angiosperm species (Merkle & Nairn, 2005), including *Populus* and *Betula*. The introduction of transgenes have included both sense and antisense strategies (referring to the orientation of the introduced gene into the plant genome) (Strauss et al., 1995; Baucher et al., 1998) and RNAi technology (Merkle & Nairn, 2005). In the antisense strategy, duplex formation between the antisense transgene and the endogenous gene transcripts is proposed to induce the degradation of duplexes and, correspondingly, lead to suppressed gene expression (Strauss et al., 1995). Regeneration system for *Leucaena* has already been established in our previous works (Shaik et al., 2009; Sirisha et al., 2008). To exploit this to produce transgenic *Leucaena* plants for reduced/altered lignin content, various phenylpropanoid pathway genes (C4H, 4CL, CCoAoMT, CCR, CAld5H, CAD and POX) were cloned and used for transformation experiments.

Fig. 5. *In vitro* shoot regeneration in L. leucocephala (cultivar K-636). A: Cotyledonary nodes, B-C: Multiple shoot induction in cotyledonary nodes on ½-MS + TDZ (0.23 µM), D: Shortened and fasciated shoots of *Leucaena* on ½-MS + TDZ (0.45 µM), E: Rooted shoots of *Leucaena* on ½-MS + NAA (0.54 µM) and F: Hardened *in vitro* propagated plant of *L. leucocephala* in Sand: soil mixture. (Shaik et al., 2009)

One day old embryo axes without cotyledons were used as explants for transformation. It was carried out by three methods: 1) Particle bombardment; 2) Particle bombardment followed by co-cultivation and 3) Agro-infusion method. The transformation efficiencies with various gene constructs are summarized in Table 3. It can be observed that maximum efficiency (100%) was noted with CAD using particle bombardment followed by co- cultivation as a means of transformation. However, CCR with the same procedure gave only 10% efficiency. In general it was concluded that a combination of particle bombardment method followed by co-cultivation was most effective in transforming the shoot meristems of *Leucaena*.

Table 3. Phenylpropanoid biosynthesis genes from *Leucaena*: C4H: Cinnamate 4-Hydroxylase; 4CL: 4 Coumarate CoA Ligase; CCoAOMT: Caffeoyl CoA O-Methyl Transferase; COMT: Caffeate *O*-Methyl Transferase; CCR: Cinnamoyl CoA Reductase; CAld5H: Coniferaldehyde 5-Hydroxylase; CAD: Cinnamyl Alcohol Dehydrogenase; POX: Peroxidase; *Partial sequence

Gene	Method	No. of embryo axis used	No. of explants survived	No. of shoots elongated	Avg. shoot length*	No. of shoots used for DNA extraction	No. PCR +ve samples	Transformation efficiency (%)#
CCR	*P B*	67	59	30	3.67	8	4	50
CCR	*PB + CC*	107	94	40	2.97	20	2	10
CCR	*A I*	43	6	6	2.07	6	1	16.66
4CL1	*P B*	231	205	154	3.72	154	92	59.74
4CL1	*PB + CC*	109	83	47	3.41	47	38	80.85
4CL1	*A I*	60	22	04	1.17	4	2	50
CAD	*P B*	438	208	42	3.06	11	7	63.63
CAD	*PB + CC*	173	114	11	2.49	2	2	100
CAD	*A I*	147	32	16	1.68	6	6	100
CAld5H	*P B*	74	54	25	3.47	7	2	28.57
CAld5H	*PB* + CC	99	74	31	2.79	14	11	78.57
CAld5H	*A I*	34	5	4	2.00	2	-	-
C4H	*P B*	-	-	-	-	-	-	-
C4H	*PB + CC*	615	61	11	3.5	10	6	60
C4H	*A I*	-	-	-	-	-	-	-
POX	*P B*	58	47	44	1.36	23	17	73.91
POX	*PB* + CC	79	64	59	1.38	31	27	87.09
POX	*A I*	32	23	19	1.15	9	5	55.56

4.4.1 Peroxidase (LlPOX): A case study

Numerous reports on peroxidase activity in lignin forming tissues have been reported, but only a few isoenzymes have been specifically associated with lignification (Sato et al.,1993; Quiroga et al., 2000; Christensen et al., 2001; Marjamaa et al., 2006). It is likely that the control of the whole lignification process requires a mechanism for the co-ordinated expression and/or activation of the monolignol biosynthetic genes/enzymes and the radical forming peroxidases. Data from transgenic plants down regulated for peroxidase activity has confirmed the role of some POX isoforms in lignin polymerization in tobacco and *Populus sieboldii* (Miq.)X *Populus grandidentata* (Michx.) (Talas-Ogras et al., 2001; Blee et al., 2003; Li et al., 2003b). Both quantitative (up to 50% reduction) and qualitative changes were reported, but no obvious growth phenotypes, other than larger xylem elements were found.

Fig. 6. Genetic transformations in *Leucaena leucocephala* by particle bombardment method using *Anti-LlPOX* construct. A: Embryos arranged for bombardment, B-C: Transient GUS expression, D. Bombarded embryos on selection, D1a and D1b. Regenerating embryos on selection, D2a and D2b. Necrosis and dying of untransformed embryos on selection, E-F: Regeneration and hardening of putative transformants

Recombinant pCAMBIA1301 binary vector harboring partial sequence of *LlPOX* in antisense orientation was used to transform the embryos isolated from aseptically germinated seeds of *Leucaena* (Fig. 6). Two fragments (one from conserved region and another from non- conserved region) of *LlPOX* cloned in anti-sense orientation were used for transformation. The *Leucaena* embryo axes were bombarded with microcarriers coated with recombinant pCAMBIA vectors using PDS-1000/He Biolistic Particle Delivery System. After growing the embryos on regeneration media without selection for one week, these embryos were subjected to three rounds of selections. The plants, which survived were shifted to ½-MS with Cytokinin, 2-ip

(2-isopentenyl adenine) 0.5 mg/ L to enhance elongation of transformed shoots. In all the above cases, the bombarded explants were subjected to transient GUS assay 48 hrs after second bombardment and the putative transgenic plants, which survived three rounds of selection, were analyzed for the gene integration into the plant genome.

As an alternative strategy, *Leucaena* embryo axes were also transformed by a combination of particle bombardment followed by co-cultivation with *Agrobacterium* (GV2260) harboring respective recombinant pCAMBIA vectors. After particle bombardment, the embryo axes were then transferred onto the respective regeneration medium, co-cultivated in the dark at 25 ± 2 °C for 3 days. After co-cultivation, the embryo axes were washed thoroughly with Cefotaxime 250 mg/ L in sterile distilled water and transferred onto the regeneration medium. Higher levels of transient GUS expression confirmed the transformation efficiency. The present study was performed using, two different antisense constructs of *Leucaena* peroxidase gene. We observed severe stunted or retarded growth in plants when transformed using constructs having conserved domain. These plants were found to grow barely up to 0.5 cm, soon followed by the death of the apical meristem and rise of a fresh axillary bud from its axis, which again dies and this process was found to be repeating. As a result, the plant attained a height of 2.5 cm on an average and even failed to produce roots when transferred to rooting medium. When non-conserved *AntiPOX* construct was used in *Leucaena* transformation, normal regeneration was noticed but the plants were thin and slow growing compared to the untransformed control plants. Comparative growth pattern of *Leucaena* are shown in Fig. 7.

Fig. 7. a. Untransformed plant not subjected to genetic transformation, b. putative transformants using construct Anti-*LlPOX (NC)* of *POX* gene and c. putative transformants using constructs *Anti-LlPOX (C)* of *POX* gene. NC-Non conserved; C-Conserved

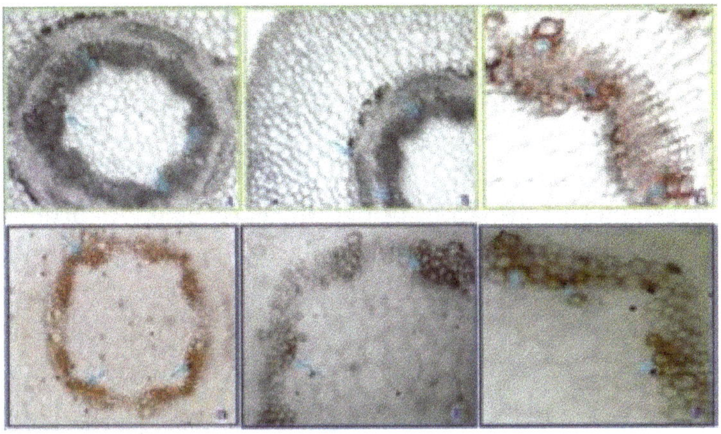

Fig. 8. Immuno-cytolocalization of POX in *Leucaena*. A, B & C stem sections of control plants showing higher levels of POX protein on xylem tissues over the transformed plants D, E & F. Control plants show a well developed vascular bundles (continuous ring) over transformants (discontinuous ring)

LlPOX was immuno-cytolocalized in the transformants generated following the above mentioned protocols. Control and transformed plants of same age group were selected. The control plants showed better growth and bio-metric parameters (height, growth and rooting) over the transformants. POX was immuno-cytolocalized in stem tissues of control untransformed plants (Fig. 8 A, B, C) and putative transformants (Fig. 8 D, E, F), with a view to find whether there exists reduction in peroxidase expression in lignifying tissues (*i.e* vascular bundle and xylem fibres). It was observed that the transformants showed reduced levels of POX near the sites of lignifications. It was also noted that *Leucaena* transformed by *AntiLlPOX* from conserved region resulted in discontinuity in vascular bundle assemblies.

Table 4. Lignin estimation of transgenic *Leucaena* plants. NC-Nonconserved; C-conserved

Genes Down-regulated	Morphological Changes	Reduction in Lignin content
4CL	No change	2-7%
CAld5H	No change	Yet to be analyzed
CCR	Stunted growth	4-13%
CAD	Stunted growth	2-8%
C4H	Stunted growth	Yet to be analyzed
POX(NC)	Stunted Growth	4-9%
POX(C)	Stunted and abnormal growth pattern	6-14%

Likewise, rest of the antisense constructs (*4CL*, *CAld5*, *CCR*, *CAD*, and *C4H*) were successfully utilized for genetic transformation of *Leucaena* and were subsequently characterized for transformation efficiency and lignin content (Table 4). Plants having antisense construct of *C4H*, *CCR*, *CAD* and *POX* showing stunted growth. But in case of *4CL* transformants no such morphological appearance were observed.

5. CONCISE SUMMARY

Extensive research on the chemistry of lignin has revealed its promise as a target for genetic engineering of plants, mostly due to its industrial and agricultural uses. Nevertheless, a significant portion of our comprehension regarding lignin biochemistry is derived from research conducted on model plants such as Arabidopsis, Tobacco, Poplar, and others. Moreover, it is important to transfer this technology to additional plant species. Leucaena, a versatile leguminous tree, is currently being studied to modify its lignin content, as it plays a crucial role in the paper and pulp industry in India. In order to meet the needs of industries, efforts were made to enhance the qualities of pulp production by genetically modifying lignin metabolism. The findings provided in this study emphasize the difficulties and constraints associated with reducing lignin levels in plants. It is crucial to strike a balance between achieving a sufficient reduction in lignin for beneficial purposes, while avoiding negative impacts on the normal growth and development of plants.

These findings may aid in the advancement of Leucaena by modifying its lignin composition and content, resulting in increased extractability of lignin. This would make the paper and pulp industry more cost-effective and environmentally benign. The plant's lignin down regulation has several advantages, including increased efficiency in saccharification for biofuel production, greater fodder digestibility, and enhanced pulping efficiency. Promising advancements in genetic engineering offer the potential to enhance lignin extraction and degradability in pulping processes. However, it is crucial to address the potential issues associated with these techniques, such as increased vulnerability to pathogens, abnormal physical traits, and undesirable metabolic functions, before implementing them on a large scale. To overcome these obstacles, it is crucial to make substantial advancements in comprehending lignin metabolism and its impact on several facets of plant biology.

However, the present genetic engineering technology offers the essential means for a thorough exploration of lignin chemistry, a task that was difficult to achieve using traditional breeding approaches.

6. REFERENCES

Axegard, P.; Jacobson, B.; Ljunggren, S. & Nilvebrant, N.O. (1992). Bleaching of kraft pulps – A research perspective. *Papier*, 46, V16–V25. ISSN 0031-1340

Bate, N. J.; Orr, J.; Ni, W.; Meromi, A.; Nadler-Hassar, T.; Doerner, P. W.; Dixon, R.A.; Lamb, C. J. & Elkind, Y. (1994). Quantitative relationship between phenylalanine ammonia-lyase levels and phenylpropanoid accumulation in transgenic tobacco identifies a ratedetermining step in natural product synthesis. *Proceedings of the National Academy of Sciences* USA, 91, 7608–7612. ISSN 00278424

Baucher, M.; Bernard-Vailhe, M. A.; Chabbert, B.; Besle, J. M.; Opsomer, C.; Van Montagu, M. & Botterman, J. (1999). Downregulation of cinnamyl alcohol dehydrogenase in transgenic alfalfa (*Medicago sativa* L.) and the effect on lignin composition and digestibility. *Plant Molecular Biology*, 39, 437–447. ISSN 1573-5028

Baucher, M.; Halpin, C.; Petit-Conil, M. & Boerjan W. (2003). Lignin: Genetic engineering and impact on pulping. *Critical Reviews in Biochemistry and Molecular Biology*, 38, 305-350. ISSN 1549-7798

Bierman, C.J. (1993). Essentials of Pulping and Papermaking. Academic Press, San Diego. Biermann, C. J. (1996). Handbook of Pulping and Papermaking, (Ed.), *Academic Press*, San Diego, CA. ISSN 0091-679X

Bell-Lelong, D. A.; Cusumano, J. C.; Meye, K. & Chapple, C. (1997). Cinnamate-4-hydroxylase expression in *Arabidopsis*. *Plant Physiology*, 113, 729–738. ISSN 1532- 2548

Blee, K. A.; Choi, J.W.; O'Connell, A.P.; Schuch, W.; Lewis, N. G. & Bolwell, G.P. (2003). A lignin-specific peroxidase in tobacco whose antisense suppression leads to vascular tissue modification. *Phytochemistry*, 64, 163-176. ISSN 0031-9422

Boerjan, W.; Ralph, J. & Baucher, M. (2003). Lignin Biosynthesis. *Annual Review of Plant Biology*, 54, 519-546. ISSN 1040-2519

Boerjan, W. (2005). Biotechnology and the domestication of forest trees. *Current Opinion in Biotechnology*, 16, 159–166. ISSN 0958-1669

Bomati, E. K. & Noel, J. P. (2005).Structural and kinetic basis for substrate selectivity in *Populus tremuloides* sinapyl alcohol dehydrogenase. *Plant Cell*, 17, 1598–1611. ISSN 1532-298X

Boniwell, J. M. & Butt, V. S. (1986). Flavin nucleotide-dependent 3-hydroxylation of 4- hydroxyphenylpropanoid carboxylic acids by particulate preparations from potato tubers. *Zeitschrift für Naturforschung*, 41, 56–60. ISSN 0939-5075

Brownleader, M. D.; Ahmed, N.; Trevan, M.; Chaplin, M. F. & Dey, P. M. (1995). Purification and partial characterization of tomato extensin peroxidase. *Plant Physiology*, 109, 1115-1123. ISSN 1532-2548

Campbell, M. M. & Sederoff, R. R. (1996). Variation in lignin content and composition. Mechanisms of control and implications for the genetic improvement of plants. *Plant Physiology*, 110, 3–13. ISSN 1532-2548

Chabannes, M.; Ruel, K.; Yoshinaga, A; Chabbert, B.; Jauneau, A.; Joseleau, J. P. & Boudet, A. M. (2001). In situ analysis of lignins in transgenic tobacco reveals a differential impact of individual transformations on the spatial patterns of lignin deposition at the cellular and subcellular levels. *Plant Journal*, 28, 271–82. ISSN 1365-313X

Chen, C.; Meyermans, H.; Burggraeve, B.; De Rycke, R. M.; Inoue, K.; Vleesschauwer, V. D.; Steenackers, M.; Van Montagu, M. C.; Engler, G. J. & Boerjan, W. A. (2000). Cell- specific and conditional expression of caffeoyl-coenzyme A-3-O-methyltransferase in poplar. *Plant Physiology*, 123, 853-867. ISSN 1532-2548

Chen, L.; Auh, C. K.; Dowling, P.; Bell, J.; Chen, F.; Hopkins, A.; Dixon, R.A. & Wang, Z. Y. (2003) .Improved forage digestibility of tall fescue (*Festuca arundinacea*) by transgenic down-regulation of cinnamyl alcohol dehydrogenase. *Plant Biotechnology Journal.*, 1, 437–449. ISSN 1863-5466

Chen, F.; Reddy, S.; M, S.; Temple, S.; Jackson, L.; Shadle, G. & Dixon, R.A. (2006). Multi-site genetic modulation of monolignol biosynthesis suggests new routes for formation of Syringyl lignin and wall-bound ferulic acid in alfalfa (*Medicago sativa* L.). *Plant Journal*, 48, 113–124. ISSN 1365-313X

Chen, F. & Dixon, R. A.(2007).Lignin modification improves fermentable sugar yields for biofuel production. *Nature Biotechnology*, 25, 759-761. ISSN 0733-222X

Chiang, V. L.; Puumala, R. J.; Takeuchi, H. & Eckert, R. E.(1988). Comparison of softwood and hardwood kraft pulping. *Tappi Journals* , 71, 173–176. ISSN 0734-1415

Christensen, J. H.; Baucher, M.; O'Connell, A. P.; Van Montagu, M. & Boerjan, W. (2000). Control of lignin biosynthesis. *Molecular Biology of Woody Plants,* Vol.1, (Ed.), No. 64, 227–67. ISBN 978-0-7923-6241-8

Christensen, J. H.; Van Montagu, M; Bauw, G. & Boerjan, W. (2001). Xylem peroxidases: purification and altered expression. *Molecular Breeding of Woody Plants,* Vol.18, 171– 176. ISBN 10- 0-444-50958-5

Croteau, R.; Kutchan, T. M. & Lewis, NG. (2000) Natural products. *Biochemistry and Molecular Biology of Plants*, 1250–1318. ISBN 0943088372

Davison, B. H.; Drescher, S. R.; Tuskan, G. A.; Davis, M. F. & Nghiem, N. P. (2006) .Variation of S/G ratio and lignin content in a *Populus* family influences the release of xylose by dilute acid hydrolysis. *Applied Biochemistry Biotechnolony* 129–132, 427–435. ISSN 1559-0291

Day, A.; Neutelings, G.; Nolin,F.; Grec, S.; Habrant, A.; Crônier, D.; Maher, B.; Rolando, C.; David, H.; Chabbert, B. & Hawkins, S.(2009). Caffeoyl coenzyme A O- methyltransferase down-regulation is associated with modifications in lignin and cell-wall architecture in flax secondary xylem. *Plant Physiology and Biochemistry*, 47, 9-19. ISSN 0981-9428

Escamilla-Trevino, L. L. & Shen, H. (2010). Switchgrass (*Panicum virgatum*) possesses a divergent family of cinnamoyl CoA reductases with distinct biochemical properties. *New Phytologist*, 185(1), 143-155. ISSN 1469-8137

Franke, R.; Humphreys, J. M.; Hemm, M. R.; Denault, J. W.; Ruegger, M. O.; Cusumano, J. C. & Chapple, C. (2002) .The *Arabidopsis REF8* gene encodes the 3-hydroxylase of phenylpropanoid metabolism. *Plant Journal*, 30, 33–45. ISSN 1365-313X

Freudenberg, K. & Neish, A. C. (1968). Constitution and Biosynthesis of Lignin. *Springer- Verlag*, pp129. ISSN 1095-9203

Goujon, T.; Sibout, R.; Pollet, B.; Maba, B.; Nussaume, L.; Bechtold, N.; Lu, F.; Ralph, J.; Mila, I.; Barri`ere,Y.; Lapierre, C., & Jouanin, L. (2003a). A new *Arabidopsis thaliana* mutant deficient in the expression of O-methyltransferase impacts lignins and sinapoyl esters. *Plant Molecular Biology*, 51, 973–989. ISSN 1573-5028

Goujon, T.; Ferret, V.; Mila, I.; Pollet, B.; Ruel, K.; Burlat, V.; Joseleau, J. P.; Barriere, Y.; Lapierre, C. & Jouanin, L. (2003b). Down-regulation of the *AtCCR1* gene in *Arabidopsis thaliana*: effects on phenotype, lignins and cell wall degradability. *Planta*, 217,218–228. ISSN 1432-2048

Goyal, Y.; Bingham, R. L. & Felker, P. (1985). Propagation of the tropical tree, *L. leucocephala* K67, by *in vitro* bud culture. *Plant Cell, Tissue and Organ Culture,* 4, 3-10.ISSN 1573- 5044

Grand, C.; B, A. & Boudet, A. M. (1983). Isoenzymes of hydroxycinnamate: CoA ligasefrom poplar stems: properties and tissue distribution. *Planta*, 158, 225-229. ISSN: 1432- 2048

Gross, G. (1985). Biosynthesis and metabolism of phenolic acids and monolignols. In Biosynthesis and Biodegradation of Wood Components. Higuchi, T. (ed), *Academic Press*, New York, 229-271. ISSN 0003-2697

Guo, D.; Chen, F.; Inoue, K.; Blount, J. W. & Dixon, R. A. (2001a). Downregulation of caffeic acid 3-O-methyltransferase and caffeoyl CoA 3-O-methyltransferase in transgenic alfalfa. impacts on lignin structure and implications for the biosynthesis of G and S lignin. *Plant Cell*, 13, 73–88. ISSN 1532-298X

Guo, D.; Chen, F.; Wheeler, J.; Winder, J.; Selman, S.; Peterson, M. & Dixon, R. A. (2001b). Improvement of in-rumen digestibility of alfalfa forage by genetic manipulation of lignin O-methyltransferases. *Transgenic Research*, 10, 457–464. ISSN 1573-9368

Halpin, C.; Knight, M. E.; Foxon, G. A.; Campbell, M. M.; Boudet, A. M.; Boon, J. J.; Chabbert, B.; Tollier, M. T. & Schuch, W. (1994). Manipulation of lignin quality by downregulation of Cinnamyl alcohol dehydrogenase. *Plant Journal.*, 6, 339–350. ISSN 1365-313X

Hammond, A. C.; Allison, M. J. & Williams, M. J. (1989). Persistence of DHP-degrading bacteria between growing seasons in subtropical Florida. *Leucaena Research Report*, 10, 66. ISSN 0254-8364

Harding, S. A.; Leshkevich, J.; Chiang, V. L. & Tsai, C. J. (2002). Differential substrate inhibition couples kinetically distinct 4-coumarate:coenzyme A ligases with spatially distinct metabolic roles in quaking aspen. *Plant Physiology,* 128, 428-438.ISSN 1532-2548

Hatton, D.; Sablowski, R.; Yung, M. H.; Smith, C.; Schuch, W. & Bevan, M. (1995). Two classes of cis sequences contribute to tissue-specific expression of a *PAL2* promoter in transgenic tobacco. *Plant Journal*, 7, 859–876. ISSN 1365-313X

He, X.; Hall, M. B.; Gallo-Meagher, M. & Smith, R. L. (2003).Improvement of forage quality by downregulation of maize O-methyltransferase. *Crop Science*, 43, 2240–2251. ISSN 0011-183X

Hegarty, M. P.; Court, R. D. & Thorne, P. M. (1964) .The determination of mimosine and 3, 4- DHP in biological material. Australia *Journal of Agricultural Research*, 15, 168. ISSN 0004-9409

Higuchi, T. (1985). Biosynthesis of lignin. In: Higuchi, T. (ed.), Biosynthesis and biodegradation of wood components. Academic Press, New York, 141-160.

Higuchi, T. (1990). Lignin biochemistry: biosynthesis and biodegradation. *Wood Science Technology*, 24, 23–63. ISSN 1432-5225

Hinman, R. L. & Lang, J. (1965) Peroxidase catalyzed oxidation of indole3acetic acid. *Biochemistry*, 4, 144-158 ISSN 1532-2548

Hoffmann, L.; Maur, S.; Martz, F.; Geoffre, P. & Legrand, M. (2002). Purification, cloning and properties of an acyltransferase controlling shikimate and quinate ester intermediates in phenylpropanoid metabolism. *J. Biological Chemistry*, 278, 95–103. ISSN 1083-351X

Hoffmann, L .; Besseau, S.; Geoffroy, P.; Ritzenthaler, C.; Meyer, D.; Lapierre, C.; Pollet, B. & Legrand, M. (2004). Silencing of hydroxycinnamoyl-coenzyme A shikimate/quinate hydroxycinnamoyltransferase affects phenylpropanoid biosynthesis. *The Plant Cell*, 16, 1446–1465. ISSN 1532-298X

Hoffmann, L.; Besseau, S.; Geoffroy, P.; Meyer, D.; Lapierre, C.; Pollet, B. & Legrand, M.(2007). Flavonoid accumulation in *Arabidopsis* repressed in lignin synthesis affects auxin transport and plant growth. *The Plant Cell*, 19, 148-162. ISSN 1532- 298X

Hu, W. J.; Kawaoka, A.; Tsai, C. J.; Lung, J.; Osakabe, K.; Ebinuma, H. & Chiang, V. L. (1998). Compartmentalized expression of two structurally and functionally distinct 4- coumarate:CoA ligase genes in aspen (*Populus tremuloides*). *Proceedings of the National Academy of Science*, 95, 5407– 5412. ISSN 0027-8424

Hu, W. J.; Harding, S. A.; Lung, J.; Popko, J. L.; Ralph, J.; Stokke, D. D.; Tsai, C. J. & Chiang, V. L. (1999). Repression of lignin biosynthesis promotes cellulose accumulation and growth in transgenic trees. *Nature Biotechnology*, 17, 808–812. ISSN 0733-222X

Humphreys,J.M.;H,M.&Chapple,C.(1999).New routes for lignin biosynthesis defined by biochemical characterization of recombinant ferulate 5-hydroxylase, a multifunctional cytochrome P450-dependent monooxygenase. *Proceedings of theNational Academy of Science* (USA), 96, 10045-10050. ISSN 0027-8424

Humphreys,J. M. & Chapple, C.(2002). Rewriting the lignin roadmap. *Current Opinion in Plant Biology*, 5, 224–229. ISSN 1369-5266

Hussain, T. M.; Thummala, C. & Ghanta, R. G. (2007). High frequency shoots regeneration of *Stercula urens* Roxb. an endangered tree species through cotyledonary node cultures. *African Journal of Biotechnology*, 6(15), 1643-1649. ISSN 1684-5315

Jones, H. D. (1984) .Phenylalanine ammonia-lyase: Regulation of its induction, and its role in plant development. *Phytochemistry*, 23, 1349–1359. ISSN 0031-9422

Jones, L.; Ennos, A. R. & Turner, S. R. (2001). Cloning and characterization of irregular xylem4 (irx4): a severely lignin deficient mutant of *Arabidopsis*. *Plant Journal*, 26, 205–216. ISSN 1365-313X

Jones, R. J. (1979). The value of *Leucaena leucocephala* as feed for ruminants in the tropics. *World Animal review*, 32, 10. ISBN 92-5-100650-4

Jouanin, L.; Goujon, T.; Nadai, V. D.; Martin, M. T.; Mila, I.; Vallet, C.; Pollet, B.; Yoshinaga, A.; Chabbert, B.; Conil, M. P. & Lapierre, C. (2000). Lignification in transgenic poplars with extremely reduced caffeic acid O methyltransferase activity. *Plant Physiology*., 123, 1363–1374. ISSN 1532-2548

Kao, Y. Y.; Harding, S. A. & Tsai, C. J. (2002). Differential expression of two distinct phenylalanine ammonia-lyase genes in condensed tannin-accumulating and lignifying cells of quaking aspen. *Plant Physiology*, 130, 796–807. ISSN 1532-2548

Kavousi, B.; Daudi, A.; Cook, C. M.; Joseleau, J. P.; Ruel, K.; Devoto, A.; Bolwell, G. P. & Blee, K. A.(2010). Consequences of antisense down-regulation of a lignification- specific peroxidase on leaf and vascular tissue in tobacco lines demonstrating enhanced enzymic saccharification. *Phytochemistry*, 71, 531-542. ISSN 0031-9422

Kojima, M. & Takeuchi, W. (1989). Detection and characterization of p-coumaric acid hydroxylase in mung bean, *Vigna mungo*, seedlings. *Biochemical Journal*, 105, 265– 270. ISSN 1470-8728

Kumar, A. & Ellis, BE. (2001). The phenylalanine ammonia-lyase gene family in raspberry. Structure, expression, and evolution. *Plant Physiology*, 127, 230– 239. ISSN 1532-2548 Lacombe, E. & Hawkins, S. (1997). Cinnamoyl CoA reductase, the first committed enzyme of the lignin branch biosynthetic pathway: cloning, expression and phylogenetic relationships. *Plant Journal*, 11(3), 429-441. ISSN 1365-313X

Lapierre, C.; Pilate, G.; Pollet, B.; Mila, I.; Leple, J.C.; Jouanin, L.; Kim, H. & Ralph, J. (2004). Signatures of cinnamyl alcohol dehydrogenase deficiency in poplar lignins. *Phytochemistry*, 65, 313–321. ISSN 0031-9422

Larsen, K. (2004). Molecular cloning and characterization of cDNAs encoding cinnamoyl CoA reductase (CCR) from barley (Hordeum vulgare) and potato (*Solanum tuberosum*). *Journal of Plant Physiology*, 161(1), 105-12. ISSN 1532-2548

Lauvergeat, V. & Lacomme, C. et al. (2001). Two cinnamoyl-CoA reductase (CCR) genes from *Arabidopsis thaliana* are differentially expressed during development and in response to infection with pathogenic bacteria. *Phytochemistry*, 57(7), 1187-1195. ISSN 0031-9422

Lee, D.; Meyer, K.; Chapple, C. & Douglas, C. J. (1997). Antisense suppression of 4-coumarate: coenzyme A ligase activity in Arabidopsis leads to altered lignin subunit composition. *Plant Cell*, 9, 985– 998. ISSN 1532-298X

Leple, J. C. & Dauwe, R. et al. (2007). Downregulation of cinnamoyl-coenzyme A reductase in poplar: multiple-level phenotyping reveals effects on cell wall polymer metabolism and structure. *Plant Cell*, 19(11), 3669-3691. ISSN 1532-298X

Lewis, N. G. & Yamamoto, E. (1990) Lignin: Occurrence, biogenesis, and biodegradation. *Annual Review Plant Physiology and Plant Molecular Biology*, 41, 455–496. ISSN 1040- 2519

Leyva, A.; Liang, X.; Pintor-Toro, J. A.; Dixon, R. A. & Lamb, C. J. (1992). cis-Element combinations determine phenylalanine ammonia-lyase gene tissue specific expression patterns. *Plant Cell*, 4, 263–271. ISSN 1532-298X

Li, L.; Popko, J. L.; Umezawa, T. & Chiang, V. L. (2000). 5-Hydroxyconiferyl aldehyde modulates enzymatic methylation for syringyl monolignol formation, a new view of monolignol biosynthesis in angiosperms. *Journal of Biological Chemistry*, 275, 6537–6545. ISSN 1083-351X

Li, L.; Cheng, X. F.; Leshkevich, J.; Umezawa, T.; Harding, S. A. & Chiang, V. L. (2001). The last step of syringyl monolignol biosynthesis in angiosperms is regulated by a novel gene encoding sinapyl alcohol dehydrogenase. *Plant Cell*, 13, 1567–1585. ISSN 1532-298X

Li, L. & Cheng, X. et al. (2005). Clarification of cinnamoyl co-enzyme A reductase catalysis in monolignol biosynthesis of Aspen. *Plant Cell Physiology*, 46(7), 1073-1082. ISSN 1471-9053

Li, X.; Weng, J. K. & Chapple, C. (2010). The Growth Reduction Associated with Repressed Lignin Biosynthesis in *Arabidopsis thaliana* Is Independent of Flavonoids. *The Plant Cell*, 22, 1620-1632. ISSN 1532-298X

Lindermayr, C.; Fliegmann, J. & Ebel, J. (2003). Deletion of a single amino acid residue from different 4-coumarate-CoA ligases from soybean results in the generation of new substrate specificities. *Journal of Biological Chemistry*, 278, 2781–2786. ISSN 1083- 351X

Logemann, E.; Parniske, M. & Hahlbrock, K. (1995). Modes of expression and common structural features of the complete phenylalanine ammonia-lyase gene family in parsley. *Proceedings of the National Academy of Science* USA, 92, 5905–5909. ISSN 0027-8424

Ma, Q. H. & Tian, B. (2005). Biochemical characterization of a cinnamoyl-CoA reductase from wheat. *Journal of Biological Chemistry*, 386(6), 553-560. ISSN 1083-351X

Ma, Q. H. (2007). Characterization of a cinnamoyl-CoA reductase that is associated with stem development in wheat. *Journal of Experimental Botany*, 58(8), 2011-2021. ISSN 0022–0957

Marita, J. M.; Vermerris, W.; Ralph, J. & Hatfield, R.D. (2003) Variations in the cell wall composition of maize brown midrib mutants. *Journal of Agricultural and Food Chemistry*, 51, 1313–1321. ISSN 0021–8561

Marjamaa, K.; Hildén, K.; Kukkola, E.; Lehtonen, M.; Holkeri, H.; Haapaniemi, P.; Koutaniemi, S.; Teeri, T. H.; Fagerstedt, K. & Lundell, T.(2006). Cloning, characterization and localization of three novel class III peroxidases in lignifying xylem of Norway spruce (*Picea abies*). *Plant Molecular Biology*, 61, 719-732. ISSN 0167–4412

Martz, F.; Maury, S.; Pincon, G. & Legrand, M. (1998). cDNA cloning, substrate specificity and expression study of tobacco caffeoyl-CoA 3-O-methyltransferase, a lignin biosynthetic enzyme. *Plant Molecular Biology*, 36, 427–437. ISSN 0167–4412

McInnes, R. & Lidgett, A. et al. (2002). Isolation and characterization of a cinnamoyl-CoA reductase gene from perennial ryegrass (*Lolium perenne*). *Journal of Plant Physiology*, 159(4), 415-422. ISSN 0176–1617

Merkle, S. & Nairn, J. (2005). Hardwood tree biotechnology. *In Vitro Cellular & Developmental Biology - Plant*, 41, 602-619. ISSN 1054-5476

Mizutani, M.; Ohta, S. & Sato, R. (1997). Isolation of a cDNA and a genomic clone encoding cinnamate 4-hydroxylase from *Arabidopsis* and its expression manner in planta. *Plant Physiology*, 113, 755-763.ISSN 0032–0889

Moore, K. J. & Jung, H. G. (2001). Lignin and fiber digestion. *Journal of Range Management*, 54, 420-430. ISSN 0022-409X

Nair, R. B.; Xia, Q.; Kartha, C. J.; Kurylo, E.; Hirji, R. N.; Datla, R. & Selvaraj, G. (2002). *Arabidopsis CYP98A3* mediating aromatic 3-hydroxylation. Developmental regulation of the gene, and expression in yeast. *Plant Physiology*, 130, 210–20. ISSN 0032–0889

Nieminen, K. M.; Kauppinen, L. & Helariutta, Y. (2004). A weed for wood? *Arabidopsis* as a genetic model for xylem development. *Plant Physiology*, 135, 653-659. ISSN 0032– 0889

O'Connell, A., Holt, K., Piquemal, J., Grima-Pettenati, J., Boudet, A., Pollet, B., Lapierre, C., Petit-Conil, M., Schuch, W. and Halpin, C. (2002). Improved paper pulp from plants with suppressed cinnamoyl- CoA reductase or cinnamyl alcohol dehydrogenase. *Transgenic Research*, 11, 495–503. ISSN 0962–8819

Odendahl, S. (1994). Environmental protection and consumer demands: a review of trends and impacts. *Pulp & Paper Canada*, 95, 144–148. ISSN 0316-4004

Ohl, S.; Hedrick, S. A.; Chory, J. & Lamb, C. J. (1990). Functional properties of a phenylalanine ammonia lyase promoter from *Arabidopsis*. *Plant Cell*, 2, 837–848. ISSN 1040–4651

Osakabe, Y.; Ohtsubo, Y.; Kawai, S.; Katayama, Y. & Morohoshi, N. (1995). Structures and tissue specific expression of genes for phenylalanine ammonialyase from a hybrid aspen. *Plant Science*, 105, 217–226.ISSN 0168–9452

Osakabe, K.; Tsao C.; Li, L.; Popko, J. L.; Umezawa, T.; Carraway, D. T.; Smeltzer, R. H.; Joshi, C. P. & Chiang, V. L. (1999). Coniferyl aldehyde 5-hydroxylation and methylation direct syringyl lignin biosynthesis in angiosperms. *Proceedings of the Natlional Academy of Science* (USA), 96, 8955-8960. ISSN 0027-8424

Pakusch, A. E.; Kneusel, R. E. & Matern, U. (1989) S-Adenosyl-lmethionine: trans-caffeoyl- coenzyme A 3-O-methyltransferase from elicitor-treated parsley cell suspension cultures. *Archives of Biochemistry and Biophysics* 271: 488–494. ISSN 0003-9861

Petersen, M.; Strack, D. & Matern, U. (1999) .Biosynthesis of phenylpropanoids and related compounds. In: Biochemistry of Plant Secondary Metabolism. Wink, M. (Ed.), Vol. 2, pp 151–222, *Sheffield Academic Press*, Sheffield, UK. ISSN 0309-0787

Petit-Conil, M.; Lapierre, C.; Pollet, B.; Mila, I.; Meyermans, H.; Pilate, G.; Leple, J. C.; Jouanin, L.; Boerjan, W. & de Choudens, C. (1999). Impact of lignin engineering on pulping processes of poplar trees. In Proceedings of the 27th EUCEPA Conference. Paris: ATIP Press, pp. 1–10.

Pichon, M. & Courbou, I. et al. (1998). Cloning and characterization of two maize cDNAs encoding cinnamoyl-CoA reductase (CCR) and differential expression of the corresponding genes. *Plant Molecular Biology*, 38(4), 671-676. ISSN 1573-5028

Piquemal, J. Lapierre, C.; Myton, K.; O'Connell, A.; Schuch, W.; Grima-Pettenati, J. & Boudet A.M. (1998). Down-regulation of cinnamoyl-CoA reductase induces significant changes of lignin profiles in transgenic tobacco plants. *Plant Journal*, 13(1), 71-83. ISSN 1365-313X

Prasad, JVNS.; Korwar, G. R.; Rao, K. V.; Mandal, U. K.; Rao, G. R.; Srinivas, I.; Venkateswarlu, B.; Rao, S. N. & Kulkarni, H. D. (2010). Optimum stand density of *Leucaena leucocephala* for wood production in Andhra Pradesh, Southern India. *Biomass Bioenergy*, 348, 1-9. ISSN 0961-9534

Quiroga, M.; Guerrero, C.; Botella, M. A.; Barcelo, A.; Amaya, I.; Medina, M. I.; Alonso, F. J.; Forchetti, S. M.; Tigier, H. & Valpuesta, V. (2000). A tomato peroxidase involved in the synthesis of lignin and suberin. *Plant Physiology*, 122, 1119-1127. ISSN 1532-2548

Raes, J.; Rohde,A.; Christensen, J. H.; Peer,Y. V. & Boerjan, W. (2003). Genome-wide characterization of the lignification toolbox in *Arabidopsis*. *Plant Physiology*, 133, 1051–1071. ISSN 1532-2548

Ralph, J.; Hatfield, R. D.; Piquemal, J.; Yahiaoui, N.; Pean, M.; Lapierre, C. & Boudet, A. M. (1998). NMR characterization of altered lignins extracted from tobacco plants down-regulated for lignification enzymes cinnamyl alcohol dehydrogenase and cinnamoyl CoA reductase. *Proceedings of the National Academy of Sciences*, 95, 12803– 12808. ISSN 0027-8424

Rastogi, S. & Dwivedi, U. N. (2003a). *Agrobacterium tumefaciens* mediated transformation of *Leucaena leucocephalas* multipurpose tree legume. *Physiology and Molecular Biology of Plants, 9*, 207-216. ISSN 0974-0430

Rastogi, S.; Rizvi, S. M. H.; Singh, R. P. & Dwivedi, U.N. (2008). *In vitro* regeneration of *Leucaena leucocephala* by organogenesis and somatic embryogenesis. *Biologia Plantarum*, 52, 743-748. ISSN 1573-8264

Reddy, M. S.; Chen, F.; Shadle, G.; Jackson, L.; Aljoe, H. & Dixon, R. A. (2005). Targeted down-regulation of cytochrome P450 enzymes for forage quality improvement in alfalfa (*Medicago sativa* L.). *Proceedings of the National Academy of Sciences* USA, 102, 16573–16578. ISSN 0027-8424

Rogers, L. A. & Campbell, M. M.(2004). The genetic control of lignin deposition during plant growth and development. *New Phytologist*, 164, 17-30. ISSN 1469-8137

Saafi, H. & Borthakur, D. (2002). In vitro plantlet regeneration from cotyledons of the tree legume of *Leucaena leucocephala. Plant Growth Regulation*, 38, 279-285. 1573-5087

Sarkanen, K. V. & Ludwig, C. H. (1971). Lignins: Occurrence, Formation, Structure, and Reactions. *New York: Wiley- Interscience*, pp 916. ISBN 0-471-75422-6

Sato, Y.; Sugiyama, M.; Gorecki, R. J.; Fukuda, H. & Komamine, A. (1993). Interrelationship between lignin deposition and the activities of peroxidase isoenzymes in differentiating tracheary elements of Zinnia. Analysis using Lα- aminoxy-β- phenylpropionic acid and 2aminoindan2phosphonic acid. *Planta*, 189, 584589. ISSN 1432-2048

Schoch, G.; Goepfert, S.; Morant, M.; Hehn, A.; Meyer, D.; Ullmann, P. & Werck-Reichhart, D. (2001). CYP98A3 from *Arabidopsis thaliana* is a 3 -hydroxylase of phenolic esters, a missing link in the phenylpropanoid pathway. *Journal of Biological Chemistry*, 276, 36566–36574. ISSN 1083-351X

Sewalt, V.; Ni, W.; Blount, J. W.; Jung, H. G.; Masoud, S. A.; Howles, P. A.; Lamb, C. & Dixon, R. A. (1997a) .Reduced lignin content and altered lignin composition in transgenic tobacco down-regulated in expression of L-phenylalanine ammonia- lyase or cinnamate 4-hydroxylase. *Plant Physiology*, 115, 41–50. ISSN 1532-2548

Sewalt, V.J.H.; Ni, W. T.; Jung, H. G. & Dixon, R. A. (1997b). Lignin impact on fiber degradation: increased enzymatic digestibility of genetically engineered tobacco (*Nicotiana tabacum*) stems reduced in lignin content. *Journal of Agricultural and Food Chemistry*, 45, 1977–1983. ISSN 0021–8561

Shadle, G.; Chen, F.; Srinivasa Reddy, M. S.; Jackson, L.; Nakashima, J. & Dixon, R. A. (2007). Down-regulation of hydroxycinnamoyl CoA:shikimate hydroxycinnamoyl transferase in transgenic *alfalfa* affects lignification, development and forage quality. *Phytochemistry*, 68, 1521–1529.

Shaik, N. M.; Arha, M.; Nookaraju, A.; Gupta, S. K.; Srivastava, S.; Yadav, A. K., Kulkarni, P. S.; Abhilash, O. U.; Vishwakarma, R. K.; Singh, S.; Tatkare, R.; Chinnathambi, K.; Rawal, S. K. & Khan, B. M. (2009). Improved method of in vitro regeneration in *Leucaena leucocephala* – a leguminous pulpwood tree species. *Physiology and Molecular Biology of Plants*, 15(4), 311-318. ISSN 0971-5894

Shanfa, L.; Yohua, Z.; Laigeng, L. & Vincent, L. C. (2006). Distinct Roles of Cinnamate 4- hydroxylase Genes in *Populus. Plant Cell Physilogy*, 47(7),905–914. ISSN 0032–0781

Sirisha,V. L.; Prashant, S.; Ranadheer, D.; Ramprasad, P.; Shaik, N. M.; Arha, M.; Gupta, S. K.; Srivastava, S.; Yadav, A. K.; Kulkarni, P. S.; Abhilash, O. U.; Khan, B. M.; Rawal, S. K. & Kavi Kishor, P. B. (2008). Direct shoot organogenesis and plant regeneration from hypocotyl explants in selected genotypes of *Leucaena leucocephala*—A leguminous pulpwood tree. *Indian Journal of Biotechnology*, 7, 388-93. ISSN 0972– 5849

Srivastava, S.; Gupta, R.; Arha, M.; Vishwakarma, R.K.; Rawal, S.K.; Kavi Kishor, P.B. & Khan, B.M. (2011). Expression analysis of cinnamoyl-CoA reductase (CCR) gene in developing seedlings of Leucaena leucocephala: A pulp yielding tree species. *Plant Physiology and Biochemistry*, 49,138-145. ISSN 0981-9428

Stafford, H. A. & Dresler, S. (1972). 4-Hydroxycinnamic acid hydroxylase and polyphenolase activities in *Sorghum vulgare. Plant Physiology*, 49, 590–595. ISSN 1532-2548

Steeves, C.; F¨orster, H.; Pommer, U. & Savidge, R. (2001). Coniferyl alcohol metabolism in conifers. I. Glucosidic turnover of cinnamyl aldehydes by UDPG: coniferyl alcohol glucosyltransferase from pine cambium. *Phytochemistry* ,57,1085–93. ISSN 0031-9422

Strauss, S. H.; Rottmann, W. H.; Brunner, A. M. & Sheppard, L.A. (1995) Genetic engineering of reproductive sterility in forest trees. *Molecular Breeding*, 1, 5-26. ISSN 1380–3743

Talas-Ogras, T.; Kazan, K. & Gözükirmizi, N. (2001). Decreased peroxidase activity in,transgenic tobacco and its effect on lignification. *Biotechnology Letter* , 23, 267-273. ISSN 0141-5492

Vailhe, M.A.B.; Besle, J. M.; Maillot, M. P.; Cornu, A.; Halpin, C. & Knight, M. (1998). Effect of down-regulation of cinnamyl alcohol dehydrogenase on cell wall composition and on degradability of tobacco stems. *Journal of the Science of Food and Agriculture*, 76, 505–514. ISSN 0022-5142

Vietmeyer, N. & Cottom, B. (1977). In Leucaena: Promising Forage and Tree Crop for the Tropics. Ruskin, F. R. (Ed.), *Proceedings of the National Academy of Sciences, Washington*, DC, 22-80. ISSN 0027-8424

Voelker, S. L.; Lachenbruch, B.; Meinzer, F. C.; Jourdes, M.; Ki, C.; Patten, A. M.; Davin, L. B.; Lewis, N. G.; Tuskan, G. A.; Gunter, L.; Decker, S. R.; Selig, M.; Sykes, R.; Himmel, M. E.; Kitin, P.; Shevchenko, O. & Strauss, S. H. (2010). Antisense down- regulation of 4CL expression alters lignification, tree growth, and saccharification potential of field-grown poplar. *Plant Physiology*, 154, 874-886. ISSN 1532-2548

Voo, K. S.; Whetten, R. W.; O'Malley, D. M. & Sederoff, R. R. (1995) 4-coumarate:coenzyme a ligase from loblolly pine xylem. Isolation, characterization, and complementary DNA cloning. *Plant Physiology*, 108, 85–97. ISSN 1532-2548

Wadenback, J.; von Arnold, S.; Egertsdotter, U.; Walter, M. H.; Grima-Pettenati, J.; Goffner, D.; Gellerstedt, G.; Gullion, T. & Clapham, D. (2007). Lignin biosynthesis in transgenic Norway spruce plants harboring an antisense construct for Cinnamoyl CoA reductase (CCR). *Transgenic Research,* 17(3), 379-92. ISSN 0962–8819

Whetten, R. & Sederoff, R. (1995). Lignin biosynthesis. *Plant Cell*, 7, 1001–1013. ISSN 1532- 298X

Whetten, R. W.; MacKay, J. J. & Sederoff, R. R. (1998). Recent advances in understanding lignin biosynthesis. *Annual Review of Plant Physiology and Plant Molecular Biology*, 49, 585–609. ISSN 1040-2519

Ye, Z. H.; Kneusel, R. E.; Matern, U. & Varner, J. E. (1994). An alternative methylation pathway in lignin biosynthesis in Zinnia. *Plant Cell*, 6, 1427–1439. ISSN 1532-298X

Ye, Z. H. & Varner, J. E. (1995). Differential expression of two O-methyltransferases in lignin biosynthesis in Zinnia elegans. *Plant Physiology* ,1084, 59–467. ISSN 1532-2548

Ye, Z. H. (1997). Association of caffeoyl coenzyme A 3-O-methyltransferase expression with lignifying tissues in several dicot plants. *Plant Physiology*, 115, 1341–1350. ISSN 1532-2548

Zhong, R.; Demura, T. & Ye, Z. H. (2006). SND1, a NAC domain transcription factor, is a key regulator of secondary wall synthesis in fibres of *Arabidopsis*. *Plant Cell*, 18, 3158- 3170. ISSN 1532-298X

Zhong, R.; Lee, C.; Zhou, J.; McCarthy, R. L. & Ye, Z. H. (2008). A battery of transcription factors involved in the regulation of secondary cell wall biosynthesis in *Arabidopsis*. *Plant Cell*, 20, 2763-2782. ISSN 1532-298X

Zhou, R. & Jackson, L. et al. (2010). Distinct cinnamoyl CoA reductases involved in parallel routes to lignin in *Medicago truncatula*. *Proceedings of the National Academy of Sciences* U S A, 107(41), 17803-17808. ISSN 0027-8424

Zimmerlin, A.; Wojtaszek, P. & Bolwell, G. P. (1994). Synthesis of dehydrogenation polymers of ferulic acid with high specificity by a purified cellwall peroxidase from French bean (*Phaseolus vulgaris* L.). *Biochemical Journal*, 299, 747-753. ISSN 1470-8728

05 PLANT GENETIC ENGINEERING FOR VIRAL RESISTANCE

1. INTRODUCTION: AN OVERVIEW

Genetic engineering is recognized as a crucial method for enhancing agricultural output and minimizing losses caused by both living organisms and environmental factors, both in the field and during storage (Sairam and Prakash, 2005; Yuan et. al., 2011). This chapter focuses mostly on the phenomenon of resistance to viral infections. Hence, it is crucial for individuals initiating a research endeavor in plant genetic engineering to possess a comprehensive comprehension of the diverse array of plant transformation techniques at their disposal, the different types of (plasmid) constructs that can be employed, and the potential consequences they may have on the safety of the end product.

Plant transformation methods encompass Agrobacterium-mediated transformation, microprojectile bombardment/biolistics, direct protoplast transformation, electroporation of cells and tissues, electro-transformation, the pollen tube pathway method, as well as infiltration, microinjection, silicon carbide mediated transformation, and liposome mediated transformation (Rakoczy-Trojanowska, 2002). Each of these strategies, which will be explained in this chapter, use a distinct strategy to transport DNA near chromosomes, where it can subsequently integrate. The utilization of markers and reporter genes in combination with various methodologies, along with the inclusion of supplementary sequences to aid integration, may have certain biosafety ramifications.

The objective of this chapter is to assess various techniques employed for plant transformation and to analyze the specific outcomes achieved in terms of virus resistance through two methods: Agrobacterium-mediated transformation and electro-transformation. The discussion will also cover the implications on biosafety.

2. PLANT TRANSFORMATION

Figure 1 depicts the overall structure of a plant cell. In order to achieve stable genetic transformation, it is necessary to transport the targeted DNA fragment through the cell wall, cell membrane, cytoplasm, and nuclear membrane into the nucleus, unless it is eliminated through pre-treatment.

Likewise, in the process of organelle transformation, the DNA needs to be conveyed over the organelle membrane in order to access the organelle's matrix. After entering the nucleus, the targeted DNA fragment needs to undergo recombination with the host chromosome in order to integrate into it. This integration ensures that the inheritance pattern of the DNA fragment matches that of the host chromosome. Currently, our understanding of the mechanics of integration is limited, and there is no specific targeting of individual chromosomes. Furthermore, there is still a significant amount of work that needs to be accomplished in regards to organelle change. The topics mentioned are extensively examined in Tinland's 1996 study, Ow's 2002 publication, Tzfira et al.'s 2004 research, Maliga's 2004 work, and Kumar et al.'s 2006 study.

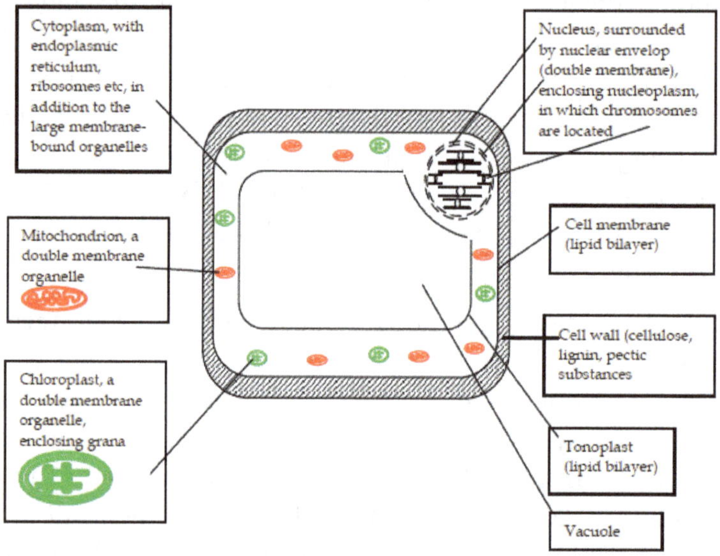

Fig. 1. Diagram to illustrate the structure of a plant cell

Genetic engineering will result in plants that carry additional genes from the same or other species, and are thus referred to as transgenic plants. Such plants may also be referred to as transformed plants, because their genotype and phenotype may have changed from one state to another, for example from disease-susceptible to disease-resistant. The term 'transformed plant' also relates to the original method of *Agrobacterium*-mediated transformation, where, after the bacterium transfers the T-DNA, the recipient plant cells become 'cancerous', and result in cankers that characterize the crown gall disease.

The term 'genetically modified plant' is much broader than 'transformed plant'. While a strict definition of 'plant transformation' may not be practical because of the varying genetics of the plants, it is generally accepted that the plant must be confirmed as transformed based on Southern DNA hybridization evidence of multiple independent transformation events showing different sized fragments correlating to different profiles of the restriction endonucleases used, and appropriate sustained phenotypic expression of the transgene exclusively in the transformed plants (Potrykus, 1991, Birch 2002).

In plant pathology, the concept of resistance and susceptibility genes is widespread. In the gene-for-gene model of pathogen incompatibility, resistance (R) genes and associated avirulence (Avr) genes have been well studied (reviewed in Belkhadir *et al.*, 2004). But one aspect that has not been well elucidated is the concept of susceptible genes. Very few susceptibility genes have been identified. However one example is the Os8N3, a host disease-susceptibility gene for bacterial blight of rice which is a vascular disease caused by *Xanthomonas oryzae* pv. *oryzae* (Yang *et al.*, 2006). Deletion of Os8N3 in rice plants by genetic engineering approaches is postulated to result in genetically engineered plants resistant to *Xanthomonas oryzae* pv. *Oryzae*. One may ask if these plants will be considered transgenic. Most susceptibility genes, however, are thought to be essential for plant growth and development, such that their deletion or mutation will result in non-viable plants.

It must be noted that 'transgenic', 'transformed' and 'genetically modified' are not equivalent terms. The definition of transformed plants should be broad enough to encompass deletions. Southern hybridization probes targeting the deletion junctions may be used to confirm the deletion event, and absence of susceptibility gene product can be demonstrated.

Conventional breeding also results in re-assortment of genes from the two genomes that are crossed, and is therefore some form of genetic modification as well. However, no genetic engineering is involved in the process, and the crosses usually involve closely related species. Genetic engineering is particularly useful when the gene/trait of interest is not present in closely related species, making conventional breeding impossible. Furthermore, conventional breeding is not precise, since extensive re-assortment of genes occurs when two species are crossed, and takes a very long time. Genetic engineering therefore becomes the approach of choice especially when there are no Biosafety issues to grapple with. The most common approach in genetic engineering involves excising the gene of interest using restriction enzymes, and cloning it into a plant transformation vector before transfer into the cells of the target species where the gene will integrate into the chromosome. This process is usually more precise and faster. In this case the resulting plants are transgenic, because they carry a gene from another species, introduced by genetic engineering.

Many transgenic plants resistant to diseases have been produced. Collinge and co-workers list the most common genes used for transgenic disease-resistant crops that have been field- tested (Collinge *et al.*, 2010). Against fungal diseases, these are the polygalacturonse inhibitor protein (grape, raspberry, tomato), proteinase (soybean), R-gene (Rpg-1, Pi9, RB2, Rps1-k) (barley, festuca, potato, soybean), cell death regulator (wheat), toxin detoxifier (barley, wheat) pathogenesis-related proteins (barley, wheat, grape, cotton, peanut, potato, rice, sweet potato, sorghum, tobacco), chitinases (alfalfa, apple, cotton, melon, onion, papaya, squash, carrot, peanut, rice, tobacco, wheat, tomato), oxalate oxidases (bean, cowpea, lettuce, sunflower, peanut, potato, soybean, tobacco), thionin (barley, potato, rice), antimicrobial peptides (cotton, grape, plum, poplar, tobacco, wheat), cecropin (cotton, maize, papaya), stilbene synthase (potato, tobacco), and antimicrobial metabolites (grape, potato, strawberry, tobacco). Against bacterial diseases, attacin (apple), cecropin (apple, papaya, pear, potato, sugarcane), hordothionin (rice, tomato), indolicidin (tobacco), lysozyme (citrus, potato, sugarcane), megainin (grape), proteinase K (rice, tomato), R-gene of pepper, tomato, rice (tomato), and transcription factors (tomato) have been field-tested. Against plant viruses, single-stranded DNA binding G5 protein (cassava), viral movement proteins (raspberry, tomato), ribonuclease (pea, potato, wheat), replicase (cassava, papaya, potato, tomato), nuclear inclusion protein (melon, potato, squash, wheat), coat protein (alfalfa, barley, beet, grape, lettuce, maize, melon, papaya, pea, peanut, pepper, pineapple, plum, potato, raspberry, soybean, squash, sugarcane, tobacco, tomato, wheat). Virus resistance will be discussed further in section 2.1.

Despite performing well in field tests, most of the transgenic plants have not been commercialized. For instance, coat protein transgenic plants make up three quarters of commercialized virus resistant plants. However, the newer and more sophisticated approaches such as RNA interference are set to become more predominant on the market.

There are still numerous obstacles to plant transformation. The effectiveness of most techniques varies depending on the plant species, and might even be exclusive to certain species or cultivars. Typically, the objective of transformation is to target a limited number of cells or an organ, with the aim of inducing growth and regeneration of an entire plant. For many agriculturally significant species, the process of regenerating entire plants in a controlled environment is not commonly practiced. Therefore, there are certain crucial crops that lack a standardized, dependable, and replicable transformation method. Hence, the endeavors to advance and enhance transformation techniques persist.

The available methods for plant transformation include Agrobacterium-mediated transformation, microprojectile bombardment/biolistics, direct protoplast transformation, electroporation, electro-transformation, as well as other methods like microinjection, silicon carbide-mediated transformation, and liposome-mediated transformation. Each of these approaches, which will be addressed in this chapter, use a distinct strategy to transport DNA toward chromosomes, where the DNA can subsequently integrate.

2.1 Plant viruses

2.1.1 Plant viral diseases

Genetic engineering in biotechnology has the capacity to enhance agricultural productivity by equipping crops with improved abilities to withstand both biotic and abiotic stress. Various research groups are investigating various aspects of both living and non-living limitations in order to enhance agricultural productivity. Nevertheless, this chapter will solely focus on biotic stress, with a specific emphasis on plant viruses. Plant viruses have a substantial negative impact on the productivity of all cultivated crops. As of the year 2000, there are approximately 675 recognized plant viruses, however the yearly financial losses in crops caused by these viruses amount to a staggering $60 billion (Fields 1996).

There are various ways of controlling viral diseases such as:

- The use of disease-free planting material. Virus-free stocks are obtained by virus elimination through heat therapy and/or meristem tissue culture. This approach is effective for seed-borne viruses, but is ineffective for viral diseases transmitted by vectors.
- Adopting cultural practices that minimize epidemics, for example by crop rotation, quarantine, rouging diseased plants and using clean implements. Pesticides may also be
- Used to control viral vectors, but the virus may be transmitted to the plant before the vector is killed.
- Classical cross protection, in which a mild strain of the virus is used to infect the crop, and protects the crop from super-infection by a more severe strain of the virus.
- Use of disease resistant planting material. Natural resistance against viruses may be bred into susceptible lines through classical breeding methods or transferred by genetic engineering.
- Engineered cross protection. This involves integration of pathogen-derived or virus- targeted sequences into DNA of potential host plants, and conveys resistance to the virus from which the sequences are derived.

Of all the methods of controlling viral diseases listed above, engineered cross protection seems to have a lot of potential that is only now beginning to be exploited. Before genetic engineering techniques were more widely accepted and applied, natural disease resistance genes bred into target cultivars by classical breeding methods constituted the major focus for introducing disease resistance into plants.

There are 139 monogenic and 40 polygenic virus resistance traits that have been described (Khetapal *et al.*, 1998; Hull 2001), but very few have been cloned, and in most cases the mechanism of resistance has not been elucidated (Ellis *et al.*, 2000; Dinesh-Kumar *et al.*, 2000). Virus-resistant crops that have been obtained by classical breeding include sugarcane resistant to Sugarcane mosaic potyvirus (SCMV) and gerkins (cucumber) resistant to Cucumber mosaic virus (CMV). The N-gene of *Nicotiana glutinosa* that is responsible for the necrotic local lesion reaction of TMV, has also been bred into some *N. tabacum* lines, resulting in the hypersensitive reaction and no systemic infection. Classical breeding has also been used to convey polygenic traits.

2.1.2 Non-viral genes

One approach to protect plants against a viral infection is by the expression of a single chain variable fragment (scFv) antibody directed against that particular virus (Tavladoraki *et al.*, 1993; Voss *et al.*, 1995). This has been demonstrated for the icosahedral Artichoke mottle crinkled tombusvirus (AMCV) and the rod-shaped Tobacco mosaic tobamovirus (TMV). However, the resistance obtained this way is not broad-spectrum resistance.

An approach that can yield broad-spectrum resistance to viral diseases is to target the inhibition of production of a product that is essential for the establishment of infection in the cell. An example is S-adenosylhomocysteine hydrolase (SAHH), an enzyme involved in the transmethylation reactions that use S- adenosyl methionine as a methyl donor (Masuta *et al.*, 1995). Lowering expression of the enzyme suppresses the 5'-capping of mRNA that is required for efficient translation. Overexpression of cytokinin in crops results in stunting. This phenotype may be due to induction of acquired resistance (Masuta *et al.*, 1995).

Expression of the pokeweed (*Phytolacca americana)* antiviral protein (PAP), a ribosome inhibiting protein (RIP), in plants protects the plants against infection by viruses (Ready *et al.*, 1986; Lodge *et al.*, 1993). In this case, expression of this single gene in the plant results in protection against a wide range of plant viruses.

2.1.3 Pathogen-derived resistance

Definition

Pathogen-derived resistance (PDR), also called parasite-derived protection is the resistance conveyed to a host organism as a result of the presence of a transgene of pathogen origin in the target host organism (Sanford & Johnson, 1985). The concept of pathogen-derived resistance predicts that a 'normal' host-pathogen relationship can be disrupted if the host organism expresses essential pathogen-derived genes. The initial hypothesis was that host organisms expressing pathogen gene products at incorrect levels, at the wrong developmental stage or in dysfunctional forms, may disrupt the normal replication cycle of the pathogen and result in an attenuated or aborted infection.

Classical cross protection

Pathogen derived resistance is an extension of the phenomenon of "cross protection" in which inoculation of a host plant with a milder strain of a pathogen can protect the plant from superinfection by more severe strains of the same or a very closely related pathogen (Wilson 1993). An example of cross protection is in tobacco where infecting tobacco plants with the U1 strain of tobacco mosaic tobamovirus (TMV) protects the plants against future infections with a more virulent strain of TMV.

In practice, the protected plants usually become superinfected, and so the definition given above is not practical. For practical purposes, cross protection is still defined by an earlier definition as "the use of a virus to protect against the economic damage by severe strains of the same virus" (Gonsalves & Garnsey, 1989). Classical cross protection, according to this practical definition, has been evaluated in the field in some countries outside Africa for the control of Citrus tristeza closterovirus (CTV), Papaya ringspot potyvirus (PRSV), Zucchini yellow mosaic potyvirus (ZYMV) and Cucumber mosaic cucumovirus (CMV) (ibid).

Engineered protection

The genetic engineering approach to cross protection was first demonstrated by Powell-Abel and co-workers who expressed the TMV coat protein gene in transgenic plants and obtained some degree of resistance against TMV (Powell-Abel *et al.*, 1986). Many viral genes and gene products have since been shown to be effective in conveying engineered PDR. Engineered PDR can be divided into protein-based PDR (coat protein-, replicase- and movement protein-mediated resistances, using these proteins in their wild type or defective forms) and nucleic acid-based PDR (antisense, sense and satellite RNA-mediated resistances, defective interfering RNA or DNA and antiviral ribozymes).

In general, when classical cross protection is incomplete, smaller lesions than in control non- protected plants are formed, indicating reduced movement and maybe reduced replication as well. On the other hand, transgenic plants engineered to confer protection to TMV show no reduction in movement or replication. However, the local lesions for PDR against PVX indicate a reduction in virus replication and movement (Hemenway *et al.*, 1988). This demonstrates the similarity between classical and engineered protection. The phenotype of PDR varies from delay in symptom development, through partial inhibition of virus replication, to complete immunity to challenge virus or inoculated viral RNA (Wilson, 1993; Baulcombe, 1996). Even a simple delay in symptom development could be useful if it allows plant biomass, seed or fruit development to outpace disease development.

Coat protein-mediated resistance

Coat protein-mediated resistance (CP-MR) is the phenomenon by which transgenic plants expressing a plant virus coat protein (CP) gene can resist infection by the same or a homologous virus. The level of protection conferred by CP genes in transgenic plants varies from immunity to delay and attenuation of symptoms. CP-MR has been reported for more than 35 viruses representing more than 15 different taxonomic groups including the tobamo-, potex-, cucumo-, tobra-, carla-, poty-, luteo-, and alfamo- virus groups. The resistance requires that the CP transgene be transcribed and translated. Hemenway and co- workers (1998) have demonstrated direct correlation between CP expression level and the level of resistance obtained. The case of CP-MR to TMV is is important because most of the earlier and more detailed work on CP-MR was done with TMV (Bevan *et al.*, 1985; Beachy *et al.*, 1986; Powell- Abel *et al.*, 1986; Register 1988 and Powell *et al.*, 1990).

2.1.4 RNA interference (RNAi)

RNA interference is the process that depends on small RNAs (sRNAs) to regulate the expression of the eukaryotic genome (Hohn and Vazquez, 2011). This newly elucidated mechanism opens up many possibilities for genetic engineering interventions due to the simplicity of the molecules involved. Small RNAs regulate many biological processes in plants, including maintenance of genome integrity, development, metabolism, abiotic stress responses and immunity to pathogens (Hohn and Vazquez, 2011; Katiya-Agarwal, 2011). The RNA molecules involved are small and of two types, micro RNAs (miRNAs) and small interfering RNAs (siRNAs). miRNAs are transcribed from miRNA genes by RNA polymerase II, as primary miRNA (pri-miRNA) that then folds into a stem loop structure (imperfectly base-paired) that is then processed in a very specific manner by a number of proteins to result in 22-24mer RNA molecules. These RNA molecules are then incorporated into AGO1 or AGO10 and guide the complex to target mRNA for cleavage or translational inhibition on the basis of sequence complementarity. siRNAs on the other hand, are derived from perfectly paired double stranded RNA (dsRNA) precursors, that are derived either from antisense or are a result of RNA-dependent RNA polymerase (RDR) transcription. Details of types of siRNAs, their origins and processing, and how this approach is used to convey virus resistance in transgenic plants are presented in Hohn and Vazquez (2011) and Katiya-Agarwal (2011).

3. *AGROBACTERIUM*-MEDIATED TRANSFORMATION

The structure of the Ti plasmid and the requirement for transfer has been established, and the natural host range of the bacterium expanded (Cheng 2004). The first reports of *in vitro* plant transformation utilised the ability of *Agrobacterium tumefaciens* to transfer a specific region of its Ti plasmid DNA into plant cells where they subsequently become integrated into the plant cell genome (Marton *et al.*, 1979; Barton *et al.*, 1983; Herrera-Estrella *et al.*, 1983). This application is based on the observation that in natural diseases of dicotyledonous plants, crown gall disease caused by *Agrobacterium tumefaciens* and hairy root disease caused by *Agrobacterium rhizogenes*, the bacterium transfers part of the DNA of its Ti or Ri plasmid DNA respectively into the host plant where it becomes integrated into the host genome (Herrera-Estrella *et al.*, 1983). The plant host cells are referred to as transformed. The transferred DNA is referred to as the T-DNA and is demarcated by conserved left and right border sequences (ibid). The integrated genes are passed on to the progeny of the initially infected cell, and their expression (using the host's transcription and translation machinery) results in the cancerous growth that characterise the crown gall or hairy root diseases that results. The tumours produce specific amino acid derivatives called opines that are utilized by the *Agrobacterium* as a carbon source (Zupan and Zambryski, 1997). Within the T-DNA is a 35 kb virulence (vir) region that includes the genes virA to virR (Zhu *et al.* 2000), flanked by imperfect 25 bp direct repeat sequences known as the left and right borders. A number of virulence genes (chv) located on the *Agrobacterium* chromosome mediate chemotaxis and attachment of the bacterium to the plant cell wall (Zupan & Zambryski, 1997).

In adapting the *Agrobacterium* system to genetic engineering, only the sequences that are essential for transfer and integration into the host genome have been retained, and DNA sequences of interest are inserted into the transferred DNA region. The first generation plasmids for *Agrobacterium*-mediated plant transformation were the disarmed Ti-plasmids. The oncogenes within the left and right borders of the naturally occurring plasmid pTiC58 were replaced with pBR322 sequences, to give pGV3850 (Zambryski *et al.*, 1983), and further improved by the addition of a selectable marker (Bevan *et al.*, 1983). Use of intermediate vectors enabled use of smaller plasmids with unique cloning sites for initial cloning experiments in *E. coli* (Matzke & Chilton 1981). The intermediate vector could be transferred from *E. coli* to *Agrobacterium* by conjugation, utilizing a helper plasmid, e.g. RK2013, to supply the requirements for conjugation (ibid). Homologous recombination between the intermediate plasmid and a resident disarmed Ti-plasmid of the *Agrobacterium* (e.g. pGV3850) resulted in a larger plasmid known as a cointegrate disarmed Ti-plasmid.

In a different approach, the virulence genes were placed in a separate plasmid such as pAL4404 where these functions would be provided in *trans* for the transfer of DNA on another smaller plasmid with only the left and right borders, markers and other sequences of interest that need to be transferred such as pBin19 in the same *Agrobacterium* cell (Zupan & Zambryski, 1997). This system is known as the binary vector system. The vectors carry a broad host range replication origin, e.g. *ori* V of pBin 19, which allows replication in *E. coli* and *Agrobacterium*. The *A. tumefaciens* is used most extensively in plant transformation because of the belief that the DNA transfer is discreet, with high proportion of integration events with single or low T-DNA copy number, compared to other methods of plant transformation (Zupan & Zambryski, 1997).

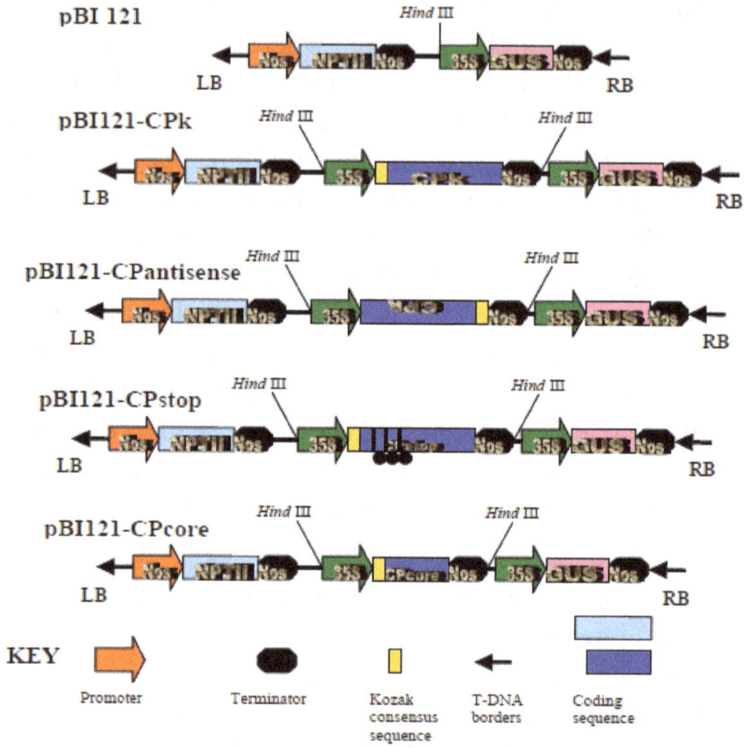

Fig. 2. Illustration of the binary plasmids used for tobacco transformation by *Agrobacterium*- mediated transformation

Plasmid origin of replication may encourage rearrangements and recombination, leading to silencing and deletion of transgene in subsequent generations. Gene disruption may occur at the site of insertion, resulting in loss of some essential functions (Birch, 1997). It is therefore important to obtain as many transformants as possible so as to be able to disregard all abnormal regenerants resulting from this or other phenomena. T-DNA transfer occurs sequentially but not always completely from the right border to the left border (Wang *et al.*, 1984).

Recently, it has also been realized that some sequences outside the borders also get transferred, and integrate into the host genome (Parmyakova *et al.*, 2008). This is undesirable in genetically modified plants for commercial release. Current efforts are to reduce or even eliminate these undesirable effects through using special vector constructs that prevent integration of vector sequences. It is thought that integration of sequences outside the borders is a result of erroneous recognition of either right or left border sequences, and Vir D proteins are central to this event. However, the transfer always starts at or adjacent to the left right borders. The reduction can be achieved by using vectors that have positive or negative selection markers, or easily identifiable markers, outside the T-DNA, or using vectors with increased numbers of terminal repeats, or with left terminal repeats surrounded by native DNA regions that serve as termination enhancers, or the so-called 'green vectors' in which the sequences outside the T DNA have been removed (Parmyakova *et al.*, 2008) Alternatively, one can use vectors in which the undesirable sequences can be removed by mechanisms such as site-specific recombination, or use vectors with sequences of plant origin only. But there still are problems associated with each approach.

Despite these limitations, *Agrobacterium*-mediated transformation is still a very useful tool in plant molecular virology. In our laboratory, *Agrobacterium*-mediated transformation was used as a tool to evaluate mechanisms of resistance to Cowpea aphid-borne mosaic virus (CABMV) in *Nicotiana benthamiana*, an experimental host of the virus. CABMV is a positive sense RNA virus that is a member of the genus *Potyvirus* (Sithole-Niang *et al.*, 1996; Mundembe *et al.*, 2009). In an experiment to evaluate the mechanisms of pathogen-derived resistance, *N. benthamiana* was transformed with recombinant pBI 121 carrying various forms of the CABMV coat protein gene, following the method of co-cultivation of leaf explants with *A. tumefaciens* described by An *et al.* (1987). The constructs used were pBI121- CPk which results in an expressed CABMV coat protein, pBI121-PC which results in antisense CP, pBI121-CPstop which results in a form of the CP mRNA that cannot be translated and CPcore which results in only the core region of the CP, together with a pBI121 control.

Evaluation of the responses of transgenic plants obtained indicate that coat protein- mediated resistance only results in delayed symptom development, while RNA mediated approaches may result in recovery or immunity. Out of 68 CP expressing transgenic plants challenged with CABMV, 19 expressed delayed symptom development; and none displayed immunity. Out of 26 CP stop lines, 3 displayed delayed symptom development, 4 tolerance, and 3 recovery phonotypes. Out of 49 antisense lines, 1 displayed delayed symptom delayed symptom development and 3 lines showed modified symptoms.

At the time of carrying out these experiments cowpea could not be transformed in a reliable, reproducible manner, and many research groups were working towards developing a suitable transformation procedure. However, the experiments with transgenic tobacco served the purpose of evaluating the effectiveness of the different approaches. Coat protein mediated resistance would only result in delayed symptom development, RNA mediated approaches are likely to give higher levels of resistance, maybe even immunity.

Therefore, as the method for cowpea transformation become available one would know which particular constructs to use to get the desired levels of resistance.

4. MICROPROJECTILE BOMBARDMENT/ BIOLISTICS

Microprojectile bombardment, also known as biolistics, is the most commonly used method falling into the category of direct gene transfer methods. In direct gene transfer methods a plasmid in which the sequences of interest are cloned is delivered across the various plant cell barriers by physical means to enter the cell where integration into the plant genome may occur. The vectors used in direct plant transformation methods usually include the gene of interest cloned between a promoter and a terminator, and the plasmid components of an origin of replication, an antibiotic resistance gene, a selectable marker for use in plants (e.g. herbicide or antibiotic resistance) or reporter gene (e.g. GUS, luciferase genes). The whole plasmid may be transferred into the plant cell and may be integrated into the plant genome as a whole or as fragments. The barriers to be crossed by the DNA in direct DNA transfer methods are the cell wall and the cell membrane before it can cross the cytoplasm and the nuclear envelop to enter the nucleoplasm where the DNA may integrate into the plant genome (Figure 1). Some direct DNA transfer procedures utilize whole plasmids, supercoiled or linear, which may ultimately integrate as a whole, or at least large parts thereof, including the gene of interest (Smith *et al.*, 2001).

Direct gene transfer methods were developed in an effort to transform economically important crops that remained recalcitrant to *Agrobacterium*-mediated transformation because of limitations such as genotype and host cell specificity. Some direct gene transfer methods may also circumvent difficult tissue culture methods.

Sanford and co-workers (1987) were the first to report of plant transformation by microprojectile bombardment. Gold or tungsten particles coated with DNA are propelled at high speed toward the plant tissue where they may penetrate the plant cell walls to introduce the DNA into the cytoplasm, vacuoles, nucleus or other structures of intact cells. A modified bullet gun or electric discharge gun is used to propel the particles (Klein *et al.*, 1987; Christou *et al.*, 1988). Inside the cell, the DNA may be expressed transiently for two or three days before being degraded, or may become integrated into the nuclear or chloroplast genome, and considered stably integrated if it is passed faithfully to subsequent generations. DNA-coated particles delivered into the nucleus are 45 times more likely to be transiently expressed than those delivered to the cytosol, and 900 times more likely to be expressed than those delivered to the vacuole (Yamashita *et al.*, 1991). Efficiency of transformation is influenced by the stage of the cell cycle (Iida *et al.*, 1991; Kartzke *et al.*, 1990). The

DNA is also likely to be expressed if it is delivered to the cell close to the time the nuclear membrane disappears at mitosis (Bower & Birch, 1990; Vasil *et al.*, 1991).

Direct DNA transfer methods seem to result in transformants with higher copy numbers than *Agrobacterium*-mediated transformation methods (Hadi *et al.*, 1996; Christou *et al.*, 1989). The multiple copies may be integrated at the same or tightly linked loci, most likely in relation to replication forks or integration hot spots resulting from initial integration events (Cooley *et al.*, 1995, Kohli *et al.*, 1998). Increasing the amount of DNA entering the cell in bombardment increases the copy number (Smith *et al.*, 2001). The DNA may undergo rearrangements (deletions, direct repetitions, inverted repetitions, ligation, concatamerization) prior to, or during integration (Cooley *et al.*, 1995). The site of integration is thought to be random. Ninety percent of T-DNA integrations are into random sites within transcriptionally active regions (Lindsey *et al.*, 1993).

Like *Agrobacterium*-mediated transformation, microprojectile bombardment also results in integration of vector sequences if they are part of the DNA molecule bombarded into the plant cell (Kohli *et al.*, 1999). However, microprojectile bombardment provides an opportunity for the introduction of minimal gene cassettes into the cells. In this approach, only the required gene expression cassettes (promoter, coding region of interest, terminator) is bombarded into the plant cells, or can be co-transformed together with marker genes to be removed before commercialization (Yao *et al.*, 2007; Zhao *et al.*, 2007). While the screening and selection might be more difficult, probably depending on detection of the gene sequence or gene product of interest, the approach is very attractive since reporter genes and selection markers are completely avoided (Zhao *et al.*, 2007).

Marker genes are unnecessary in established transgenic plants, and also limit options when additional transgenes are to be added (stacking) to the original transgenic line. Herbicide resistance genes may potentially be transferred to weeds by outcrossing. Consumers may also worry about the possibility of antibiotic resistance genes spreading to gut microflora, even though there is no scientific evidence for this.

A variation of the microprojectile bombardment method designed to increase the chances of integration is the **Agrolistic transformation method.** In this method, the transforming plasmid is transferred to the plant cell by a direct mechanism together with a second plasmid coding for *A. tumefaciens* proteins involved in the integration process (Zupan & Zambryski, 1997). Transient expression of the *A. tumefaciens* proteins will direct integration of the plasmid into the plant cell genome. As a result, entry of the plasmid into the cell is by a direct/physical mechanism, but integration into the genome is by a mechanism similar to *Agrobacterium*-mediated transformation. The agrolistic transformation method was expected to address one of the main drawbacks of the microprojectile bombardment method which is that there seem to be a high incidence of high copy number. However, a second drawback that the gene gun accessories are very expensive is still valid.

5. ELECTROPORATION AND PEG-MEDIATED TRANSFORMATION OF PROTOPLASTS

Plant cell walls can be removed by enzymatic degradation to produce protoplasts. Polyethylene glycol (PEG) causes permeabilization of the plasma membrane, allowing the passage of macromolecules into the cell. Pazkowski and co-workers were the first to produce transgenic plants after PEG transformation of protoplasts, and many more monocotyledonous and dicotyledonous species have now been transformed using this method (Pazkowski *et al.* 1984). In electroporation, the protoplasts are subjected to an electric pulse that renders the plasma membrane of the protoplasts permeable to macromolecules. The cell wall and whole plants can be regenerated, if procedures exist.

Thetransgenicplantsgeneratedusingthesemethodsseemtohavecharacteristics similar to those of plants derived from all other direct transformation methods. However, it is important to note that carrier DNA (usually ~500 bp fragments of calf thymus DNA) is usually included in the transformation mixture to increase transformation efficiency. This may have some consequences in terms of prevalence of transgene rearrangements and integration of superfluous sequences (Smith *et al.*, 2001).

The cell cycle stage of the protoplasts at the time of transformation influence the transgene integration pattern. Non-synchronized protoplasts produce predominantly non-rearranged single copy transgenes in contrast to M phase protoplasts that give multiple copies usually at separate loci (Kartzke *et al.*, 1990). The S phase protoplasts give high copy numbers, usually with rearrangements. Irradiation of protoplasts shortly before or after addition of DNA in direct transformation procedures increases both the frequency of transformation and number of integration sites (Koehler *et al.*, 1989, 1990, Gharti-Chhertri *et al.*, 1990). This is consistent with a mechanism of integration that is partly mediated by DNA repair mechanisms.

The main drawbacks of these methods are that protoplast cultures are not easy to establish and maintain, and regeneration of whole plants from the protoplasts is often unreliable for some important species.

6. ELECTROPORATION OF INTACT CELLS AND TISSUES

DNA can be introduced into intact cells and tissues in a manner similar to electroporation of protoplasts. Thus pollen, microspores, leaf fragments, embryos, callus, seeds and buds can be used as targets for transformation (Rakoczy-

Trojanowska 2002). Protocols for efficient electroporation of cell suspensions of tobacco, rice and wheat (Abdul-Baki, *et al.*, 1990; De la Pena, *et al.*, 1987; Zaghmout and Trolinder, 1993), and protocols for regeneration of transgenic plants are available. For maize in particular, the transformation efficiencies are comparable to those obtained by bombardment (Dashayes *et al.*, 1985; D'Halluin *et al.*, 1992).

7. ELECTRO-TRANSFORMATION

DNA can also be delivered into cells, tissues and organs by electrophoresis (Ahokas 1989; Griesbach and Hammond, 1994; Songstad *et al.*, 1995). This method is known as transformation by electrophoresis or electro-transformation. The tissue to be transformed is placed between the cathode and anode. The anode is placed in a pipette tip containing agarose mixed with the DNA to be used for transformation. The assembly is illustrated in Figure 3.

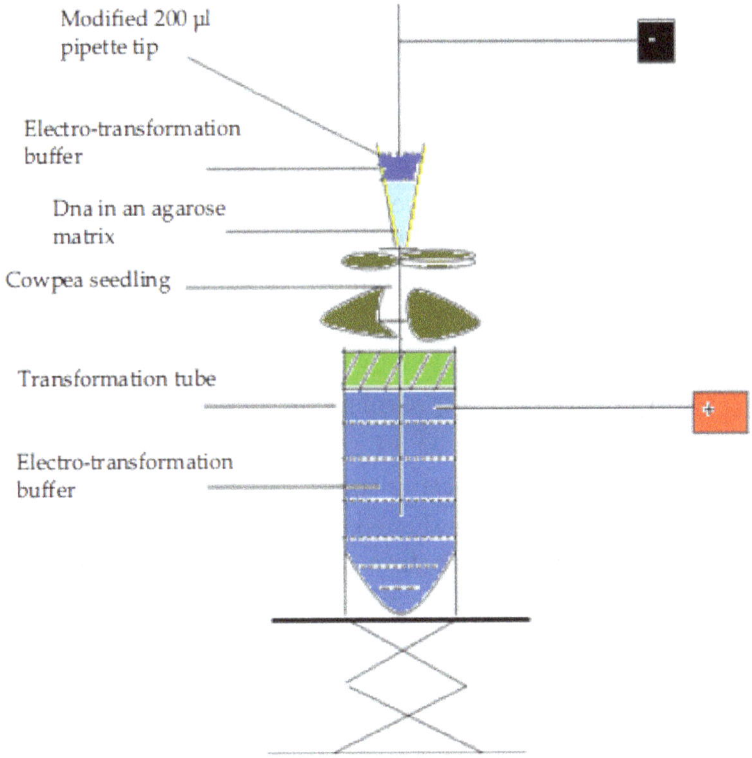

Fig. 3. Diagrammatic illustration of the electro-transformation equipment and experimental set-up

We used this method of transformation on cowpea seedlings, at a time when there was no efficient, reliable, reproducible method for cowpea transformation. The main obstacles to cowpea transformation were that the tissues into which DNA could be introduced failed to regenerate whole plants. We therefore decided to target apical meristems for transformation. In the event of successful transformation, the seeds from transgenic branches of the cowpea plants would be transgenic, and could be screened for desired transformation events.

We had previously made constructs based on CABMV coat protein gene designed to confer various levels of resistance to the virus in transgenic plants (Figure 2). Circular or linearised binary plasmid constructs were electrophoresed into the apical meristematic region of cowpea seedling of various ages and lengths, untreated or pre-treated with acid or alkali, under various conditions of current and voltage as summarized in Table 1.

7.1 Electrotransformation of cowpea

Cowpea (*Vigna unguiculata* C until the seeds germinated (7 – 12 d). Variety 475/89) seeds were sterilized by shaking in 10% (v/v) bleach for 10 min at room temperature, and washed with double distilled water for 5 min. The seeds were then rolled on a moistened paper towel and placed in a beaker with water and incubated in the growth room at 28 For each transformation attempt, a seedling was removed from the paper towel, pre-treated (where applicable) and placed in the transformation tube. About 1 µl of DNA (0.5µg/µl, circular, or linearized by *Nhe*I or *Nhe*I/*Nde*I digestion) was mixed with about 9 µl of 2% (v/v) low melting point agarose (made up in transformation buffer) and allowed to set at the tip of a 200 µl pipette whose tip had been widened by cutting. Both the pipette tip and the transformation tube (Figure 2) were filled with transformation buffer (0.12 M LiCl, 1 mM Hepes, 0.54 mM MgCl2, 0.005% L-ascorbic acid, pH 7.2). The setup (Figure 3) was connected to a power source and allowed to run under the various current and voltage settings.

The aspects of the seedlings that were noted include the height and age of the plant on the day of manipulation, whether the cotyledons were still attached to the plant or had fallen off, and whether the first true leaves were open or closed. The pretreatments were: none, punched meristem, seedling were exposed to temperatures of 35 °C for 1 hour before manipulation, the manipulations were carried out at increased temperatures of >30 °C, meristems and leaves pretreated with 0.1M HCl, or 0.1 M CaCl2, or 2,4-D + kinetin, NAA + BAP. The voltage settings used were DC or AC, at 30, 40, 125 or 250 V; the current was either 1.0 or 0.15 mA), the duration was kept constant at 15 min. The distance between the electrodes varied with the length of the seedling, and was recorded.

A common feature of the GUS positive plants in Table 1 is that the manipulations were carried out on plants that had straight stems, first true leaves open and cotyledons still attached to the seedling. No pre-treatment other than maybe punching the meristem appear to be necessary. The pre-treatments except punching the meristem do not seem to increase transformation efficiency. Both DC and AC are effective in delivery DNA to the plant cells.

Table 1. below summarizes the potentially transgenic events that were obtained in the experiment

Plant ID at screening	DNA construct	Current/ Time/ Distance	Age (days)/ Size	Stem	First true leaves	Cotyledons	Notes
		between	**(cm)**				
		electrodes					
217	pBI121-	0.15 V	7 d	Straight	Open	On	No
	CPcore,	15 min	8 cm				pretreatment
	circular	7 cm					
301	pBI121-	0.15 V	8 d	Straight	Open	On	No
	CPk, *Nhe*I	15 min	6 cm				pretreatment,
	linearized	1.5 cm					AC 30 sec
309	pBI121-	0.15 V	3 d	Straight	Open	On	No
	CPk, *Nhe*I	15 min	5 cm				pretreatment
	linearized	7 cm					
398	pBI121-	0.15 V	8 d	Straight	Open	On	Punched
	CPk, *Nhe*I	15 min	9 cm				meristem
	linearized	6 cm					

The leaves of GUS positive plants had a sectored appearance; this was not unexpected since the transformation procedure targets the general apical meristem area of the cowpea seedling. As a result, both meristematic and somatic cells may become transformed, to result in a chimeric plant. Such a chimeric plant appears as a mosaic of transformed and non- transformed sectors, and poses a challenge in terms of sampling especially in this particular case where a destructive GUS assay was used. Since PCR is very sensitive and amplifies any signal present, the CP transgene could be detected in some GUS positive plants. However, the signal detected by both the GUS assay and PCR could be transient, and Southern analysis is the standard way of determining whether integration has occurred. Southern and other analyses of these lines through subsequent generations, if fertile, would be necessary. There is need to ensure that the germline is transformed to enable the transgene to be passed to subsequent generations.

GUS positive sectors were obtained only from plants that had cotyledons attached, open first true leaves and had developed straight stems at the time of manipulation. The electrotransformation procedure stresses the seedling, and only those seedlings that have developed sufficiently will take up exogenous DNA, survive and develop using the food reserve of cotyledons as well as the photosynthate from first true leaves. The pBI121 binary constructs used in this experiment have a gene for kanamycin resistance.

However, kanamycin resistance is not an effective assay against germinating cowpea seedlings since the germinating cowpea seedlings were not affected by kanamycin. This is probably because of the large food reserves of the seedlings.

The various seedling pre-treatments except punching the meristem did not appear to improve transformation efficiency. Punching the meristem wounds the seedling and may make the meristematic cells more accessible to the exogenous DNA since the epidermal cells will have been removed. Acid and calcium chloride pretreatments were expected to make the cell wall and cell membrane respectively more permeable to DNA. Besides chemically weakening the cell wall, acid pretreatment may also induce the production of expansins that may result in further weakening of cell walls (Cosgrove, 2001). The heat and plant growth substance pretreatments were expected to induce other chemical messengers and heat shock proteins that may increase the chances of integration events in the cell (Hong & Verling, 2001). However, no improvement in transformation efficiency was observed.

The mechanism of DNA integration after uptake by electrophoresis is not known, but is likely to occur by non-homologous recombination into sites on the genome that are undergoing repair or replication, as is the case for other direct DNA transfer methods (Smith *et al.*, 2001). Not all GUS-positive lines tested CP-positive possibly because of incomplete transfer. This also means that it is possible that some transformants were GUS-negative but CP-positive, and these would not detected in this screening procedure.

Transformation by electrophoresis, if successful, is a procedure that can be used to avert one of the major concerns of GMOs. The procedure does not necessarily require the use of selectable markers such as antibiotic or herbicide resistance genes, and only the exact sequence required for a particular characteristic in the transgene may be used. It is not understood how integration would occur, but T-DNA borders do not seem to be required. DNA integration by direct transformation methods appears to be random. In this experiment, transformation is not enhanced by pre-treatment with high temperature, hydrochloric acid, calcium chloride, kinetin, BAP or NAA. Both circular and linearised DNA seemed to be effective. However, the seedling must have developed a straight stem with the first true leaves open, but the cotyledons must be intact. This may be important in ensuring survival of the seedling after the rather harsh handling and subjection to electrophoresis that stresses the plant.

8. OTHER METHODS OF PLANT TRANSFORMATION

8.1 Microinjection

DNA can also be delivered to the plant cell nucleus or cytoplasm by microinjection. This approach is more widely used for large animal cells such as frog egg cells or cells of mammalian embryo. Animal cells are usually immobilised with a holding pipette and gentle suction. For plant cells, the cell wall which contains a thick layer of cellulose and lignins is a barrier to the glass microtools. Removal of the cell wall to form protoplasts might allow use of the microtools, but the plant cells might release hydrolases and other toxic compounds from the vacuole, leading to rapid death of the cells (Lorz *et al.*, 1981). Protoplasts may also be attached to glass slides by coating with polyL- lysine, or by or agarose. Poly-L-lysine is toxic to some cells. Agarose reduces visibility around the cells to be manipulated. Microinjection has been used for the transformation of tobacco (Schnorf *et al.*, 1991), petunia (Griesbach, 1987), rape (Neuhaus *et al.*, 1987) and barley (Holm *et al.*, 2000), with the transgenic plants being recovered at very low frequencies. Microinjection therefore remains of limited use for plant transformation, even though it would be very attractive for introduction of whole chromosomes into plant cells.

8.2 Silicon carbide whisker-mediated transformation

In this method of plant transformation, silicon carbide crystals (average dimensions of 0.6 μm diameter, 10 – 80 μm long) are mixed with DNA and plant cells by vortexing, enabling the crystals to pierce the cell walls (Kaeppler *et al.*, 1990, Songstad *et al.*, 1995). The method appears to be widely adaptable, and can be used with as little as

0.1 μg DNA. It appears as if there is a lot of scope for further development of this method of plant transformation (Thompson *et al.*, 1995).

The method is simple and easy to adapt to new crops, but the transformation efficiencies are low, and the fibres must be handled with care since they pose a health risk to the experimenter. Success has however been reported with maize (Bullock *et al.*, 2001; Frame *et al.*, 1994; Kaepler *et al.*, 1992; Petolino *et al.*, 2000; Wang *et al.*, 1995), rice (Nagatani, 1997), wheat (Brisibe, *et al.*, 2000; Serik, *et al.*, 1996), tobacco (Kaeppler *et al.*, 1990), *Lolium multiflorum*, *L. perenne*, *Festuca arundinacea*, and *Agrostis stolonifera* (Dalton *et al.*, 1998).

8.3 The pollen tube pathway

DNA is applied to the cut styles shortly after pollination, and flows down the pollen tube to reach the ovules. This approach has been used to transform rice (Luo an Wa, 1988), wheat (Mu *et al.*, 1999), soybean (Hu and Wang 1999), *Petunia hybrida* (Tjokrokusumo *et al.*, 2000) and watermelon (Chen *et al.*, 1998). Relatively high transformation efficiencies have been reported. A modification of the procedure is to inject plasmid DNA or *A. tumefaciens* carrying the plasmid DNA into inflorescences in the premeiotic stage, without removing the stigma, as was done for rye (De la Pena *et al.*, 1987), to result in high transformation efficiencies.

8.4 Liposome mediated transformation

Liposomes are microscopic spherical vesicles that form when phospholipids are hydrated. They can be loaded with a variety of molecules, including DNA. Liposomes loaded with DNA can be made to fuse with protoplast membrane and deliver their contents into the cytoplasm by endocytosis. Liposomes can also be carried through the pores of pollen grains to fuse with the membrane of the pollen grain. Transgenic plants have been reported by liposome-mediated transformation only from tobacco (Dekeyser *et al.*, 1990) and wheat (Zhu *et al.*, 1993). The process is inexpensive, but is laborious and inefficient, and so has not been widely adopted. It might be worthwhile to consider delivering the liposomes through the pollen tube pathway.

8.5 Infiltration

Infiltration (vacuum infiltration) is a method for plant transformation almost exclusively used for the transformation of *Arabidopsis*. Inflorescences of plants in early generative phase (5 – 15 cm) are immersed in *A. tumefaciens* and 5% sucrose. The inflorescences are then placed under vacuum for several minutes. Typically 0.5 to 4% of the seeds harvested from the inflorescences will be transgenic (Chung *et al.*, 2000; Clough *et al.*, 1998; Ye *et al.*, 1999). This method is highly optimized and works well for *Arabidopsis*.

Table 2. Summary of plant transformation methods

	Transformation Method	Short Description	Pros	Cons	Main Results Achieved
Indirect transfer methods	*Agrobacterium*-mediated	T-DNA mobilized from *Agrobacterium* into the plant cell	Based on a naturally occurring	Marker and reporter genes required Vector back-borne often	Mono- and dicotyledonous plants
		under the direction	process	integrated into the	Field-tested and
		of *Agrobacterium*-		plant genome	commercialized.
		encoded virulence			Very successful
		proteins			
Direct transfer	Microprojectile	Tungsten or gold	Not cultivar	Multiple copies often	Non-homologous
methods	bombardment/	microprojectiles	or genotype	reported	recombination.
	Biolistics	coated with DNA are	dependent		Also organelle
		propelled at high			transformation
		speed across the cell			
		barriers into the			
		nucleus			
	Direct protoplast	With cell wall	Introduction	Dependent on ability to	Can also be used
	transformation –	removed, DNA can	of DNA into	regenerate whole plants	for organelle
	electroporation or	be moved into the	protoplasts is	from protoplast	transformation
	PEG-mediated	cell by methods	easy		
		similar to those used			
		on bacteria			

	Transformation Method	Short Description	Pros	Cons	Main Results Achieved
	Electroporation of	High voltage	Higher	Protocol for	Maize, rice,
	cells and tissues	discharge is used to	regeneration	regeneration required	tobacco and wheat
		open pores on the	success than		
		cell membrane and	with		
		carry DNA into the	protoplasts		
		cell			
	Electro-	Electric current is	Circumvents	Low success rates.	Experimental
	transformation	used to carry DNA	problems	Needs further	
		cells or tissues of	associated	investigation of factors	
		intact plants	with	to improve success	
			regeneration,		
	Microinjection	DNA delivered	Can	Practical only for	Tobacco, *Petunia*,
		through a needle	potentially be	protoplasts.	rape and barley
		into cells	used for the		
		immobilized by	introduction		
		microtools	of whole		
			chromosomes		
	Silicon carbide	Silicon carbide	The method is	Low transformation	Tobacco, maize,
	mediated	whiskers coated with	widely	efficiencies. Silicon	rice, other grasses.
	transformation	DNA pierce and	adaptable,	carbide whiskers are a	

	Transformation Method	Short Description	Pros	Cons	Main Results Achieved
		enter the cells	and requires	health risk to the	
			little DNA	experimenter.	
	The pollen tube	DNA delivered to	Apparently	Apparently widely	Successful for rice,
	pathway	ovule via cut end of	widely	applicable, but	wheat, soybean,
		pollen tube	applicable.	particular protocols	water melon and
				need to be developed	*Petunia hybrida*
	Liposome	Liposomes loaded	Uptake	Effective only for	Success for tobacco
	mediated	with DNA are made	depends on	protoplasts	and wheat
	transformation	to fuse with	the natural		
		protoplast	process of		
		membrane	endocytosis		
	Infiltration	A suspension of	Simple	Not generally	Very efficient for
		Agrobacterium cells	procedure	applicable to most	*Arabidopsis*
		habouring the DNA		species	
		construct of interest			
		is vacuum-infiltrated			
		into inflorescences			

9. CONCISE SUMMARY

Currently, there is a diverse range of plant transformation techniques available for producing plants that are resistant to viruses. These methods are listed in Table 2. Agrobacterium-mediated transformation and microprojectile bombardment techniques have been employed to generate virus-resistant plants that have undergone field trials and, in some cases, have been made available for commercial use. Transgenic plants serve as valuable study material for gaining deeper insights into the techniques of plant transformation. Nevertheless, the ever-increasing expectations of consumers necessitate the ongoing enhancement of these techniques, with the aspiration that certain procedures will progress towards being marker-free and vector-free approaches for transforming plants.

10. REFERENCES

Abdul-Baki, A.A., Saunders, J.A., Matthews, B.F. and Pittarelli, G.W. (1990). DNA uptake during electroporation of germinating pollen grains. Plant Sci. 70:181-190.

Ahokas, H. (1989). Transfection of germinating barley seed electrophoretically with exogenous DNA. *Theor. Appl. Genet.* 77: 469-472.

An, G., Watson, B.D., Stachel, S., Gordon, M.P. and Nester, E.W. (1987). New cloning vehicles for transformation of higher plants. *EMBO Journal*, 4: 277 – 284.

Baulcombe, D.C. (1996). Mechanisms of pathogen-derived resistance to viruses in transgenic plants. *The Plant Cell*, 8: 1833 – 1844.

Barton, K.A., Binns, A.N., Matzke, A.J.M. and Chilton, M.D. (1983). Regeneration of intact tobacco plants containing full-length copies od genetically engineered T-DNA, and transmission of T-DNA to R1 progeny. *Cell*, 32: 1033 – 1043.

Beachy, R.N., Abel, P, Oliver, M.J., (1986). Potential for applying genetic information to studies of virus pathogenesis and cross protection. In: M. Zaitlin, P. Day and A. Hollaender (eds). *Biotechnology in Plant Science: Relevance to Agriculture in the Eighties*, pp 265 – 275.

Beavan, M.W., Flavell, R.B., and Chilton, M.D. (1983). A chimaeric antibiotic resistance gene as a selectable marker for plant cell transformation. Nature, 304: 184 – 187.

Bevan, M.W., Mason, S.E. and Goelet, P. (1985). Expression of tobacco mosaic virus coat protein by a cauliflower mosaic virus promoter in plants transformed by *Agrobacterium*. *EMBO J.*, 4: 1921 – 1926

Belkhadir, Y., Subramaniam, R. and Dangl, J.L. (2004). *Current Opinions in Plant Biology* 7: 391 – 399.

Birch, R.G. (1997). Plant transformation: Problems and strategies for practical application. *Annual Reviews of Plant Physiology Plant Molecular Biology*, 48: 297 – 326.

Brisibe, E.A, Gajdosava, A., Olsen, A. and Andersen, S.B. (2000). Cytodifferentiation and transformation of embryogenic callus lines derived from anther culture of wheat. *J. Exp. Bot.* 51:187-196.

Bower, R. and Birch, R.G. (1990). Competence for gene transfer by electroporation in a subpopulation of protoplasts from a uniform carrot cell suspensions. *Plant Cell Reports*, 9: 386 – 389

Bullock, W., Dias, D., Bagnal, S., Cook, K., Teronde, S., Ritland, J., Spielbauer, D., Abbaraju, R., Christensen, J. and Heideman, N. A high effuciency maize "whisker" transformation system. *Plant and Animal Genomes IX Conference*, San Diego, CA, Jan 13-17, 2001. Abstr. 148.

Chen, W.S., Chiu, C.C., Liu, H.Y., Lee, T.L., Cheng, J.T., Lin, C.C., Wu, Y.Y. and Chang, H.Y. (2002). Gene transfer via pollen-tube pathway for anti-fusarium wilt in watermelon. *Bioch. Mol. Biol. Intern.* (46) 1201-1209.

Cheng, M., Lowe, B.A., Spenser, T.M, Ye, X.D. and Armstrong, C.L. (2004). Factors influencing Agrobacterium-mediated transformation of monocotyledonous species. *In Vitro Cellular & Developmental Biology – Plant* (40): 31 – 45.

Chung, M.H., Chen, M.K. and Pan, S.M. (2000). Floral spray transformation can efficiently generate *Arabidopsis* transgenic plants. *Transgen. Res.* 9: 471-476.

Christou, P., McCabe, D.E., and Swain, W.F. (1988). Stable transformation of soybean callus by DNA-coated gold particles. *Plant Physiology*, 87: 671 – 674.

Christou, P., Swain, W.F., Yang, N-S., and McCabe, D.E. (1989). Inheritance and expression of foreign genes in transgenic soybean plants. *Proceeding of the National Academy of the Sciences*, U.S.A. 86: 7500 – 7504.

Clough, S.J. and Bent, A.F. Floral dip: a simplified method for *Agrobacterium*-mediated transformation of *Arabidopsis thaliana*. *Plant J.* 16 (1998) 735-743.

Cooley, J., Ford, T., and Christou, P. (1995). Molecular and genetic characterisation of elite transgenic rice plants produced by electric discharge particle acceleration. *Theoretical and Applied Genetics*, 90: 744 – 104.

Cosgrove, D.J. (2001). Loosening plant cell walls by expansins. *Nature*, 407: 321 – 326.

Dalton, S.J., Bettany, A.J.E., Timms, E., and Morris, P. (1998). Transgenic plants of *Lolium multiflorum, Lolium perenne, Festuca arundinacea*, and *Agrostis solonifera* by silicon carbide whisker-mediated transformation of cell suspension cultures. *Plant Science*, 132: 31 – 43.

Dekeyser, R.A., Claes, B., De Rycke, R.M.U., Habets, M.E., Van Montagu, M.C. and Caplan, A.B. (1990). Transient gene expression in intact and organized rice tissues. *The Plant Cell* 2: 591-601.

Deshayes, A., Herrera-Estrella, L. and Caboche, M. (1985). Liposome-mediated transformation of tobacco mesophyll protoplasts by an Escherichia coli plasmids. *EMBO J.* 4:2731-2737.

Dinesh-Kumar, S.P., Tham, W-H. and Baker, B.J. (2000). Structure-function analysis of the tobacco mosaic virus resistance gene N. *Proceedings of the National Academy of the Sciences, USA*, 97 (26): 14789 – 14794.

Ellis, J., Dodds, P. and Pryor, T. (2000). The generation of plant disease resistance gene specificities. *Trends in Plant Sciences*, 5: 373 – 379.

Ellis, J., Dodds, P. and Pryor, T. (2000). Structure, function and evolution of plant disease resistance genes. *Current Opinions in Plant Biology*, 3: 278 – 284.

Frame, B.R., Drayton, P.R., Bagnall, S.V., Lewnau, C.J., Bullock, W.P., Wilson, H.M., Dunwell, J.M., Thompson, J.A. and Wang, K. (1994). Production of fertile transgenic maize plants by silicon carbide whisker-mediated transformation. *Plant J.* 6 941-948.

Frame, B.R., Drayton, P.R., Bagnall, S.V., Lewnau, C.J., Bullock, P., Wilson, H.M., Dunwell, J.M., Thompson, J.A., and Wang, K. (1994). Production of fertile transgenic maize plants by silicon carbide whisker-mediated transformation. *Plant Journal*, 6: 941 – 948.

Gharti-Chhetri, G.B., Cherdshewasart, W., Dewulf, J., Paszkowski, J., Jocabs, M. and Negrutiu, I. (1990). Hybrid genes in the analysis of transformation conditions. III. Temporal-spatial fate of NPTII gene integration, its inheritance and actors affecting these processes in *Nicotiana plumbaginifolia*. Plant Molecular Biology, 14: 687 – 696.

Gonsalves, D and Gansey, S.M. (1989). Cross protection techniques for control of plant virus diseases in the tropics. *Plant Diseases*, 73: 592 - 597.

Griesbach, R.J. (1987). Chromosome-mediated transformation via microinjection. *Plant Sci.*50:69-77.

Griesbach, R.J. and Hammond, J. An improved method for transforming plants through electrophoresis. *Plant Sci.* 102 (1994) 81-89.

Hadi, M.Z., McMullen, M.D., and Finer, J.J. (1996). Transformation of 12 different plasmids into soybean via particle bombardment. *Plant Cell Reports*, 15: 500 – 505.

Hemmenway, C., Fang, R.-X., Kaniewski, W.K., Chua, N.-H. and Turner, N.E. (1988). Analysis of the mechanism of protection in transgenic plants expressing the potato virus X coat protein or its antisense RNA. *EMBO J.*, 7: 1273 – 1280.

Herrera-Estrella, L., Depicker, A., Van Montagu, M. and Schell, J. (1983). Expression of chimeric genes transferred into plant cells using a Ti-plasmid-derived vector. *Nature*, 303: 209 – 213.

Hohn, T. and Vasquez, F. (2011). RNA silencing pathway of plants: Silencing and its suppression by plant DNA viruses. *Biochemica et Biophysica Acta* (xx): x – xx. Doi: 10:1016/j.bbagrm.2011.06.002

Holm, P.B., Olsen, O., Schnorf, M., Brinch-Pederse, H. and Knudsen, S. Transformation of barley by microinjection into isolated zygote protoplasts. *Transgen. Res.* 9 (2000) 21- 32.

Hong, S.W. and Verling, E. (2001). Hsp101 is necessary for heat tolerance but dispensable for development and germination in the absence of stress. *Plant Journal*, 27: 25 – 35

Hu, C.Y. and Wang, L. (1999). In planta soybean transformation technologies developed in China: procedure, confirmation and field performance. *In Vitro Cell. Dev. Biol.-Plant* 35: 417-420.

Hull, R. (2001). Matthews' Plant Virology (Fourth Edition). Academic Press, San Diego.

Kaeppler, H.F., Gu, W., Somers, D.A., Rines, H.W. and Cockburn, A.F. (1990). Silicon carbide fiber-mediated DNA delivery into plant cells. *Plant Cell Reports*, 9: 415 – 418.

Kaeppler, H.F., Somers, D.A., Rines, H.W. and Cockburn, A.F. (1992). Silicon carbide fibre mediated stable transformation of plant cells. *Theoretical and Applied Genetics*, 84: 560 – 566.

Kartzke, S., Saedler, H., and Meyer, P. (1990). Molecular analysis of transgenic plants derived from transformations of protoplasts at various stages of the cell cycle. *Plant Science*, 67: 63 –72.

Katiyar-Agarwal, S. and Jin, H. (2010). Role of small RNAs in host-microbe interactions. *Annual Review of Phytopathology*, 48: 225 – 246.

Khetarpal , R.K., Maisonneuve, B., Maury *et al.*, (1998). Breeding for resistance to plant viruses. In: A. Hadidi, R.K. Khetarpal and H. Koganezawa (eds) *Plant Virus Disease Control.* pp. 14 – 32. APS Press, St Paul, MN.

Klein, T.M., Wolf, E.D. and Wu, R. (1987). High velocity microprojectiles for delivering nucleic acids into living cells. *Nature*, 327: 70 –73.

Kohli, A., Ghareyazie, B., Kim, H.S., Khush, G.S., Bennett, J. and Khush, G.S. (1996). Cytosine methylation implicated in silencing of β-glucuronidase gene in transgenic rice. *Rice Genetics Symposium-III*, pages 825 – 828. IRRI, Manilla, Philippines.

Koehler, F., Cardon, G., Poehlman, M., Gill, R., and Schieder, O. (1989). Enhancement of transformation rate in higher plants by low-dose irradiation – Are DNA repair systems involved in the incorporation of exogenous DNA into the plant genome? *Plant Molecular Biology*, 12: 189 – 200.

Koehler, F., Benediktsson, I., cardon, G., Andreo, C.S., and Schieder, O.(1990). Effect of various irradiation treatments of plant protoplasts on the transformation rates after direct gene transfer. *Theoretical and Applied Genetics*, 79: 679 – 685.

Kohli, A., Leech, M., Vain, P., Laurie, D.A. and Christou, P. (1998). Transgene organisation in rice engineered though direct DNA transfer supports a two-phase integration mechanism mediated by the establishment of integration hot spots. *Proceedings of the National Academy of the Sciences*, USA, 95: 7203 – 7208.

Kumar, S., Allen, G.C. and William, F.T. (2006). Gene targeting in plants: fingers on the move. Trends in Plant Science, 11(4): 159 – 161.

Lindsey, K., Wei, W., Clarke, M.C., McArdale, H.F. and Rooke, L.M. (1993). Tagging genomic sequences that direct transgene expression by activation of a promoter trap in plants. *Transgenic Research*, 2: 33 – 47.

Lodge, J.K., Silva-Rosales, L., Proebsting, W.M. and Dougherty, W.G. (1993). Induction of a highly specific antiviral state in transgenic plants: Implications for regulation of gene expression and virus resistance. *Plant Cell*, 5 (12): 1749 - 1759.

Lörz, H., Paszkowski, J., Dierks-Ventling, C. and Potrykus, I. (.1981) Isolation and characterization of cytoplasts and miniprotoplasts derived from protoplasts of cultured cells. *Physiol. Plant* 53:385-391.

Luo, Z.X. and Wa, R. (1998). A simple method for the transformation of rice via pollen-tube pathway. *Plant Mol. Biol. Report* 6 :165-174.

Marton, L., Wullems, G.J., Molendijk, L. and Schilperoort, R.A. (1979). *In vitro* transformation of cultured cells from *Nicotiana tabacum* by *Agrobacterium tumefaciens*. *Nature*, 277: 129 – 131.

Masuta, C., Tanaka, H., Vehara, K., Kuwata, S., Koiwai, A. and Noma, I. (1995). Broad resistance to plant viruses in transgenic plants conferred by antisense inhibition of a host gene essential in S-adenosyl methionine-dependent transmethlation reactions. *Proceedings of the National Academy of the Sciences*, USA, 92 (13): 6117 - 6121.

Matzke, A.J.M. and Chilton, M. (1981). Site specific insertion of genes into T-DNA of the *Agrobacterium* tumour-inducing plasmid: an approach to genetic engineering of higher plant cells. *Journal of Molecular and Applied Genetics*, 1: 39 – 49.

Maliga, P. (2004). Plastid transformation in Higher Plants. *Annual Reviews of Plant Biology* 55: 289 – 313.

Mundembe, R., Matibiri, A. and Sithole-Niang, I. (2009). Transgenic plants expressing the coat protein gene of cowpea aphid-borne mosaic potyvirus predominantly convey the delayed symptom development phenotype. African Journal of Biotechnology, 8(12): 2682 – 2690.

Nagatani, N., Honda, H., Shimada, T. and Kobayashi, T. (1997). DNA delivery into rice cells and transformation using silicon carbide whiskers. *Biotechnol. Techniq.* 11:781-786.

Ow, D.W. (2002). Recombinase-directed plant transformation for the post-genomic era. *Plant Molecular Biology* 48: 183 – 200.

Parmyakova, N.V., Shumnyi, V.K. and Deineko, E.V. (2008). Agrobacterium-mediated transformation of plants: Transfer of vector DNA fragments in the plant genome. *Russian Journal of Genetics* 45 (3): 266 – 275.

Pazkowski, J., Shillito, R.D., Saul, M., Mandak, V., Holn, T., Holn, B., and Potrykus, I. (1984). Direct gene transfer to plants. *EMBO Journal*, 3: 2717 – 2722.

Petolino, J.F., Hopkins, N.L., Kosegi, B.D. and Skokut, M. (2000). Whisker mediated transformation of embryogenic callus of maize. *Plant Cell Rep.* 19 (2000) 781-786.

Powell-Abel, P., Nelson, R.S., De B., Hoffmann, N., Rodger, S.G., Fraley, R.T. and Beachy, R.N. (1986). Delay of disease development in transgenic plants that express the tobacco mosaic virus coat protein gene. *Science*, 232: 738 - 743.

Powell-Abel, P., Saunders, P.R., Turner, N., Frayley, R.T. and Beachy, R.N. (1990). Protection against tobacco mosaic virus infection in transgenic plants requires accumulation of coat protein rather than coat protein RNA sequences. *Virology*, 175: 124 – 130.

Rakoczy-Trojanowska, M. (2002). Alternative methods of plant transformation – a short review. *Cellular and Molecular Biology Letters* 7: 849 – 858.

Ready, M.P., Brown, D.T. and Robertus, J.D. (1986). Extracellular localisation of pokeweed antiviral protein. *Proceedings of the National Academy of the Sciences*, (USA), 83: 5053 – 5056.

Register, J.C. and Beachy, R.N. (1988). Resistance to TMV in transgenic plants results from interference with an early event in infection. *Virology*, 166: 524 – 532.

Sairam, R.V. and Prakash, C.S. (2005). OBPC Symposium: Maize 2004 and Beyond – Can agricultural biotechnology contribute to food security? *In Vitro Cellular & Developmental Biology - Plant* 41: 424 – 430.

Sanford, J.C. and Johnson, S.A. (1985). The concept of parasite-derived resistance: deriving resistance genes from the parasite's own genome. *Journal of Theoretical Biology*, 115: 395 – 405.

Sanford, J.C., Klein, T.M., Wolf, E.D., and Allen, N. (1987). Delivery of substances into cells and tissues using a particle bombardment process. *Journal of Particulate Science and Technology*, 5: 27 – 37.

Sithole-Niang, I., Nyathi, T., Maxwell, D.P. and Candresse, T. (1996). Sequence of the 3'- terminal region of Cowpea aphid-borne mosaic virus (CABMV). *Archieves of Virology*, 141: 935 – 943

Smith, N., Kilpatrick, J.B. and Whitelam, G.C. (2001). Superfluous transgene integration in plants. *Critical Reviews in Plant Sciences*, 20(3): 215 – 249.

Songstad, D.D., Somers, D.A., and Griesbach, R.J. (1995). Advances in alternative DNA delivery techniques. *Plant Cell Tissue and Organ Culture*, 40: 1 –15.

Tavladoraki, P., Benvenuto, E., Trinca, S., De Martinis, D., Cattanco, A. and Galeffi, P. (1993). Transgenic plants expressing a functional single chain Fv antibody are specifically protected from virus attack. *Nature* (London), 366 (6454): 469 – 472).

Thompson, J.A., Drayton, P.R., Frame, B.R., Wang, K. and Dunwell, J.M. (1995). Maize transformation utilizing silicon carbide whiskers: a review. *Euphytica*, 85: 75 – 80.

Thottappilly, G. and Rossel, H.W. (1992). Virus diseases of cowpea in tropical Africa. *Tropical Pest Management* 38(4): 337 – 348.

Tinland, B. (1996). The integration of T-DNA into plant genomes. *Trends in Plant Sciences* 1 (6): 178 – 183.

Tjokrokusumo, D., Heinrich, T., Wylie, S., Potter, R. and McComb, J. (2000). Vacuum infiltration of Petunia hybrida pollen with *Agrobacterium tumefaciens* to achieve plant transformation. *Plant Cell Rep.* 19:792-797.

Tzfira, T, Jianxiong, L., Lacroix, B. and Citovsky, V. (2004). *Agrobacterium* T-DNA integration: molecules and models. *Trends in Genetics* 20 (8): 375 - 378

Vasil, V., Brown, S.M., Re, D., Fromm, M.E., and Vasil, I.K. (1991). Stably transformed callus lines from microprojectile bombardment of cell suspension cultures of wheat. *Bio/Technology*, 9: 743 – 747.

Voss, A., Niersbach, M., Hain, R., Hirsch, H.J., Liao, Y.C., Kreuzaler, F. and Fischer, R. (1995). Reduced virus infectivity in *N. tabacum* secreting a TMV-specific full size antibody. *Molecular Breeding*, 1 (1): 39 - 50.

Wang, K., Herrera-Estrella, L., Van Montagu, M. and Zambryski, P. (1984). Right 25 bp terminus sequence of the nopaline T-DNA is essential for and determines direction of DNA transfer from *Agrobacterium* to the plant genome. *Cell* 38: 455 – 462.

Wang, H.L., Yeh, S.D., Chiu, R.J. and Gonsalves, D. (1987). Effectiveness of cross-protection by mild mutants of papaya ringspot virus for control of ringspot disease of papaya in Taiwan. *Plant Diseases*, 71: 491 – 497.

Wang, K., Drayton, P., Frame, B., Dunwell, J., Thompson, J.A. (1995). Whisker mediated plant transformation: an alternative technology. *In Vitro Cell Dev. Biol.* 31 (1995) 101-104.

Wilson, T.M.A. (1993). Strategies to protect crop plants against viruses – pathogen derived resistance blossoms. *Proceedings of the National Academy of the Sciences*, USA, 90: 3134– 3141.

Yamashita, T., Iida, A. and Morikawa, H. (1991). Evidence that more than 90% of β-glucuronidase-expressing cells after particle bombardment directly receive the foreign gene in their nucleus. Plant Physiology, 97: 829 – 831.

Yang, B., Sugio, A. and White, F.F. (2006). Os8N3 is a host disease-susceptibility gene for bacterial blight of rice. *Proceedings of the National Academy of the Sciences*, USA, 103 (27): 10503 -10508.

Ye, G.N., Stone, D., Pang, S.Z., Creely, W., Gonzalez, K. and Hinchee, M.(1999), *Arabidopsis* ovule is the target for *Agrobacterium in planta* vacuum infiltration transformation. *Plant J.* 19:249-257.

Yuan, D., Bassie, L., Sabalza, M., Miralpeix, B. *et. al* (2011). The potential impact of plant biotechnology on Millenium Development Goals. *Plant Cell Reports* 30: 249 – 265.

Zaghmout, O.M.F. and Trolinder, N. (1993). Simple and efficient method for directly electroporating plasmid DNA into wheat callus cells. *Nucl. Acid Res.* 21:1048.

Zambryski, P., Joos, H., Genetello, C., Leemans, J., Van Montagu, M., and Schell, J. (1983). Ti plasmid vector for the introduction of DNA into plant cells without alteration of normal regeneration capacity. *EMBO Journal*, 2: 2143 – 2150.

Zhu, J., Oger, P.M., Schrammeiler, B., Hooykaas, P.J.J., Farrand, S.K. and Winans, S.C. (2000). The basis of crown gall tumerigenesis. *Journal of Bacteriology*, 182: 3885 – 3895

Zhu, Z., Sun, B., Liu, C., Xiao, G. and Li, X. (1993). Transformation of wheat protoplasts mediated by cationic liposome and regeneration of transgenic plantlets. *Chin. J. Biotech.* 9:257-261.

Zupan, J. and Zambryski, P. (1997). The Agrobacterium DNA transfer complex. *Critical Reviews in Plant Science*, 16: 279 – 295.

06 ENHANCING SOYBEAN REGENERATION VIA SOMATIC EMBRYOGENESIS AND GENETIC TRANSFORMATION: A STRATEGIC APPROACH

1. INTRODUCTION: AN OVERVIEW

The seed, which houses the embryo, serves as the principal reproductive unit in angiosperms. In flowering plants, like in other eukaryotes, the embryo originates from the zygote resulting from the fusing of gametes. Throughout history, numerous plant species have developed diverse mechanisms of asexual embryogenesis to overcome various environmental and genetic obstacles that hinder fertilization (Sharma & Thorpe, 1995; Raghavan, 1997).

Somatic embryogenesis (SE) is a type of asexual reproduction that involves the development of embryos from somatic or gametic cells, without the fusing of gametes (Williams & Maheswaran, 1986). This phenomenon, which is specific to plants, can occur either spontaneously or in vitro through experimental induction (Dodemam et al., 1997). The procedure is possible because plants have the ability to regenerate individual somatic cells into a complete plant, known as cellular totipotency (Reinert, 1959).

Somatic embryogenesis (SE) has been documented in tissue cultures of many angiosperm and gymnosperm plant species. This process entails a sequence of morphological alterations that bear resemblance, in multiple respects, to the developmental changes observed in zygotic embryos. Embryogenic structures in soybean (Glycine max (L.) Merrill) have been observed in histological sections in several studies (Barwale et al., 1986; Finer & McMullen, 1991; Kiss et al., 1991; Liu et al., 1992; Sato et al., 1993). Christou & Yang (1989), Fernando et al. (2002), Rodrigues et al. (2005), and Santos et al. (2006) conducted a study to describe the different stages of development in soybean somatic embryos. The phases of the pro-embryo, globular, heart-shaped, torpedo, and cotyledonary embryos were observed, closely mimicking the development of zygotic embryos. Nevertheless, the primary distinctions between zygotic and somatic embryogenic processes were the lack of a distinctive suspensor and the delayed development of inner organization (Santos et al., 2006).

2. SOYBEAN SOMATIC EMBRYOGENIC PROCESS

In general, the *in vitro* soybean somatic embryogenic process can be divided into different phases: induction, proliferation, histodifferentiation, maturation, germination and conversion into plants.

2.1 Somatic embryo induction

According to Sharp et al. (1982), the induction of somatic embryogenesis (Fig. 1 A1, B1, C1) can be considered as termination of the existing gene expression pattern in the explant tissue, and its replacement for an embryogenic gene expression program in those cells of the explant tissue which will give rise to somatic embryos. These authors used the term "induced embryogenic determined cell" (IEDC) to describe an embryogenic cell that has been originated from a non-embryogenic cell. Cells from very immature zygotic embryos, which already have their embryogenic gene expression program activated, were termed "pre-embryogenic determined cells" (PEDCs). For the purposes of regeneration, both terms may be referred to simply as "embryogenic cells" (ECs) (Carman, 1990; Merkle et al., 1995).

There is a major developmental difference among explants with respect to the ontogeny of somatic embryos. The obtainment of somatic embryogenesis in legumes depends on whether the explant tissue consists of PEDCs (for example, very immature zygotic embryos) or non-ECs (for example, differentiated plant tissues). In the first case, a stimulus to the explant may be sufficient to induce cell division for the formation of somatic embryos, which appear to arise directly from the explant tissue in a process referred to as **direct embryogenesis** (Fig. 1 A1, B1). In contrast, non-EC tissue must undergo several mitotic divisions in the presence of an exogenous auxin for induction of the ECs. Cells resulting from these mitotic divisions are manifested as a callus, and the term **indirect embryogenesis** is used to indicate that a callus phase intervenes between the original explant and the appearance of somatic embryos (Fig. 1 C1) (Merkle et al., 1995).

Thus, the somatic embryo induction process can be achieved using different approaches, as illustrated in Figure 1. Somatic embryos induced from very immature zygotic embryos (torpedo-stage) upon exposure to cytokinins were only obtained in clovers (*Trifolium* ssp.) (Maheswaran & Williams, 1984) (Fig. 1 A). In soybean, somatic embryos can be induced in response to auxins, and regenerated directly from cotyledonary-stage zygotic immature embryos without an intervening callus phase (Lazzeri et al., 1985; Finer, 1988; Bailey et al., 1993; Santarém et al., 1997) (Fig. 1 B). Finally, some legumes, notably alfalfa (*Medicago sativa*), can be regenerated from leaf-derived callus (Bingham et al., 1988). In this case, the tissue responds to combinations of auxins and cytokinins (Fig. 1 C).

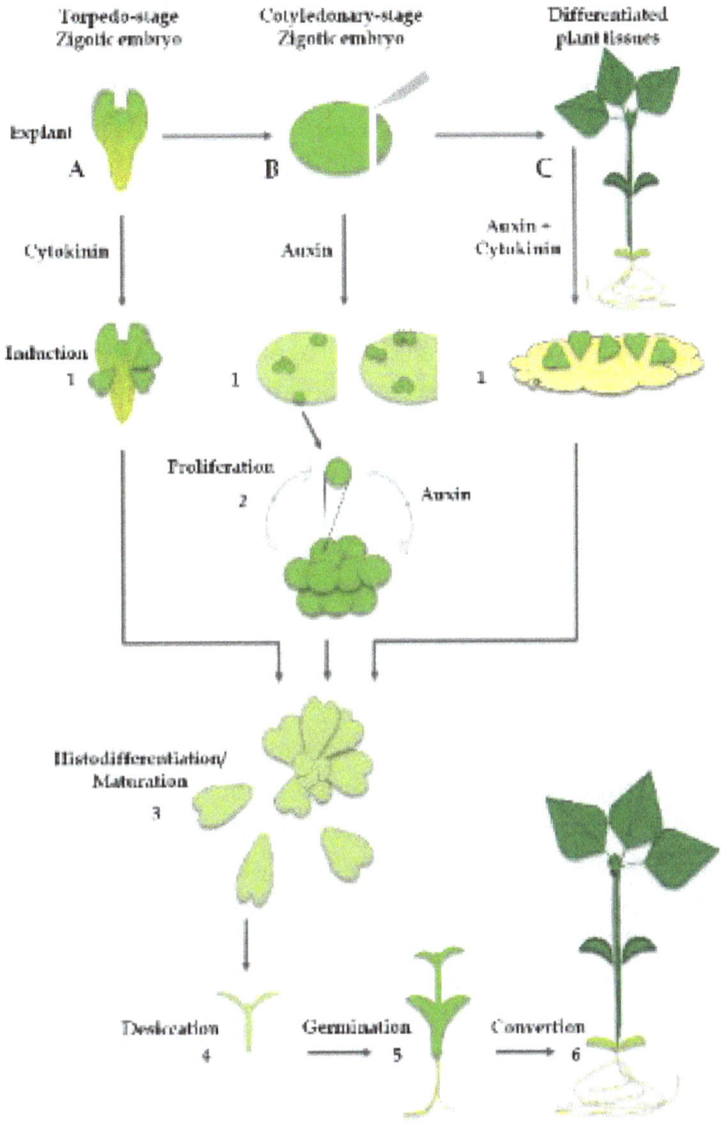

Fig. 1. Embryogenic processes in legumes. Somatic embryos may be induced (1), histodifferentiated/matured (3), desiccated (4), germinated (5), and converted into plants (6). Alternatively, auxin can be used to maintain repetitive embryogenesis – embryo proliferation (2), which continues until auxin is withdrawn from the medium, allowing somatic embryos to resume their development. (A) The youngest zygotic embryos respond to cytokinin; (B) older zygotic embryos respond to auxin, and (C) differentiated plant tissues respond to combination of auxin and cytokinins by forming callus. (Adapted from Parrott et al., 1995. Drawing by S. N. C. Richter)

The type of growth regulator and explant, as well as genotype ability to respond to *in vitro* stimulus, are the main factors affecting somatic embryogenesis induction.

The role of exogenous cytokinins during the induction phase depends on whether somatic embryogenesis is direct or indirect. When SE is originated from callus, the frequency of somatic embryo formation is enhanced by cytokinins. However, in direct systems, such as in soybean, in which somatic embryos are formed directly from immature zygotic embryos, addition of a cytokinin reduces the frequency of embryo formation (Merkle et al., 1995). Soybean SE is induced by two auxins: α-naphthaleneacetic acid (NAA) and 2,4-dichlorophenoxyacetic acid (2,4-D), but the most commonly used is 2,4-D. The exact mechanism underlying the auxin-induced somatic embryo formation is not understood, but some studies with other legumes suggested certain auxin-induced cellular processes such as embryo-specific DNA methylation (Vergara et al., 1990), disruption of tissue integrity by interrupting cell–cell interaction (Smith & Krikorian, 1989) and establishment of cell polarity (Merkle et al., 1995). However, auxins are not the only substances able to induce embryogenesis. Several other factors that alter gene expression programs (e.g., stress) or disrupt cell-cell interaction (physical disruption of the tissue) can also direct this transition (Gharyal & Maheshwari, 1983; Dhanalakshmi & Lakshmanan, 1992).

The choice of explant is a critical factor that determines the success of most tissue culture experiments. Immature, meristematic tissues proved to be the most suitable explant for somatic embryogenesis in legumes (Lakshmanan & Taji, 2000). For instance, cotyledons of immature zygotic embryos have been the most used explants for the induction of SE in soybean (Lazzeri et al., 1985; Finer, 1988; Bailey et al., 1993; Santarém et al., 1997; Droste et al., 2002). However, in this species, somatic embryos have also been obtained from leaf and stem (Ghazi et al., 1986), cotyledonary node (Kerns et al., 1986), anther (Santos et al., 1997; Rodrigues et al., 2005) and embryonic axes (Kumari et al., 2006).

The last but not least important factor affecting somatic embryo induction is plant genotype (Merkle et al., 1995). In soybean, considerable variation in embryogenic capacity was found to exist between individual genotypes (Komatsuda et al., 1991; Bailey et al., 1993a, b; Santos et al., 1997; Droste et al., 2001; Meurer et al., 2001; Tomlin et al., 2002; Hiraga et al., 2007; Yang et al., 2009; Droste et al., 2010) as will be discussed below (Genotype-dependent response and screening of highly responsive cultivars section).

2.2 Embryo proliferation

An inherent trait of embryogenic tissue is its ability to maintain its embryogenic state indefinitely. This type of cell proliferation has been referred to by different names, including secondary, recurrent, or repetitive embryogenesis (Fig. 1 B2). In soybean, primary somatic embryos can originate from multiple cells, while secondary somatic embryos (i.e. those coming from another somatic embryo) likely to originate from a single cell (Merkle et al., 1995). Hartweck et al. (1988) discovered that somatic embryos in soybean zygotic cotyledons arise from clusters of cells, but Sato et al. (1993) observed that embryos in globular-stage soybean somatic embryos proliferate from individual cells. The proliferation of embryogenic cells is influenced by various factors, some of which are regulated during the culture phase, while others remain unidentified. Factors that have been examined include plant genotype and growth regulators, which are also linked to the induction phase (Merkle et al., 1995).

Auxin is the most well established factor linked to the continual growth of embryogenic cells. Soybean can undergo secondary somatic embryo growth when it is kept in a media that contains the auxin 2,4-D, as demonstrated by Finer and Nagasawa in 1988. Sato et al. (1993) demonstrated that individual epidermal cells can generate secondary somatic embryos in soybeans. The precise mechanism by which auxin stimulates cell division remains unclear. Additionally, the amount of auxin needed to sustain repeated embryogenesis is contingent upon the specific culture regimen utilized.

2.3 Embryo histodifferentiation and maturation

After induction, somatic embryos start an ontogenetic development process similar to that of their zygotic counterparts (Merkle et al., 1995). The process of organ formation through which a globular-stage embryo develops into a cotyledon-stage embryo has been termed histodifferentiation (Fig. 1 B3) (Carman, 1990).

In general, continued embryo histodifferentiation beyond the globular stage and subsequent maturation requires the removal of growth regulators from the medium - or at least a decrease in growth regulator concentrations associated with induction and proliferation - down to levels that enable proper embryo development (Merkle et al., 1995). Auxins suppress the development of the apical meristem, probably governed by the same mechanism involved in the establishment of apical dominance (Parrott et al., 1995).

The role of cytokinins during the histodifferentiation is more difficult to assess. Aung et al. (1982) observed a decrease in endogenous cytokinin levels during soybean zygotic embryo development. On the other hand, the inclusion of a cytokinin during the histodifferentiation phase can compensate for auxin-induced detrimental effects on meristem development. Globular-stage soybean somatic embryos, when exposed to a cytokinin, decrease the development rates while the apical meristem elongates so as to form multiple shoots (Wright et al., 1991).

Due to the fact that traditional embryogenesis protocols have typically used the same medium for histodifferentiation and the subsequent maturation phases, it is difficult to review the literature and determine if a given treatment affects the histodifferentiation stage or the maturation stage (Merkle et al., 1995).

Following histodifferentiation, the period of embryo development in which cell expansion and reserve deposition occur is considered the maturation phase (Fig. 1 B3) (Bewley et al., 1985). The time required for somatic embryos to achieve physiological maturity is species- specific, mirroring the maturation period of zygotic embryos *in planta* (Parrott et al., 1995). Analyses of reserve accumulations in developing somatic embryos have revealed both similarities and differences in comparison to zygotic embryos. These differences may primarily be attributed to the *in vitro* maturation conditions used. Many reports on somatic embryos reserve accumulations have not been carried out with optimal maturation protocols. Hence, manipulation of culture conditions to prolong and improve embryo maturation, and to prevent precocious germination, will probably add to the similarities observed between zygotic and somatic embryos (Merkle et al., 1995).

Currently, a trend is observed in the specification of numerous protocols towards providing various growth regulators in the medium during embryo maturation. Yet, there is ample information suggesting that neither exogenous auxin nor cytokinin is actually required for normal embryo maturation, as evidenced by the normal development of soybean zygotic (Hsu & Obendorf, 1982) or somatic embryos (Parrott et al., 1988) in media devoid of any growth regulators. Indeed, poorly developed meristem or swollen hypocotyls may be an undesired outcome of the application of exogenous auxins and cytokinins, respectively. Consequently, a treatment that binds and removes auxin, by adding activated charcoal to the maturation medium (Ebert & Taylor, 1990), for instance, may improve embryo development and enhance germination (Buchheim et al., 1989). Differently from auxins and cytokinins, abscisic acid (ABA) may be necessary during embryogenesis to initiate the synthesis of storage proteins and proteins involved in desiccation tolerance (Galau et al., 1990).

2.4 Embryo germination and conversion into plants

Embryos removed from maturation conditions for further development often display poor or aberrant subsequent germination, growth and vigor (Parrott et al., 1988). These observations suggest that further post-maturation treatments are required. One of the fundamental aspects of zygotic embryo development not normally encountered during somatic embryo development is desiccation (Fig. 1 B4), which leads to embryo quiescence. It has been proposed that desiccation is required for the correct transition from an embryo maturation program to a germination program (Kermode, 1990). The desiccation period has been linked with the synthesis of proteins associated with the ability to germinate (Rosenberg & Rinne, 1986, 1988). Partial desiccation has been shown to enhance conversion of soybean somatic embryos (Hammatt & Davey, 1987; Parrott et al., 1988; Buchheim et al., 1989).

While most studies report the development of roots and the germination of somatic embryos, little distinction is made between germination and conversion. According to Ranch et al. (1985), soybean germination refers to both root and shoot development in an embryo with intact hypocotyl (Fig. 1 B5). Walker & Parrott (2001) described soybean apparent conversion as the development of expanded trifoliolates and branched roots under *in vitro* conditions (Fig. 1 B6). However, according to the authors, the definition of actual conversion refers to survival following transfer to soil.

While germination capacity may be affected by several culture medium components or environmental manipulations, one aspect that cannot be manipulated so readily is the genetic background of the embryo (Merkle et al., 1995). Studies with soybean plants indicated that germination/conversion capacity is also greatly influenced by genotype (Komatsuda & Ohyama, 1988; Bailey et al., 1993; Santos et al., 1997; Droste et al., 2001, 2010).

3. GENOTYPE-DEPENDENT RESPONSE AND SCREENING OF HIGHLY RESPONSIVE CULTIVARS

The more closely the pattern of somatic embryo gene expression matches that of zygotic embryos, the greater the chance of obtaining highly efficient regeneration systems. Such normalization of gene expression patterns will be achieved through the optimization of media and culture protocols for each individual stage of embryo development. According to Meurer et al. (2001), there are two ways to optimize soybean regeneration via embryogenic cultures. The first is screening a large number of new cultivars in order to identify those with embryogenic potential. Since SE is a heritable trait (Parrott et al., 1989), the potential for embryogenesis can be also improved through conventional crossing between non- responsive cultivars and highly competent cultivars (Kita et al., 2007). An alternative and additive approach is to optimize embryogenesis protocols for improve results of an interesting cultivar.

Several studies have revealed the differences among soybean genotypes in their capacity to respond to the different steps of somatic embryogenesis (Bailey et al., 1993; Santos et al., 1997; Simmonds & Donaldson, 2000; Droste et al., 2001; Meurer et al., 2001; Tomlin et al., 2002; Hiraga et al., 2007; Yang et al., 2009; Droste et al., 2010). The efficiencies were shown to be different among cultivars at each SE phase: induction, proliferation, histodifferentiation/maturation, germination and conversion (Bailey et al., 1993; Santos et al., 1997). As each phase must be under independent genetic control, the best cultivar performance in one stage not necessarily is the best in another one. In this context, Bailey et al. (1993a) reported that PI 417138 was among the least inducible genotypes studied, though it did have the highest germination and conversion capacity. In the same way, IAS-5 cultivar had the lowest embryo yield but the highest conversion capacity (Santos et al., 1997). Tomlin et al. (2002) and Hiraga et al. (2007) suggested that differences in somatic embryo induction efficiency among soybean cultivars are likely to be attributed to differences in endogenous auxin levels or in the auxin sensitivity.

One of the major challenges in soybean is to identify genotypes highly responsive to SE induction, embryo proliferation and conversion into plants. A direct comparison among previous reports is difficult because each research group has adopted different protocols for SE and evaluation methods. In some studies only the first phases such as induction and/or proliferation were analyzed, while in other studies genotypes were screened upon plant regeneration.

Simmonds & Donaldson (2002) screened 18 of 20 short-season soybean genotypes in Canada for proliferative embryogenic capacity. Only five genotypes produced embryogenic cultures which were proliferative for at least six months. Nine soybean cultivars representing different US growing regions were evaluated at each of three locations using uniform embryogenic induction and proliferation protocols. Several cultivars were identified as uniformly embryogenic at the primary induction phase at all locations, among which Jack was the best (Meurer et al., 2001).

Twenty-six genotypes including soybean wild relatives and Japanese cultivars were screened for differences of competence in both somatic embryogenesis and subsequent shoot formation. All genotypes were able to induce somatic embryos but with wide variation. *Glycine gracilis*, *G. gracilis* T34, Masshokutou (kou 502) and Masshokutou (kou 503) varieties presented high competence of germination from somatic embryos (Komatsuda & Ohyama, 1988). Hiraga et al. (2007) examined the capacity for plant regeneration through somatic embryogenesis in Japanese soybean cultivars. Induction of somatic embryos from immature cotyledons, embryo proliferation in liquid medium and differentiation into cotyledon-stage embryos were evaluated. Yuuzuru and Yumeyutaka cultivars were found to have high potentials of plant regeneration through SE, being superior or comparable to North American cultivar Jack.

From a preliminary experiment using 98 Chinese soybean genotypes, 12 varieties were selected for further study in order to enhance the efficiency of somatic embryogenesis and plant regeneration. Significant differences in somatic embryogenesis were found among genotypes. N25281, N25263, and N06499 varieties were shown to have the highest somatic embryogenic capacities. The greatest average number of plantlets regenerated per explant was observed in N25281 (Yang et al., 2009).

Regarding Brazilian genotypes, Bonacin et al. (2000) demonstrated the influence of genotype influence in somatic embryogenic capability of five cultivars, of which

BR-16, FT-Cometa and IAS-5 were the most embryogenic ones. That study only assessed somatic embryo induction stage. Other studies also reported the high capacity of IAS-5 to produce somatic embryos (Santos et al., 1997; Di Mauro et al., 2000; Droste et al., 2001). More recently, Droste et al. (2010) identified Brazilian soybean genotypes with potential to respond to *in vitro* culture stimuli for somatic embryo induction, embryo proliferation and plant regeneration. Somatic embryos were induced in all eight tested genotypes, but differences were observed at each stage. IAS-5 and BRSMG 68 Vencedora had high embryo induction frequencies, repetitive embryogenic proliferation, low precocious embryo germination, better embryo differentiation and plant regeneration. Thus, this work identified BRSMG 68 Vencedora and confirmed IAS-5 as genotypes with high potential for somatic embryogenesis and plant regeneration.

4. OPTIMIZATION OF SOYBEAN SOMATIC EMBRYOGENESIS PROTOCOL

Although somatic embryogenesis was described long ago and may be considered a routine procedure for other plant species, the first record of the event in soybean was made by Beversdorf & Bingham (1977), when no more than a few embryos were produced. It was only in 1983 that Christianson et al. (1983) regenerated, for the first time, soybean plants via SE. Subsequently, several papers have reported the development of somatic embryos from cotyledons of immature embryos. But until today the measured frequencies at which these embryos convert into plants are significantly lower than that expected (Parrott et al., 1988; Meurer et al., 2001; Walker & ning protocols for initiation, proliferation, and histodifferentiation/ maturation of soybean somatic embryogenesis & Parrott, 2001; Droste et al., 2001, 2010). Therefore, efforts have been directed towards developing soybean somatic embryogenesis protocol.

4.1 Induction and proliferation of somatic embryos

Somatic embryos induced on 2,4-D are friable, translucent, yellowish-green in color, and globular to torpedo in shape. Somatic embryos induced on NAA are compact, opaque, pale-green in color, with an advanced morphology, forming cotyledon-like structures (Lazzeri et al., 1987; Hoffmann et al., 2004). Furthermore, Lazzeri et al. (1987) described that somatic embryos initiated on NAA had more normal embryo morphology. Significantly higher than that produced in the presence of NAA (Hoffmann et al., 2004). It is noteworthy to mention that the type of auxin used in the medium also in most protocols described in the early studies failed to promote the satisfactory induction of somatic embryos. This limitation was overcome when embryos were induced from immature soybean cotyledons by placing the explant on high levels of 2,4-D (40 mg/l) (Finer, 1988). Comparing the capacity to induce soybean SE, the mean number of embryos produced on 2,4-D was signi. The synergistic effect of pH, solidifying agent, 2,4-D concentration, explants orientation and wounding have also been reported to improve efficiency of somatic embryo induction (Santarém et al., 1997). Embryo initiation was higher when explants were cultured with the abaxial side facing the induction medium containing high concentration of 2,4-D, pH adjusted to 7.0 and solidified with GelriteTM. Effect of pH was not observed or was slighter when somatic embryos were induced in the presence of other growth regulators (Lazzeri et al., 1987; Komatsuda & Ko, 1990; Hoffman et al., 2004; Bonacin et al., 2000). It has been suggested that the effect of pH on somatic embryo initiation may be related to auxin uptake into cultured explants, and that a pH of 7.0 may facilitate slower and more gradual uptake of 2,4-D at relatively high level. Wounding treatment did not increase the number of embryos, although in wounded explants somatic embryos were induced earlier than in non- wounded counterparts (Santarém et al., 1997). In addition, efficiency on SE induction was highest when in a medium containing 2-3% sucrose. Cultures initiated on lower sucrose concentrations tended to produce higher amount of friable embryos, while increased concentrations of this sugar impaired embryo induction (Lippmann & Lippmann, 1984; Lazzeri et al., 1987, 1988; Hoffmann et al., 2004).

After induced, the early-staged somatic embryos can be maintained and proliferated by subculturing the tissue on either semi-solid medium (Finer, 1988) or liquid suspension culture medium (Finer & Nagasawa, 1988). As liquid medium allows greater contact of plant tissue with medium components, proliferation in liquid is usually more efficient than on a semi-solid medium (Samoylov et al., 1998a). On the other hand, soybean embryogenic suspension cultures can be very difficult to establish and maintain (Santarém et al., 1999). Although maintenance of soybean embryogenic cultures in liquid medium was facilitated by the development of FN medium (Finer & Nagasawa, 1988), its efficiency remains low. Changes in individual medium components, such as carbohydrate type and concentration, total nitrogen, ammonium, nitrate, and other macronutrients, improved proliferation of soybean suspension cultures in liquid medium and resulted in the development of an optimized medium, referred to as FN Lite (Samoylov et al., 1998a).

Somatic embryos incubated in a medium containing NAA do not proliferate so well as those produced on a medium containing 2,4-D (Liu et al., 1992). As discussed above, somatic embryos initiated on NAA are more advanced in embryo morphology than those induced on 2,4-D. As a consequence, they are also not suitable for use in establishing repetitive cultures (Lazzeri et al., 1987; Hoffmann et al., 2004). Embryogenic cultures induced from immature cotyledons on a medium containing 40 mg/l required lower levels of 2,4-D for efficient proliferation (Bailey et al., 1993; Santarém & Finer, 1999). Twenty mg/l 2,4-D are necessary to efficiently maintain repetitive embryogenesis on semi-solid medium (Santarém & Finer, 1999), while 5 mg/l 2,4-D are sufficient on liquid suspension culture medium (Samoylov et al., 1998a). Furthermore, in contrast to the effect of pH on embryo initiation, cultures maintained on semisolid medium at pH 5.8 or 7.0 showed no difference in proliferation rates, suggesting that once embryogenesis has been induced, the tissue does not necessary have to be maintained under the same conditions (Santarém et al., 1997).

4.2 Histodifferentiation/maturation of somatic embryos and recovery of plants

Even after three decades of intense research, soybean regeneration via somatic embryogenesis remains low if compared to other crops. Developmental limitations in somatic embryos are usually related to inadequate culture media composition (Merkle et al., 1994). So, efforts to better mimic the developmental environment of zygotic embryos were made to further improve SE. Deficiency in maturation stage has been identified as the main obstacle to somatic embryo conversion into plant. Somatic and zygotic embryos have been shown to diverge in sugars, protein, and total lipid accumulation, indicating that the somatic embryos did not develop properly and, as consequence, bring about difficulties for plant regeneration (Chanprame et al., 1998). In addition to embryo nutrition, appropriate sugar accumulation has been shown to be involved in soybean seed desiccation tolerance (Blackman et al., 1992).

Histodifferentiation and maturation of soybean somatic embryos has been basicallyachieved through the use of two protocols. The first one is a two-step process, whereby embryos are First histodifferentiated on MSM6AC medium (Bailey et al., 1993), which consists of solidied MS basal salts supplemented with 6% maltose and 0.5% activated charcoal. After 30 days, embryos are transferred to the same medium, but without charcoal, in order to allow growth and maturation to proceed. The second protocol is based on a liquid medium termed FNLS3, which consists of basal Finer & Nagasawa "Lite" (FNL) salts histodifferentiated on MSM6AC medium (Bailey et al., 1993), which consists of solidified MS basal salts supplemented with 6% maltose and 0.5% activated charcoal. After 30 days, embryos are transferred to the same medium, but without charcoal, in order to allow growth and maturation to proceed. The second protocol is based on a liquid medium termed FNLS3, which consists of basal Finer & Nagasawa "Lite" (FNL) salts supplemented with 3% sucrose (Samoylov et al., 1998b). The main advantage of the liquid FNLS3 medium over the semi-solid MSM6AC/MSM6 medium is its ability to produce larger numbers of mature somatic embryos in a short period of time. On the other hand, liquid based protocols require greater care during handling.

Successful plant recovery has been reported using both protocols, though numerous studies suggested modifications to these basic media in order to optimize histodifferentiation and maturation of soybean somatic embryos, with the final goal of obtaining higher conversion rates. The main adaptations were: the type and concentration of carbon source, addition of osmotic agents, growth regulator agents and/or amino acids, type of basal salts, as well as the medium supplementation with activated charcoal (Tables 1 and 2). Modifications in other culture media are not considered in the present review.

4.2.1 Carbon source and osmotic agents

Carbon source is critical for embryo nutritional health and improves somatic embryo maturation. Carbohydrates are commonly used as carbon sources for *in vitro* development of tissues, and maltose and sucrose have been usually added to soybean tissue culture media in an optimal concentration range of 3% to 6% (Samoylov et al., 1998b; Körbes & Droste, 2005; Schmidt et al., 2005). Carbon source type and concentration required for histodifferentiation/maturation appear to differ between solid and liquid media.

The effect of carbohydrates on embryo histodifferentiation and maturation on liquid medium was analyzed by Samoylov et al. (1998b). FNL medium supplemented with 3% sucrose (FNL0**S3**) or 3% maltose (FNL0**M3**) were compared. Data indicated that sucrose promotes embryo growth and significantly increases the number of cotyledon-stage embryos recovered during histodifferentiation and maturation. However, the percentages of plants recovered from embryos differentiated and matured in FNL0**S3** was lower than those grown in FNL0**M3** (Samoylov et al., 1998b). This limitation was partially solved by adding 3% sorbitol to the medium, which resulted in an increment in germination and conversion frequencies (Walker & Parrott, 2001). The modified medium was named FNLS3S3 medium. It was suggested that sorbitol acts as an osmotic agent and/or promotes accumulation of triglycerides in somatic embryos. After addition of sorbitol, the effects of using 3% sucrose (FNL0**S3**S3) or 3% maltose (FNL0**M3**S3) were compared again (Schmidt et al., 2005). Conversion rate of embryos differentiated on maltose was higher when compared to those obtained in sucrose. However, sucrose was considered the most appropriate carbon source, since its use in liquid maturation media resulted in a higher number of larger embryos, which required less time to reach physiological maturity.

The media that have been used in the two-step histodifferentiation/maturation process (MS**M6**AC/MS**M6**) contain 6% maltose (Bailey et al., 1993). Körbes & Droste (2005) compared conversion frequencies when 6% maltose was replaced by 3% maltose (MS**M3**) or 6% sucrose (MS**S6**). Results showed that maturation in MSS6 medium leads to an increment in rates of histodifferentiated embryos with normal morphology, as well as in plant recovery. The quality of somatic embryos can be positively influenced by a low osmotic potential in maturation medium (Walker & Parrott, 2001; Körbes & Droste, 2005). Carbohydrates can act as an osmotic agent (Li et al., 1998). Since molecular weights of maltose and sucrose are very cant effect on the osmotic potential of the medium could be expected (Schmidt et al., 2005). Nevertheless, differences observed in embryos matured in media supplemented with maltose or sucrose have been related to these compounds' osmotic potentials (Samoylov et al., 1998b; Körbes similar, no signi & Droste, 2005). Influence of other osmotic agents (sorbitol, mannitol and polyethylene glycol) has also been tested in soybean somatic embryogenesis (Walker & Parrott, 2001; Körbes & Droste, 2005; Schmidt et al., 2005), but positive effects were only reported for sorbitol (Walker & Parrott, 2001), as described above.

4.2.2 Amino acids

The nitrogen source is also a critical component for proper embryo maturation. With a view to soybean embryo proliferation, 1 g/l asparagine was included in the formulation of the original FN liquid medium (Finer & Nagasawa, 1988), and it was also used in the FN Lite proliferation medium (Samoylov et al., 1998a) and FNL his to differentiation/maturation medium (Samoylov et al., 1998b). On the other hand, culture media supplemented with glutamine were shown to be beneficial to zygotic soybean embryos (Thompson et al., 1977), by increasing embryo size (Lippmann & Lippmann, 1993; Dyer et al., 1987), and inducing storage oil and protein synthesis (Saravitz & lter-sterilized glutamine were larger and exhibited an overall higher quality. The modified medium was named FNLS3S3G30 medium. Raper, 1995). Schmidt et al. (2005) compared the effect of FNLS3S3 supplemented with asparagine or glutamine on embryo histodifferentiation/maturation. The results showed that cotyledonary-stage embryos obtained on 30 mM on 30 mM flter-sterilized glutamine were larger and exhibited an overall higher quality. The modified medium was named FNLS3S3G30 medium.

Methionine supplementation has also been reported to be helpful for growth stimulation (Coker et al., 1987). Thus, addition of 1 mM methionine to FNL0S3S3 medium was tested and a clear positive effect on soybean somatic embryo histodifferentiation and maturation, manifested mainly by conversion percentages, was demonstrated (Schmidt et al., 2005). Again, this medium was renamed FNLS3S3G30M1 medium.

Improvements in the efficiency of solid MSM6 histodifferentiation/maturation medium were also obtained by using the ingredients from the optimized FNLS3S3G30M1 liquid medium recipe, specifically glutamine and methionine (Schmidt et al., 2005).

4.2.3 Basal salts

A comparison of embryo development showed that embryos differentiated into yellow- green cotyledon-stage faster when cultured on FNLS3 than when maintained on MSM3 liquid media (Samoylov et al., 1998b). On the other hand, the number of histodifferentiated embryos and the frequency of germinated embryos recovered from MSM3 were higher than those obtained on FNLS3. Further studies showed that germination rates were increased when embryos histodifferentiated on FNLS3 supplemented with sorbitol, glutamine and methionine (Walker & Parrott, 2001; Schmidt et al., 2005).

4.2.4 Abscisic acid

Abscisic acid (ABA) is a growth regulator involved in plant development, especially during embryo development and maturation and in response to abiotic stresses. In seeds, ABA induces storage protein synthesis and affects dormancy induction and maintenance (Rock & Quatrano, 1995). ABA prevents precocious seed germination and is thought to play a major role in the sequence of events leading to desiccation tolerance (Hoekstra et al., 2001). ABA supplementation in culture media was shown to affect somatic embryogenesis of a variety of plants. Having this in mind, it was suggested that medium containing low ABA concentrations (0.38 or 1 μM) helps soybean immature somatic embryos to achieve maturity and further develop apical meristems (Ranch et al., 1885; Lazzeri et al., 1987). However, independently of the protocol used, no significant effect was observed when ABA was added to histodifferentiation or maturation medium (Tian & Brown, 2000; Schmidt et al.,

2005; Weber et al., 2007). Tian & Brown (2000) investigated the effect of ABA addition to culture media in different embryogenic stages: proliferation, histodifferentiation and maturation. The positive effects of ABA were observed only when embryos at a globular stage (proliferation) were treated prior to histodifferentiation induction. Addition of 50 μM ABA promoted growth of histodifferentiated embryos, increased proportion of morphological normal histodifferentiated embryos, improved embryo viability after desiccation and increased germination frequency. In agreement, Weber et al. (2007) demonstrated that presence of ABA during proliferation or during both proliferation and maturation stages increased percentage of converted plants.

Table 1. Optimization of MSM6AC medium. D20 - MS salts (Murashige & Skoog, 1962), B5 vitamins (Gamborg et al., 1968), 3% sucrose, 20 mg/l 2,4-D, and 0.2% Gelrite or 0.3% Phytagel (pH 5.8) (Wright et al., 1991). MSM6AC - MS salts, B5 vitamins, 6% maltose, 0.5% activated charcoal and 0.25% Gelrite or 0.3% Phytagel (pH 5.8) (Bailey et al., 1993). MSM6 - MS salts, B5 vitamins, 6% maltose and 0.25% Gelrite or 0.3% Phytagel (pH 5.8) (Finer & McMullen, 1991). MSS6 - MS salts, B5 vitamins, 6% sucrose, 0.3% Phytagel (pH 5.8) (Körbes & Droste, 2005). MSO - MS basal salts, B5 vitamins, 3% sucrose, and 0.2% Gelrite, at pH 5.8

Reference	Main modification	Culture tissue stage				Beneficial effects
		Proliferation	Histodifferentiation	Maturation	Regeneration / conversion	
Tian & Brown, 2000	Abscisic acid (ABA) addition	FN + 50 µM ABA	MSM6AC	MSM6AC	MSO	• promote embryo • growth • increase • histodifferentiation of • embryo with normal • morphology • improve embryo • viability • enhance embryo • germination
Weber et. al., 2007	ABA addition	D20 + 50 µM ABA	MSM6 + 50 µM ABA	MSM6	MSO	• increase embryo • conversion into plants
Körbes & Droste, 2005	Maltose replacement with sucrose	D20	MSS6	MSS6	MSO	• increase • histodifferentiation of • embryo with normal • morphology

Reference	Main modification	Culture tissue stage				Beneficial effects
		Proliferation	Histodifferentiation	Maturation	Regeneration / conversion	
						• increase embryo • conversion into plants
Schmidt et al., 2005	Glutamine and methionine addition		MSM6 + 30 mM glutamine + 1 mM Methionine			• increase the • number of • histodifferentiated • embryos
						• increase embryo • conversion into plants
Droste et al., 2010	Activated charcoal (AC) addition	D20	MSS6 + 0.5 % AC	MSS6	MSO	• increase the • number of • histodifferentiated • embryos

Table 2. Optimization of FNLS3 medium. FN Lite - FN Lite macro salts (Samoylov et al., 1998a), MS micro salts (Murashige & Skoog, 1962), B5 vitamins (Gamborg et al., 1968), 6.7 mM L-asparagine, 1% sucrose , 146.1 mM mannitol and 0.5 mg/l 2,4-D, at pH 5.8 (Samoylov et al., 1998a). FNL0S3 – FN Lite macro salts, MS micro salts, B5 vitamins, 6.7 mM L- asparagine and 3% sucrose, at pH 5.8 (Samoylov et al., 1998b). MSO - MS basal salts, B5 vitamins, 1.5% sucrose, and 0.2% Gelrite, at pH 5.8.

Reference	Main Modification	Culture tissue stage: Proliferation	Histodifferentiation/ maturation	Regeneration / conversion	Beneficial effects
Walker & Parrott, 2001	Sorbitol addition	D20/ FN Lite	FNL0S3 + 1.5-3% sorbitol	MSO	• reduce fresh weight of mature embryos
					• increase embryo • regeneration and • conversion into plants
Schmidt et al., 2005	Standard medium	FN Lite	FNL0S3S3	MSO	• increase number of • matured embryos
					• give rise to larger • embryos
					• need shorter time to • reach physiological • maturity
					• reduce embryo • conversion into plants
	Sucrose replacement by maltose	FN Lite	FNL0M3S3	MSO	• reduce number of • matured embryos,
					• give rise to smaller • embryos

Reference	Main Modification	Culture tissue stage			Beneficial effects
		Proliferation	Histodifferentiation/ maturation	Regeneration / conversion	
					• need longer time to • reach physiological • maturity
					• increase embryo • conversion into plants
	Glutamine addition	FN Lite	FNL0S3S3 + 30 mM glutamine	MSO	• give rise to larger • embryos
	Methionine addition	FN Lite	FNL0S3S3G30 + 1 mM Methionine	MSO	• increase embryo • conversion into plants

4.2.5 Activated charcoal

Activated charcoal (AC) is a porous material composed of carbon. AC has a unique adsorption capacity and is often used in plant tissue culture to improve cell growth and development. Applicability of AC is credited mainly to its capacity as adsorbent of inhibitory substances in the culture medium (Thomas, 2008). As previously discussed (Induction and proliferation of somatic embryos section), high concentrations of 2,4-D are necessary to stimulate soybean somatic embryo induction and proliferation (Finer, 1988; Samoylov et al., 1998a; Santarém & Finer, 1999). On the other hand, this growth regulator can lead to abnormal embryo histodifferentiation or development of apical meristem, especially in long-term cultures under exposure to 2,4-D. AC is presumably able to adsorb 2,4-D or other auxins released from developing tissues, promoting a more normal embryo morphology and increasing germination ability (Merkle et al., 1995).

AC was present in a medium originally described by Bailey et al. (1993) for embryo histodifferentiation/maturation, but it has not been applied in further studies (Samoylov et al., 1998b; Walker & Parrott, 2001; Körbes & Droste, 2005; Schmidt et al., 2005; Weber et. al., 2007). Recently, it has been observed that addition of activated charcoal and ABA to the first step histodifferentiation/ maturation medium increased the number of histodifferentiated embryos (Droste et al., 2010). Since positive effects were not identified when ABA was added in histodifferentiation/maturation medium (Schmidt et al., 2005; Weber et. al., 2007) (discussed in the section about abscisic acid), the benefits described by Droste et al. (2010) must be related to AC supplementation. It is important to stress that studies were carried out with different genotypes and that soybean response to *in vitro* culture conditions is genotype-dependent. Further studies, specially focusing on embryo histodifferentiation and maturation, are important and necessary to improve soybean embryogenesis of highly responsive cultivars.

After appropriate maturation, a desiccation stage is required. Conversion of partial- desiccated soybean somatic embryos was shown to proceed more vigorously than non- desiccated ones (Buchheim et al., 1989). The partial desiccation has been adopted by most research teams (Bailey et al., 1993; Samoylov et al., 1998b; Tian & Brown, 2000; Walker & Parrott, 2001; Droste et al., 2002, 2010; Körbes & Droste, 2005; Schmidt et al., 2005; Yang et al., 2009; Wiebke-Strohm et al., 2011). Although there is a broad number of media available for histodifferentiation/maturation, MSO (MS basal salts, B5 vitamins, 3% sucrose, and 0.2% Gelrite™, at pH 5.8) medium has been almost exclusively used for soybean somatic embryos germination/ conversion stage (Finer & McMullen, 1991; Bailey et al., 1993; Samoylov et al., 1998b; Tian & Brown, 2000; Walker & Parrott, 2001; Droste et al., 2002, 2010; Körbes & Droste, 2005; Schmidt et al., 2005; Wiebke-Strohm et al., 2011).

5. GENETIC TRANSFORMATION OF SOYBEAN SOMATIC EMBRYOS VIA PARTICLE BOMBARDMENT AND BOMBARDMENT/ AGROBACTERIUM INTEGRATED SYSTEM

Plant genetic transformation is described as the introduction of recombinant DNA in plant cells using genetic engineering methods. Transgenic plants represent a priceless tool for molecular, genetic, biochemical and physiological studies. Plant genetic transformation also offers a significant advancement for soybean breeding programs, in terms of allowing the production of novel and genetically diverse plant materials.

Advancements in the use of plant transgenesis have been reported for a large number of species. Such progress entails the adoption of different protocols, which include the genetic transformation mediated by polyethyleneglycol (PEG) and liposomes (known under the heading of *chemical methods*), microinjection, electroporation and particle bombardment (called *physical methods*), as well as the use of viral and/or bacterial vectors, as in agroinfection and the *Agrobacterium* system (named *biological methods*).

Soybean transformation was first reported in 1988 by two independent groups using different methods (Hinchee et al., 1988; Christou et al., 1988). Even after more than two decades, the stable transformation of soybean cannot yet be considered as routine because it depends on the ability to bring together efficient transformation and regeneration techniques. Two methods have been successfuly used: **particle bombardment** (McCabe et al., 1988; Christou et al., 1989; Finer & McMullen, 1991; Christou & McCabe, 1992; Finer et al., 1992; Stewart et al., 1996; Aragão et al., 2000; Droste et al., 2002; Homrich et al., 2008; Wu et al., 2008; Li et al., 2009; Hernandez-Garcia et al., 2009; Xing et al., 2010; Viana et al., 2011) and ***Agrobacterium tumefaciens* system** (Parrott et al., 1989; Trick et al., 1997; Trick & Finer, 1998; Aragão et al., 2000; Yan et al., 2000; Ko et al., 2003, 2004; Paz et al., 2006; Hong et al., 2007; Miklos et al., 2007; Liu et al., 2008; Wang & Xu, 2008; Wiebke-Strohm et al., 2011).

Regardless of the method used, the unicellular origin of soybean secondary somatic embryos makes them a useful target tissue for transformation, allowing the production of fully transformed plants. The first target used for transformation was primary somatic embryos, but chimerical plants were obtained (Parrott et al., 1989) due to the multicellular nature of primary embryos (Sato et al., 1993). Finer (1988) showed that the secondary somatic embryos proliferated directly from the apical or terminal portions of the older primary somatic embryos. Sato et al. (1993) proved that somatic embryo proliferation occurred from single epidermal cells of existing somatic embryos. Using proliferative embryos as target tissue many studies succeed in regenerating completely transformed plants (Finer & McMullen,

1991; Finer et al., 1992; Stewart et al., 1996; Trick et al., 1997; Trick & Finer, 1998; Droste et al., 2002; Homrich et al., 2008; Schmidt et al., 2008; Wu et al., 2008; Li et al., 2009; Hernandez-Garcia et al., 2009; Xing et al., 2010; Wiebke-Strohm et al., 2011).

5.1 Agrobacterium system

Agrobacterium tumefaciens is a soil-borne Gram-negative phytopatogenic bacterium that naturally infects different plants causing the crown gall disease (DeCleene & DeLay, 1976). The origin of these sicknesses is interkingdom horizontal gene transfer. When virulent strains of *Agrobacterium* infect plant cells, they transfer one or more segments of DNA (transferred DNA or T-DNA) from Ti (**T**umor **i**nducing) plasmids into host plant cells (recently reviewed by Gelvin, 2010 a,b; Pitzschke & Hirt, 2010). In recent decades, disarmed (non-tumorigenic) *A. tumefaciens* ed plants. In order to obtain engineered binary vectors derived from Ti plasmids, oncogenes present in T-DNA region are replaced by any foreign DNA of interest (Gelvin, 2010b). Strains have also provided a means to produce genetically modi The advantages of *Agrobacterium*-mediated gene transfer include the possibility of transferring relatively large segments of DNA, lower number of transgene copies integration into plant genomes, rare transgene rearrangement, lower frequency of genomic DNA interspersion and reduced abnormal transgene expression (Gelvin, 2003; Kohli et al., 2003). Moreover, this system involves low operating cost and simplicity of transformation protocols. On the other hand, plants differ greatly in their susceptibility to *Agrobacterium*- mediated transformation. These differences occur among species, cultivars or tissues (Droste et al., 1994; Gelvin, 2010b). In addition, in our laboratory this transformation system usually results in lower transformation rates, if compared to particle bombardment.

Agrobacterium-mediated transformation system is a growing trend in crop transformation programs (Somers et al., 2003). Soybean has long been considered recalcitrant to *Agrobacterium* (DeCleene & DeLey, 1976), especially due to the low success rate in recovering transgenic plants. Studies with tumorigenic *Agrobacterium* strains (Pedersen et al., 1983; Droste et al., 1994; Mauro et al., 1995) and transient assays (Meurer et al., 1998; Droste et al., 2000) showed that soybean can be readily transformed by this bacterium. Currently, it is known that addition of acetosyringone during bacterial infection, combination of appropriate *Agrobacterium* strain and soybean cultivar, as well as development of super virulent bacterium strain and suitable plasmids increase efficiency of soybean transformation (Somers et al., 2003; Ko et al., 2003, 2004; Wiebke-Strohm et al., 2011).

Transgenic plants regenerated via somatic embryogenesis were obtained using two different target tissues: immature zygotic cotyledons and secondary somatic embryos. In the first case, wounded immature zygotic cotyledons are co-cultivated with *Agrobacterium* suspension, after which embryogenic tissue formation is induced from the surface of these cotyledons. In the second case, proliferative secondary somatic embryos are first obtained and then submitted to transformation experiments.

Transformation of zygotic cotyledons by *A. tumefaciens* and subsequent regeneration of transformed soybean plants was first reported by Parrott et al. (1989). Plants were regenerated from primary somatic embryos and, due to embryo multi-cellular origin, plants were chimeric. As a consequence, transgenes were not present in the germ line and transmitted to the progeny. Still using zygotic cotyledons as target, a new approach was developed in which formation of secondary somatic embryos was allowed under continued selection system. The unicellular origin of secondary somatic embryos permitted recovery of complete, stable and fertile transgenic plants, whose progeny also displayed these characteristics (Yan et al., 2000; Ko et al., 2003).

Transformation of proliferative secondary somatic embryos via *A. tumefaciens* has proven to be challenging and only succeeded when combined to physical methods that generate an entry point to bacteria penetration. Instead of the conventional transformation system, two alternative methods have been proposed for this target tissue: the Sonication-Assisted *Agrobacterium*-mediated Transformation (SAAT) (Trick et al., 1997; Trick & Finer, 1998), and the combined DNA-free particle bombardment and *Agrobacterium* system (bombardment/ *Agrobacterium* integrated system) (Droste et al., 2000; Wiebke et al., 2006, Wiebke-Strohm et al., 2011). The difference between these methods lies in the technique used to induce tissue wounding: while the first one used sonication, the second relied upon bombardment. Although these methods proved to be feasible, both systems are time- consuming, laborious, and depend on the availability of specific equipments for routine application.

It is important to stress that the success in soybean *Agrobacterium*-mediated transformation and regeneration via somatic embryogenesis depends on the availability of cultivars with superior response to *in vitro* culture stimuli and high susceptibility to this bacterium. So far, regeneration of transgenic plants was achieved using immature zygotic cotyledons of Jack, Williams and Dwight cultivars (Yan et al., 2000; Ko et al., 2003) or proliferative secondary somatic embryos of Chapman, Bragg, IAS5 & BRMG 68 Vencedora cultivars (Trick et al., 1997; Trick & Finer, 1998; Droste et al., 2000; Wiebke-Strohm et al., 2011).

5.2 Particle bombardment

Genetic transformation by particle bombardment, also known as particle or projectile acceleration, biolistics or biobalistics, involves the insertion of DNA into intact cells and tissues using accelerated microprojectiles propelled at high velocities. These projectiles have the capability to penetrate the cell and nucleus walls, allowing for the release of DNA fragments (Trick & Finer, 1997). Within this cellular structure, foreign DNA can be incorporated into the chromosomal DNA through either illegitimate or homologous recombination processes, which rely only on cell components (Sanford, 1990; Kohli et al., 2003). Particle bombardment allows for the delivery of DNA into plant cells using plasmids (Hadi et al., 1996; Homrich et al., 2008) or gene cassettes (Fu et al., 2000; Breitler et al., 2002). This technique involves the attachment of DNA to metallic particles known as microcarriers. The metals utilized in the procedure must be chemically unreactive, such as gold or tungsten, in order to avoid any interaction with DNA or cellular constituents (Christou et al., 1990). The DNA-particle complex is propelled towards the target cells utilizing several devices that rely on distinct acceleration processes.

Particle bombardment can be accomplished using gene guns that operate at either high or low helium pressure. Precise control over tissue penetration allows for accurate targeting of the bulk of DNA-carrying particles to a single cell layer. This aspect is crucial because various explants may necessitate distinct acceleration conditions to achieve optimal particle penetration (Christou et al., 1990). According to Sato et al. (1993), the beginning of secondary somatic embryos is carried out by individual epidermal cells. To achieve effective transformation, a shallow penetration is achieved by bombarding with low helium pressure. Finer et al. (1992) created a particle accelerator dubbed Particle Inflow Gun (PIG) that operates at low helium pressure.

Particle bombardment offers a significant benefit in its ability to transfer genes to any cell or tissue type, regardless of genotype, without the need to consider compatibility between host and bacteria, as is necessary with the Agrobacterium method. Alternatively, by employing this method, several duplicates of DNA are introduced, which have the potential to undergo recombination or fragmentation (Hadi et al., 1996; Kohli et al., 2003). Multiple studies have been published on the use of particle bombardment to transform soybeans using embryogenic tissues. These studies include works by Finer & McMullen (1991), Finer et al. (1992), Stewart et al. (1996), Droste et al. (2002), Homrich et al. (2008), Schmidt et al. (2008), Wu et al. (2008), Li et al. (2009), Hernandez-Garcia et al. (2009), and Xing et al. (2010).

6. CONCISE SUMMARY

Proliferative somatic embryos are highly suited and convenient for soybean genetic transformation. Nevertheless, even after a span of twenty years, the consistent alteration of somatic embryos cannot yet be regarded as a customary practice. The lack of a highly effective regeneration technique is the primary constraining constraint. Progress has been achieved in identifying cultivars that have a high likelihood of embryogenesis, as well as in improving the media and techniques used for culture. Significant endeavors are underway to enhance the efficiency of soybean plant regeneration by somatic embryogenesis. By identifying the crucial elements, the procedures for each specific phase of the somatic embryogenic process can be enhanced to better mimic the formation of zygotic embryos in plants.

7. REFERENCES

Aragão, F.J.L.; Sarokin, L.; Vianna, G.R. & Rech, E.L. (2000). Selection of transgenic meristematic cells utilizing a herbicidal molecule results in the recovery of fertile transgenic soybean [*Glycine max* (L) Merrill] plants at a high frequency. *Tag Theoretical and Applied Genetics*, Vol.101, No.1-2, (2000), pp. 1-6, ISSN 0040-5752

Aung, L.H.; Buss, G.R.; Crosby, K.E. & Brown, S.S. (1982). Changes in the hormonal levels of soybean fruit during ontogeny. *Phyton*, Vol.45, (1982), pp. 182-185, ISSN 0031-9457

Bailey, M.A.; Boerma, H.R. & Parrott, W.A. (1993). Genotype effects on proliferative embryogenesis and plant regeneration of soybean. *In Vitro Cellular and Developmental Biology-Plant*, Vol.29, No.3, (July 1993) pp. 102–108, ISSN 1054-5476

Barwale, U.B.; Kerns, H.R. & Widholm, J.M. (1986). Plant regeneration from callus cultures of several soybean genotypes via embryogenesis and organogenesis. *Planta*, Vol.167, No.4, (1986), pp. 473-481, ISSN 0032-0935

Beversdorf, W.D. & Bingham, E.T. (1977). Degrees of differentiation obtained in tissue cultures of *Glycine* species. *Crop Science*, Vol.17, No.2, (1977), pp. 307-311, ISSN 0011-183X

Bingham, E.T; McCoy T.J. & Walker K.A. (1988). Alfalfa tissue culture. In: *Alfalfa and alfalfa improvement* , A.A. Hanson, D.K. Barnes, R.R. Hill Jr, (Eds), 903-929, American Society of Agronomy, ASA-CSSA-SSSA, ISBN 0-89118-094-X, Madison WI

Blackman, S.A.; Obendorf, R.L. & Leopold, A.C. (1992). Maturation proteins and sugars in desiccation tolerance of developing soybean seeds. *Plant Physiology*, Vol.100, No.1, (September 1992), pp. 225–230, ISSN 0032-0889

Bonacin, G.A.; DiMauro, A.O.; Oliveira, R.C. & Perecin, D. (2000). Induction of somatic embryogenesis in soybean: physicochemical factors influencing the development of somatic embryos. *Genetics and Molecular Biology*, Vol.23, No.4, (December 2000), pp. 865-868, ISSN 1415-4757

Breitler, J.C.; Labeyrie, A.; Meynard, D.; Levavre, T. & Guiderdoni, E. (2002). Efficient microprojectile bombardment-mediated transformation of rice using gene cassettes. *Tag Theoretical and Applied Genetics*, Vol.104, No.4, (2002), pp. 709-719, ISSN 0040- 5752

Buchheim, J.A.; Colburn, S.M. & Ranch, J.P. (1989). Maturation of soybean somatic embryos and the transition to plantlet grown. *Plant Physiology*, Vol.89, (1989), pp. 768-77, ISSN 0032-0889

Carman, J.G. (1990). Embryogenic cells in plant tissue cultures: occurrence and behavior. *In Vitro Cellular and Developmental Biology-Plant*, Vol.26, (August 1990), pp. 743–756, ISSN 0883-8364

Chanprame, S.; Kuo, T.M. & Widholm, A.M. (1998). Soluble carbohydrate content of soybean [*Glycine max* (L.) Merr.] somatic and zygotic embryos during development. *In Vitro Cellular and Developmental Biology-Plant*, Vol.34, (January-March 1998), pp. 64-68, ISSN 1054-5476

Christianson, M.L.; Warnick, D.A. & Carlson, P.S. (1983). A morphogenetically competent soybean suspension culture. *Science*, Vol.222, No.4624 (November 1983), pp.632- 634, ISSN 0036-8075

Christou, P. & McCabe, D.E. (1992). Prediction of germ-line transformation events in chimeric R0 transgenic soybean plantlets using tissue-specific expression patterns. *The Plant Journal*, Vol.2, No.3, (May 1992), pp. 283–290, ISSN 0960-7412.

Christou, P. & Yang, N.S. (1989). Developmental aspects of soybean (*Glycine max*) somatic embryogenesis. *Annals of Botany*, Vol.64, No.2, (1989), pp. 225-234, ISSN 0305- 7364

Christou, P.; McCabe, D.E. & Swain, W.F. (1988). Stable transformation of soybean callus by DNA-coated gold particles. *Plant Physiology*, Vol.87, (1988), pp. 671–674, ISSN 0032- 0889

Christou, P.; McCabe, D.E.; Martinell, B.J. & Swain, W.F. (1990). Soybean genetic engeneering-commercial products of transgenic plants. *Trends Biotechnology*, Vol.18, (1990), pp. 145-151, ISSN 0167-7799

Christou, P.; Swain, W.F.; Yang, N.S. & McCabe, D.E. (1989). Inheritance and expression of foreign genes in transgenic soybean plants. *Proceedings of the National Academy of Sciences of the United States of America*, Vol.88, (October 1989), pp. 7500-7504, ISSN 0027-8424

Coker, G.T.I.; Garbow, J.R. & Schaefer, J. (1987). 15N and 13C NMR determination of methionine metabolism in developing soybean cotyledons. *Plant Physiology*, Vol.83, (1987), pp. 698–702, ISSN 0032-0889

DeCleene, M. & DeLey, J. (1976). The host range of crown gall. *Botanical Gazette*, Vol.42, (1976), pp.389-466, ISSN 0006-8071

Dhanalakshmi, S. & Lakshmanan, K.K. (1992). *In vitro* somatic embryogenesis and plant regeneration in *Clitoria ternatea. Journal of Experimental Botany*, Vol.43, No.2, (1992), pp. 213–219, ISSN 0022-0957

Di Mauro, A.O.; de Oliveira, R.C. & de Oliveira, J.A. (2001). Capacidade embriogênica da cultivar IAS-5 de soja. *Pesquisa Agropecuária Brasileira*, Vol.36, No.11, (November 2001), pp. 1381-1385, ISSN 0100-204X

Dodeman, V.L.; Ducreux, G. & Kreis, M. (1997). Zygotic embryogenesis *versus* somatic embryogenesis. *Journal of Experimental Botany,* Vol. 48, No.313, (August 1997), pp. 1493-1509, ISSN 0022-0957

Droste, A.; Bodanese-Zanettini, M.H.; Mundstock, E. & Hu, C.Y. (1994). Susceptibility of Brazilian soybean cultivars to *Agrobacterium tumefaciens. Brazilian Journal of Genetics*, Vol.17, (1994), pp. 83–88, ISSN 0100-8455

Droste, A.; Pasquali, G. & Bodanese-Zanettini, M.H. (2000). Integrated bombardment and *Agrobacterium* transformation system: an alternative method for soybean transformation. *Plant Molecular Biology Reports*, Vol.18, No.1, (2000), pp. 51-59, ISSN 0735-9640

Droste, A.; Pasquali, G. & Bodanese-Zanettini, M.H. (2002). Transgenic fertile plants of soybean [*Glycine max* (L) Merrill] obtained from bombarded embryogenic tissue. *Euphytica*, Vol. 127, No.3, (2002), pp. 367-376, ISSN 0014-2336

Droste, A.; Silva, A.M.; Souza, I.F.; Wiebke-Strohm, B.; Bücker-Neto, L.; Bencke, M.; Sauner, M.V. & Bodanese-Zanettini, M.H. (2010). Screening of Brazilian soybean genotypes with high potential for somatic embryogenesis and plant regeneration. *Pesquisa Agropecuária Brasileira*, Vol.45, No.7, (July 2010), pp.715-720, ISSN 0100-204X

Droste, A.; Leite, P.C.P.; Pasquali, G.; Mundstock, E.C. & Bodanese-Zanettini, M.H. (2001). Regeneration of soybean via embryogenic suspension culture. *Scientia Agricola*, Vol.58, No.4, (October-Deyember 2001), pp.753-758, ISSN 0103-9016

Dyer, D.J.; Cotterman, C.D. & Cotterman, J.C. (1987). Comparison of *in situ* and *in vitro* regulation of soybean seed growth and development. *Plant Physiololgy*, Vol.84, No.2, (Juny 1987), pp. 298–303, ISSN 0032-0889

Ebert, A. & Taylor, H.F. (1990). Assessment of the changes of 2,4-dichlorophenoxyacetic acid concentrations in plant tissue culture media in the presence of activated charcoal. *Plant Cell, Tissue and Organ Culture*, Vol.20, No.3, (1990), pp. 165-172, ISSN 0167- 6857

Fernando, J.A.; Vieira, M.L.C.; Geraldi, I.O. & Appezzato-da-Gloria, B. (2002). Anatomical study of somatic embryogenesis in *Glycine max* (L.) Merrill. *Brazilian Archives of Biology and Technology*, Vol.45, No.3, (September 2002), pp. 277-286, ISSN 1516- 8913

Finer, J. J. & Nagasawa, A. (1988). Development of an embryogenic suspension culture of soybean (*Glycine max* Merrill). *Plant Cell, Tissue and Organ Culture*, Vol.15, (1988), pp. 125—136, ISSN 0167-6857

Finer, J.J. & McMullen, M.D. (1991). Transformation of soybean via particle bombardment of embryogenic suspension culture tissue. *In vitro Cellular and Developmental Biology- Plant*, Vol.27, No.4, (October 1991), pp. 175-182, ISSN 1054-5476

Finer, J.J. (1988). Apical proliferation of embryogenic tissue of soybean (*Glycine max* (L.) Merrill). *Plant Cell Reports*, Vol.7, No.4, (1988), pp. 238-241, ISSN 0721-7714

Finer, J.J.; Vain, P.; Jones, M.W. & McMullen, M.D. (1992). Development of the particle inflow gun for DNA delivery to plant cells. *Plant Cell Reports*, Vol.11, No.7, (1992), pp.323–328, ISSN 0721-7714

Fu, X.; Due, L.T.; Fontana, S.; Bong, B.B.; Tinjuangjun, P.; Sudhakar, D.; Twyamn, R. M.; Christou, P. & Kohli, A. (2000). Linear transgene constructs lacking vector backbone sequences generate low-copy-number transgenic plants with simple integration patterns. *Transgenic Research*, Vol.9, (2000), pp. 11-19, ISSN 0962- 8819

Galau, G.A.; Jakobsen, K.S. & Hughes, D.W. (1991). The controls of late dicot embryogenesis and early germination. *Physiologia Plantarum*, vol.81, No.2, (February 1991), pp. 280- 288, ISSN 0031-9317

Gamborg, O.L.; Miller, R.A. & Ojima, K. (1968). Nutrient requirements of suspension cultures of soybean root cells. *Experimental Cell Research*, Vol.50, No.1, (April 1968), pp.151-158, ISSN 0014-4827

Gelvin, S.B. (2003). *Agrobacterium* and plant transformation: the biology behind the "gene- jockeying" tool. *Microbiology and Molecular Biology Reviews*, Vol.67, No.1, (March 2003), pp. 16-37, ISSN 1092-2172

Gelvin, S.B. (2010a). Finding a way to the nucleus. *Current Opinion in Microbiology*, Vol.13, No.1, (February 2010), pp. 53–58, ISSN 1369-5274

Gelvin, S.B. (2010b). Plant proteins involved in *Agrobacterium*-mediated genetic transformation. *Annual Review Phytopathology*, Vol.48, (March 2010), pp. 45-68, ISSN 0066-4286

Gharyal, P.K. & Maheswari, S.C. (1983). *In vitro* differentiation of somatic embryos in a leguminous tree, *Albizia lebbeck* L. *Naturwissenschaften*, Vol.68, (1983), pp. 379–380, ISSN 0028-1042

Ghazi T.D.; Cheema H.V. & Nabors M.W. (1986). Somatic embryogenesis and plant regeneration from embryonic callus of soybean [*Glycine max* (L.) Merr.]. *Plant Physiology*, Vol.77, (1986), p. 863-868, ISSN 0032-0889

Hadi, M.Z.; McMullen, M.D. & Finer, J.J. (1996). Transformation of 12 different plasmids into soybean via particle bombardment. *Plant Cell Reports*, Vol.15, No., (1996), pp. 500-505, ISSN 0721-7714

Hammatt, N. & Davey, M.R. (1987). Somatic embryogenesis and plant regeneration from cultured zygotic embryos of soybean (*Glycine max* L. Merr.). *Journal of Plant Physiology*, Vol.128, No.3, (1987), pp. 219-226, ISSN 0176-1617

Hartweck, L.M.; Lazzeri, P.A.; Cui, D.; Collins, G.B. & Williams, E.G. (1988). Auxin-orientation effects on somatic embryogenesis from immature soybean cotyledons. *In Vitro Cellular and Developmental Biology-Plant*, vol.24, No.8, (August 1988), pp. 821-828, ISSN 1054-5476

Hernandez-Garcia, C.M.; Martinelli, A.P.; Bouchard, R.A. & Finer, J.J. (2009). A soybean (*Glycine max*) polyubiquitin promoter gives strong constitutive expression in transgenic soybean. *Plant Cell Reports*, Vol.28, No.5, (2009), pp. 837–849, ISSN 0721- 7714

Hinchee, M.A.; Connor-Ward, D.V.; Newell, C.A.; McDonell, R.E.; Sato, S.J.; Gasser, C.S.; Fishhoff, D.A.; Re, D.B.; Fraley, R.T. & Horsch, R.B. (1988). Production of transgenic soybean plants using *Agrobacterium*-mediated DNA transfer. *Nature Biotechnology*, Vol.6, (1988), pp. 915-922, ISSN 1087-0156

Hiraga, S.; Minakawa, H.; Takahashi, K.; Takahashi, R.; Hajika, M.; Harada, K. & Ohtsubo, N. (2007). Evaluation of somatic embryogenesis from immature cotyledons of Japanese soybean cultivars. *Plant Biotechnology*, v.24, No. 4, (September, 2007), p.435-440, ISSN 1342-4580

Hoffmann, N.; Nelson, R.L. & uence of media components and pH on somatic embryo induction in three genotypes of soybean.□Korban S.S. (2004). In *Plant Cell, Tissue and Organ Culture*, Vol.77, No.2, (2004), pp. 157–163, ISSN 0167-6857

Homrich, M.S.; Passaglia, L.M.P.; Pereira, J.F.; Bertagnolli, P.F.; Pasquali, G.; Zaidi, M.A.; Altosaar, I. & Bodanese-Zanettini, M.H. (2008). Resistance to *Anticarsia gemmatalis* Hübner (Lepidoptera, Noctuidae) in transgenic soybean (*Glycine max* (L) Merrill Fabales, Fabaceae) cultivar IAS5 expressing a modified Cry1Ac endotoxin. *Genetics and Molecular Biology*, Vol.31, No.2, (2008), pp. 522–531, ISSN 1415-4757

Hong, H.P.; Zhang, H.; Olhoft, P.; Hill, S.; Wiley, H.; Toren, E.; Hillebrand, H.; Jones, T. & Cheng, M. (2007). Organogenic callus as the target for plant regeneration and transformation via *Agrobacterium* in soybean (*Glycine max* (L) Merr). *In Vitro Cellular and Developmental Biology–Plant*, Vol.43, No.6, (2007), pp. 558-568, ISSN 1054-5476

Hsu, F.C. & Obendorf, R.L. (1982). Compositional analysis of in vitro matured soybean seeds. *Plant Science Letters*, Vol.27, No.2, (October 1982), pp. 129-135, ISSN 0168- 9452

Kermode, A.R. (1990). Regulatory mechanism involved in the transition from seed development to maturation *Critical Reviews in Plant Sciences*, Vol.9, (1990), pp. 155- 195, ISSN 0735-2689

Kerns, H.R.; Barwale, V.B. & Meyer, M.M. (1986). Correlation of cotyledonary node shoot proliferation and somatic embryoid development in suspension cultures of soybean [*Glycine max* (L.) Merr.]. *Plant Cell Reports*, Vol.5, No.2, (1986), p. 140-143, ISSN 0721-7714

Kiss, E.; Heszky, L.E.; Gyulai, G.; Horváth, H.S. & Csillag, A. (1991). Neomorph and leaf differentiation as alternative morphogenetic pathways in soybean tissue culture. *Acta Biologica Hungarica*, Vol. 42, No.4, (1991), pp. 313-321, ISSN 0236-5383

Kita Y.; Nishizawa K,; Takahashi M.; Kitayama M. & Ishimoto M. (2007). Genetic improvement of the somatic embryogenesis and regeneration in soybean and transformation of the improved breeding lines. *Plant Cell Reports*, vol.26, No. 4, (April, 2007), pp. 439-447, ISSN 1432-203X

Ko, T.S.; Lee, S.; Farrand, S.K. & Korban, S.S. (2004). A partially disarmed vir helper plasmid, pKYRT1, in conjunction with 2,4-dichlorophenoxyacetic acid promotes emergence of regenerable transgenic somatic embryos from immature cotyledons of soybean. *Planta*, Vol.218, No.4, (February), pp. 536-541, ISSN 0032-0935

Ko, T.S.; Lee, S.; Krasnyanski, S. & Korban, S.S. (2003). Two critical factors are required for efficient transformation of multiple soybean cultivars: *Agrobacterium* strain and orientation of immature cotyledonary explant. *Tag Theoretical and Applied Genetics*, Vol.107, No.3, (2003), pp. 439-447, ISSN 0040-5752

Kohli, A.; Twyman, R.M.; Abranches, R.; Weget, E.; Stoger, E. & Christou, P. (2003). Transgene integration, organization and interaction in plants. Plant Molecular Biology, Vol.52, No.2, (2003), pp. 247-258, ISSN 0167-4412

Komatsuda, T. & Ko, S.W. (1990). Screening of soybean (*Glycine max* (L.) Merrill) genotypes for somatic embryo production from immature embryo. *Japanese Journal of Breeding*, Vol.40, (1990), pp. 249-251, ISSN 0536-3683

Komatsuda, T. & Ohyama, K. (1988). Genotype of high competence for somatic embryogenesis and plant regeneration in soybean *Glycine max*. *Theoretical and Applied Genetics*, Vol. 75, No. 5, (1998), pp. 695-700, ISSN 0040-5752

Komatsuda, T., Kanebo, K., & Oka, S. (1991). Genotype × sucrose interactions for somatic embryogenesis in soybean. *Crop Science*. Vol.31, No.2, (1991), pp. 333–337, ISSN 0011-183X

Körbes, A.P. & Droste, A. (2005). Carbon sources and polyethylene glycol on soybean somatic embryo conversion. *Pesquisa Agropecuária Brasileira*, Vol.40, No.3, (March 2005), pp. 211-216, ISSN 0100-204X

Kumari, B.D.R.; Settu, A. & Sujatha, G. (2006). Somatic embryogenesis and plant regeneration in soybean. *Indian Journal of Biotechnology*, Vol.5, (April 2006), p. 243- 245, ISSN 0972-5849

Lakshmanan P. & Taji A. (2000). Somatic Embryogenesis in Leguminous Plants. *Plant Biology*, Vol.2, No.2, (March 2000), pp. 136–148, ISSN 1435-8603

Lazzeri, P.A., Hildebrand, D.F., Sunega, J., Williams, E.G. & Collins, G.B. (1988). Soybean somatic embryogenesis: interactions between sucrose and auxin. *Plant Cell Reports*, Vol.7, (1988), pp. 517–520, ISSN 0721-7714

Lazzeri, P.A.; Hildebrand, D.F. & Collins, G.B. (1987). Soybean somatic embryogenesis: effects of hormones and culture manipulations. *Plant Cell, Tissue and Organ Culture*, Vol.10, (1987), pp. 197–208, ISSN 0167-6857

Lazzeri, P.A.; Hilderbrand, D.F. & Collins, G.B. (1985). A procedure for plant regeneration from immature cotyledon tissue of soybean. *Plant Molecular Biology Reporter*, Vol.3, No.4, (Winter 1985), pp. 160-167, ISSN 0735-9640

Li, X.Y.; Huang, F.H.; Murphy, B. & Gbur Junior, E.E. (1998). Polyethylene glycol and maltose enhance somatic embryo maturation in loblolly pine (*Pinus taeda* L.). *In Vitro Cellular and Developmental Biology-Plant*, Vol.34, (1998), pp. 22-26, ISSN 1071-2690

Li, Z.; Xing, A.; Moon, B.P.; McCardell, R.P.; Mills, K. & Falco, S.C. (2009). Site-Specific Integration of Transgenes in Soybean via Recombinase-Mediated DNA Cassette Exchange. *Plant Physiology*, Vol.151, No.3, (2009), pp. 1087–1095, ISSN 0032- 0889

Lippmann, B. & Lippmann, G. (1984). Induction of somatic embryos in cotyledonary tissue of soybean, *Glycine max* L. Merr. *Plant Cell Reports*, Vol.185, No.3, (September 1984), pp. 215–218, ISSN 0721-7714

Lippmann, B. & uencing plant recovery from isolated embryos. Lippmann, G. (1993). Soybean embryo culture: factors in *Plant Cell, Tissue and Organ Culture*, Vol.32, No.1, (1993), pp. 83–90, ISSN 0167-6857

Liu, S.J.; Wei, Z.M. & Huang, J.Q. (2008). The effect of co-cultivation and selection parameters on *Agrobacterium*-mediated transformation of Chinese soybean varieties. *Plant Cell Reports*, Vol.27, No.3, (2008), pp. 489-498, ISSN 0721-7714

Liu, W.; Moore, P.J. & Collins, G.B. (1992). Somatic embryogenesis in soybean via somatic embryo cycling. *In Vitro Cellular and Developmental Biology-Plant*, Vol.28, No.3, (July 1992), pp. 153-160, ISSN 1054-5476

Maheswaran, G. & Williams, E.G. (1984). Direct somatic embryoid formation on immature embryos of *Trifolium repens, T. pretense* and *Medicago sativa*, and rapid clonal propagation of *T. repens*. *Annals of Botany*, Vol.54, (1984), pp. 201-211, ISSN 0305- 7364

Mauro, A.O.O.; Pfeiffer, T.W. & Collins, G.B. (1995). Inheritance of soybean susceptibility to *Agrobacterium tumefaciens* and its relationship to transformation. *Crop Science*, Vol.35, No.4, (1995), pp. 1152-1156, ISSN 0011-183X

McCabe, D.E.; Swain, W.F.; Martinell, B.J. & Christou, P. (1988). Stable transformation of soybean (*Glycine max*) by particle acceleration. *Nature Biotechnology*, Vol.6, (1988), pp. 923–926, ISSN 1087-0156

Merkle, S.A.; Parrott, W.A. & Flinn, B.S. (1995). Morphogenic aspects of somatic embryogenesis. In: *In vitro embryogenesis in plants*, T.A. Thorpe, (Ed.), 155–203, Kluwer Academic, ISBN 0-7923-3149-4, Dordrecht, Netherlands

Meurer, C.A.; Dinkins, R.D.; Redmond, C.T.; Mcallister, K.P.; Tucker, D.T.; Walker, D.R.; Parrott, W.A.; Trick, H.N.; Essig, J.S.; Frantz, H.M.; Finer, J.J. & Collins, G.B. (2001). Embryogenic response of multiple soybean [*Glycine max* (L.) Merr.] cultivars across three locations. *In Vitro Cellular and Developmental Biology-Plant*, Vol.37, No.67, (January-February 2001), pp. 62-67, ISSN 1054-5476

Miklos, J.A.; Alibhai, M.F.; Bledig, S.A.; Connor-Ward, D.C.; Gao, A.G.; Holmes, B.A.; Kolacz, K.H.; Kabuye, V.T.; MacRae, T.C.; Paradise, M.S.; Toedebusch, A.S. & Harrison, L.A. (2007). Characterization of soybean exhibiting high expression of a synthetic *Bacillus thuringiensis* cry1A transgene that confers a high degree of resistance to Lepidopteran pests. *Crop Science*, Vol.47, (2007), pp. 148-157, ISSN 0011-183X

Murashige, T. & Skoog, F. (1962). A revised medium for rapid growth and bio assays with tobacco tissue cultures. *Physiologia Plantarum*, Vol. 15, (1962), pp. 473–497, ISSN 0031-9317

Parrott, W.A.; Dryden, G.; Vogt, S.; Hildebrand, D.F.; Collins, G.B. & Williams, E.G. (1988). Optimization of somatic embryogenesis and embryo germination in soybean. *In Vitro Cellular and Development Biology-Plant*, Vol.24, (1988), pp. 817-820, ISSN 1071-2690

Parrott, W.A.; Durham, R.E. & Bailey, M.A. (1995). Somatic embryogenesis in legumes. In: *Biotechnology in Agriculture and Forestry, Somatic Embryogenesis and Synthetic Seed II*, Y. P. S. Bajaj, (Ed.), 199–227, Springer-Verlag, ISBN 0-387-57449-2-X, Berlin

Parrott, W.A.; Hoffman, L.M.; Hildebrand, D.F.; Williams, E.G. & Collins, G.B. (1989). Recovery of primary transformants of soybean. *Plant Cell Reports*, Vol.7, No.8, (1989), pp. 615-617, ISSN 0721-7714

Parrott, W.A.; Williams E.G.; Hildebrand, D.F. & Collins, G.B. (1989). Effect of genotype on somatic embryogenesis from immature cotyledons of soybean. *Plant Cell, Tissue and Organ Culture*, Vol. 16, No. 1, (1989) pp. 15-21, ISSN 0167-6857

Paz, M.M.; Martinez, J.C.; Kalvig, A.B.; Fonger, T.M. & Wang, K. (2006). Improved cotyledonary node method using an alternative explant derived from mature seed for efficient *Agrobacterium*-mediated soybean transformation. *Plant Cell Reports*, Vol.25, No.3, (2006), pp. 206-213, ISSN 0721-7714

Pitzschke, A. & Hirt, H. (2010). New insights into an old story: *Agrobacterium*-induced tumour formation in plants by plant transformation. *The EMBO Journal*, Vol.29, No.6, (February 2010), pp. 1021-1032, ISSN 0261-4189

Raghavan, V. (1986). *Embryogenesis in Angiosperms. A Developmental and Experimental Study.* Raghavan, V. ISBN 0-521-26771-4, Cambridge, U.K., Cambridge University Press

Ranch, J.P.; Oglesby, L. & Zielinski, A.C. (1985). Plant regeneration from embryo-derived tissue culture of soybeans. *In Vitro Cellular and Developmental Biology-Plant*, Vol.21, No.11, (1985), pp. 653-658, ISSN 1071-2690

Reinert, J. (1958). Morphogenese und ihre Kontrolle an Gewebekulturen aus Karotten. Naturwissenschaften, Vol.45, (1958), pp. 344–349, ISSN 0028-1042

Rock, C.D. & Quatrano, R.S. (1995). The role of hormones during seed development. In: *Plant hormones: physiology, biochemistry and molecular biology,* P.J. DAVIES, (Ed.), pp. 671-697, Kluwer Academic, ISBN 0-7923-2985-6, Dordrecht, Netherlands

Rodrigues, L.R.; Oliveira, J.M.S.; Mariath, J.E.A. & Bodanese-Zanettini, M.H. (2005). Histology of embryogenic responses in soybean anther culture. *Plant Cell, Tissue and Organ Culture*, Vol.80, No.2, (February 2005), pp. 129-137, ISSN 0167-6857

Rosenberg, L.A. & Rinne, R.W. (1986). Moisture loss as a prerequisite for seedling growth in soybean seeds (*Glycine max* L. Merr.). *Journal of Experimental Botany*, Vol.37, No.11, (1986), pp. 1663-1674, ISSN 0022-0957

Rosenberg, L.A. & Rinne, R.W. (1988). Protein synthesis during natural and precocious soybean seed (*Glycine max* L. Merr.) maturation. *Plant Physiology*, Vol.87, No.2, (June 1988), pp. 474-478, ISSN 0032-0889

Samoylov, V.M.; Tucker, D.M. & Parrott, W.A. (1998b). A liquid medium-based protocol for rapid regeneration from embryogenic soybean cultures. *Plant Cell Reports*, Vol.18, (1998), pp. 49–54, ISSN 0721-7714

Samoylov, V.M; Tucker, D.M. & Parrott, W.A. (1998a). Soybean [*Glycine max* (L.) Merrill] embryogenic cultures: the role of sucrose and total nitrogen content on proliferation. *In Vitro Cellular and Developmental Biology-Plant*, Vol.34, (January- March 1998), pp. 8–13, ISSN 1071-2690

Sanford, J.C. (1988). The biolistic process. *Trends Biotechnology*, Vol.6, No.12, (December 1988), pp. 299-302, ISSN 0167-7799

Sanford, J.C. (1990). Biolistic plant transformation. *Physiologia Plantarum*, Vol.79, No.1, (May 1990), pp. 206-209, ISSN 0031-9317

Santarém, E.R. & Finer, J.J. (1998). Transformation of soybean [*Glycine max* (L.) Merrill] using proliferative embryogenic tissue maintained on semi-solid medium. *In Vitro Cellular and Developmental Biology-Plant*, Vol.35, (November-December 1998), pp. 451–455, ISSN 1071-2690

Santarém, E.R.; Pelissier, B. & Finer, J.J. (1997). Effect of explant orientation, pH, solidifying agent and wounding on initiation of soybean somatic embryos. *In Vitro Cellular and Developmental Biology-Plant*, Vol.33, No.1, (January 1997), pp. 13–19, ISSN 1071-2690

Santos, K.G.B.; Mariath, J.E.A.; Moço, M.C.C. & Bodanese-Zanettini, M.H. (2006). Somatic Embryogenesis from Immature Cotyledons of Soybean (*Glycine max* (L.) Merr.): Ontogeny of Somatic Embryos. *Brazilian Archives of Biology and Technology*, Vol. 49, No.1, (January 2006), pp. 49-55, ISSN 1516-8913

Santos, K.G.B.; Mundstock, E.; Bodanese-Zanettini, M.H. (1997). Genotype-specific normalization of soybean somatic embryogenesis through the use of an ethylene inhibitor. *Plant Cell Reports*, Vol. 16, No. 12, (June, 1997), pp. 859-864, ISSN 0721-7714

Saravitz, C.H. & Raper, C.D.Jr. (1995). Responses to sucrose and glutamine by soybean embryos grown *in vitro*. *Physiologia Plantarum*, Vol.93, (1995), pp. 799–805, ISSN 0031-9317

Sato, S.; Newell, C.; Kolacz, K.; Tredo, L.; Finer, J. & Hinchee, M. (1993). Stable transformation via particle bombardment in two different soybean regeneration systems. *Plant Cell Reports*, Vol.12, No.7-8, (1993), pp. 408- 413, ISSN 0721-7714

Schmidt, M.A.; LaFayette, P.R.; Artelt, B.A. & Parrott, W.A. (2008). A comparison of strategies for transformation with multiple genes via microprojectile-mediated bombardment. *In Vitro Cellular and Developmental Biology-Plant*, Vol. 44, No.3, (2008), pp. 162-168, ISSN 1071-2690

Schmidt, M.A.; Tucker, D.M.; Cahoon, E.B. & Parrott, W.A. (2005). Towards normalization of soybean somatic embryo maturation. *Plant Cell Reports*, Vol.24, (2005), pp. 383- 391, ISSN 1432-203X

Sharma, K. K. & Thorpe, T. A. (1995). Asexual Embryogenesis in Vascular Plants in Nature, In: *In Vitro Embryogenesis in Plants*, T.A.Thorpe, (Ed.), 17–72, Kluwer Academic, ISBN 0-7923-3149-4, Dordrecht, Netherlands

Sharp, W.R., Evans D.A. & Sondahl, MR. (1982). Application of somatic embryogenesis to crop improvement. In: *Plant tissue culture 1982.* Proceedings of the Fifth International Congress of Plant Tissue Culture, A. Fujiwara, (Ed), 759-762, Japanese Association for Plant Tissue Culture, Maruzen, Tokyo

Simmonds, D.H. & Donaldson, P.A. (2000). Genotype screening for proliferative embryogenesis and biolistic transformation of short-season soybean genotypes. *Plant Cell Reports*, Vol. 19, No. 5, (2000), pp. 485-490, ISSN 0721-7714

Smith, D.L. & Krikorian, A.D. (1989). Release of somatic embryogenic potential from excised zygotic embryos of carrot and maintenance of proembryonic cultures in hormone- free medium. *American Journal of Botany*, Vol.76, No.12, (1989), pp. 1832–1840, ISSN 0002-9122

Somers, D.A.; Samac, D.A. & Olhoft, P.M. (2003). Recent Advances in Legume Transformation. *Plant Physiology*, Vol. 131, No.3, (March 2003), pp. 892–899, ISSN 0032-0889

Stewart Jr, C.N.; Adang, M.J.; All, J.N.; Boerma, H.R.; Cardineau, G.; Tucker, D. & Parrott, W.A. (1996). Genetic transformation, recovery, and characterization of soybean (*Glycine max* [L.] Merrill) transgenic for a synthetic *Bacillus thuringiensis* CRY1A(c) gene. *Plant Physiology*, Vol.112, No.1, (September 1996), pp. 121-129, ISSN 0032- 0889

Thomas, T.D. (2008). The role of activated charcoal in plant tissue culture. *Biotechnology Advances*, Vol.26, (August 2008), pp. 618–631, ISSN 0734-9750

Thompson, J.F.; Madison, J.T. & Muenster, A.-M.E. (1977). *In vitro* culture of immature cotyledons of soya bean (*Glycine max* L. Merr.). *Annals of Botany*, Vol.41, No.1, (1977), pp. 29–39, ISSN 0305-7364

Tian, L.N. & Brown, D.C.W. (2000). Improvement of soybean somatic embryo development and maturation by abscisic acid treatment. *Canadian Journal of Plant Science*, Vol.80, (2000), pp. 721-276, ISSN 0008-4220

Tomlin, E.S.; Branch, S.R.; Chamberlain, D.; Gabe, H., Wright, M.S. & Stewart, C.N. (2002). Screening of soybean, *Glycine max* (L.) Merrill, lines for somatic embryo induction and maturation capability from immature cotyledons. *In Vitro Cellular and Developmental Biology-Plant*, Vol.38, (November-December 2002), pp. 543–548, ISSN 1071-2690

Trick, H.N. & Finer, J.J. (1997). SAAT: sonication-assisted *Agrobacterium*-mediated transformation. *Transgenic Research*, Vol.6, (1997), pp. 329-336, ISSN 0962-8819

Trick, H.N. & Finer, J.J. (1998). Sonication-assisted *Agrobacterium*-mediated transformation of soybean (*Glycine max*) embryogenic suspension culture tissue. *Plant Cell Reports*, Vol.17, (1998), pp. 482-488, ISSN 0721-7714

Trick, H.N.; Dinkins, R.D.; Santarém, E.R.; Di, R.; Samoylov, V.; Meurer, C.A.; Walker, D.R.; Parrott, W.A.; Finer, J.J. & Collins, G.B. (1997). Recent advances in soybean transformation. *Plant Tissue Culture and Biotechnology*, Vol.3, No.1, (March 1997), pp. 9–24, ISSN 1817-3721

Vergara, R.; Verde, F.; Pitto, L.; LoSchiavo, F. & Terzi, M. (1990). Reversible variations in the methylation pattern of carrot DNA during somatic embryogenesis. *Plant Cell Reports*, Vol.8, No.12, (1990), pp. 697–701, ISSN 0721-7714

Vianna, G.R.; Aragão; F.J.L. & Rech, E.L. (2011). A minimal DNA cassette as a vector for genetic transformation of soybean (*Glycine max*). *Genetics and Molecular Research*, Vol.10, No.1, (March 2011), pp. 382-390, ISSN 1676-5680

Walker, D.R. & Parrott, W.A. (2001). Effect of polyethylene glycol and sugar alcohols on soybean somatic embryo germination and conversion. *Plant Cell, Tissue and Organ Culture*, Vol.64, No.1, (January 2001), pp. 55–62, ISSN 0167-6857

Wang, G. & Xu, Y. (2008). Hypocotyl-based *Agrobacterium*-mediated transformation of soybean (*Glycine max*) and application for RNA interference. *Plant Cell Reports*, Vol.27, No.7, (2008), pp. 1177-1184, ISSN 0721-7714

Weber, R.L.M.; Körber, A.P.; Baldasso, D.A.; Callegari-Jacques, S.M.; Bodanese-Zanettini, M.H. & Droste, A. (2007). Beneficial effect of abscisic acid on soybean somatic embryo maturation and conversion into plants. *Plant Cell Culture and Micropropagation*, Vol. 3, No. 1, (2007), pp. 1-9, ISSN 1808-9909

Wiebke, B.; Ferreira, F.; Pasquali, G.; Bodanese-Zanettini, M.H. & Droste, A. (2006). Influence of antibiotics on embryogenic tissue and *Agrobacterium tumefaciens* suppression in soybean genetic transformation. *Bragantia*, Vol.65, No.4, (2006), pp. 543-551, ISSN 0006-8705

Wiebke-Strohm, B.; Droste, A.; Pasquali, G.; Osorio, M.B.; Bücker-Neto, L.; Passaglia, L.M.P.; Bencke, M.; Homrich, M.S.; Margis-Pinheiro, M. & Bodanese-Zanettini, M.H. (2011). Transgenic fertile soybean plants derived from somatic embryos transformed via the combined DNA-free particle bombardment and Agrobacterium system. *Euphytica*, Vol.177, No.3, (2011), pp. 343-354, ISSN 0014- 2336

Williams, E. G. & Maheswaran, G. (1986). Somatic embryogenesis: factors influencing coordinated behavior of cells as an embryogenic group. *Annals of Botany*, Vol.57, No.4, (April 1986), pp. 443-462, ISSN 0305-7364

Wright, M.S.; Launis, K.L.; Novitzky, R.; Duesiing, J.H. & Harms, C.T. (1991). A simple method for the recovery of multiple fertile plants from individual somatic embryos of soybean [*Glycine max* (L.) Merrill]. *In Vitro Cellular and Developmental Biology- Plant*, Vol.27, (1991), pp. 153-157, ISSN 1071-2690

Wu, C.; Chiera, J.M.; Ling, P.P. & Finer, J.J. (2008). Isoxaflutole treatment leads to reversible tissue bleaching and allows for more effective detection of GFP in transgenic soybean tissues. *In Vitro Cellular and Developmental Biology-Plant*, Vol.44, No.6, (2008), pp. 540–547, ISSN 1054-5476

Xing, A.; Moon, B.P.; Mills, K.M.; Falco, S.C. & Li, Z. (2010). Revealing frequent alternative polyadenylation and widespread low-level transcription read-through of novel plant transcription terminators. *Plant Biotechnology Journal*, Vol.8, No.7, (September 2010), pp. 772–782, ISSN 1467-7644

Yan, B.; Reddy, M.S.S.; Collins, G.B. & Dinkins, R.D. (2000). *Agrobacterium tumefaciens* mediated transformation of soybean [*Glycine max* (L) Merrill] using immature zygotic cotyledon explants. *Plant Cell Reports*, Vol.19, No.11, (2000), pp. 1090-1097, ISSN 0721-7714

Yang, C.; Zhao, T.; Yu, D. & Gai J. (2009). Somatic embryogenesis and plant regeneration in Chinese soybean (*Glycine max* (L.) Merr.)-impacts of mannitol, abscisic acid, and explants age. *In Vitro Cellular and Developmental Biology-Plant*, Vol. 45, No. 2, (April 2009), pp. 180-181, ISSN 1054-5476

07 GENETIC ENGINEERING AND BIOTECHNOLOGY ARE USED UO MANIPULATE GROWTH HORMONES

1. INTRODUCTION: AN OVERVIEW

Biotechnology, in its contemporary form, refers to the application of genetic engineering methods to change microbes, plants, and animals for the purpose of creating profitable products and processes that are advantageous to humans. The techniques that emerged in the mid-1970s as part of the biotechnological revolution have enabled the isolation and manipulation of specific genes. These techniques have also facilitated the development of transgenic microorganisms that can produce therapeutic proteins, including vaccines, enzymes, and hormones, primarily of eukaryotic origin.

Biotechnology is utilized in various domains including food production, industrial waste degradation, mining, and medical. Notable accomplishments include the creation of drugs in genetically modified animals and plants, as well as the profitable utilization of gene sequences derived from the human genome project and other programs focused on economically valuable plants and animals, both ongoing and forthcoming.

Human growth hormone was the second product of this new technology, following insulin. Genentech initially developed and commercialized this medication, which was subsequently employed in clinical settings to address growth issues and dwarfism (1). In addition, transgenic organisms have been utilized to manufacture growth hormones from various animal species, which have found applications in both the aquatic animal and livestock industries.

2. THE GROWTH HORMONE (GH) FAMILY

Growth hormones (GHs) are a group of proteins that share structural similarities and have specific functions. These functions include prolactin (Prl), somatolactin (SL), chorionic somatomammotropin (CS), proliferin (PLF), and proteins linked to Prl (PLP) (2). This family exemplifies one of the most varied protein families in terms of physiological characteristics, which have emerged through the process of gene duplication. The family has extensively researched two of its members, GH and Prl, which have been observed in a wide range of species from primordial fish to mammals. However, the distribution and study of other members of the family are not as extensive.

2.1 Structure of growth hormones

GHs (see Figure 1), in general, have a molecular weight of around 22,000 Daltons (22 kDa or simply 22k) and do not require post-translational modifications. They are synthesized in somatotrophs in the hypophysis, intervening as an important endocrine factor in postnatal somatic growth and lactation.

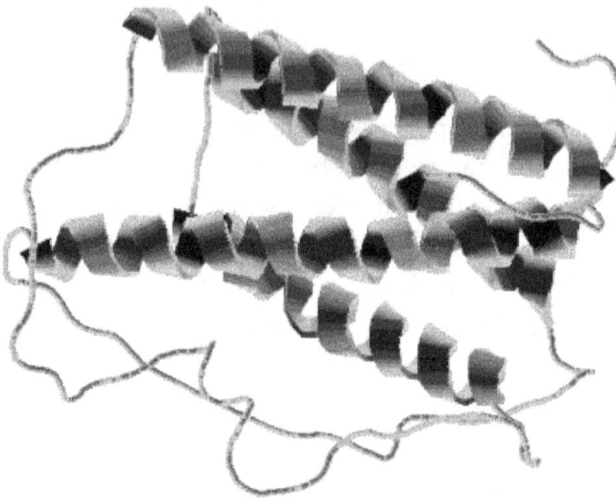

Fig. 1. Growth hormones' consensus tridimensional structure. The GHs have in general 190 aminoacidic residues, four alpha helixes, and two sulphide bonds

2.2 Hormones of the human growth hormone family

HGH22k

HGH22k (or HGHN) is the main product of the GH gene (hGH-N) active in the hypophysis and it is responsible for postnatal growth as well as being an important modulator of carbohydrate, lipid, nitrogen and mineral metabolism. It is the best known hormone and the only one of the HGH family that has been commercialized.

As mentioned, besides being the cure for hypophyseal dwarfism, HGH22k postulated benefits are as an anabolic in athletics and for the treatment of trauma because of its postulated regenerative properties (3).

HGH20k

In addition to the mRNA of HGH22k, an alternative processing pathway of the primary transcript of the hGH-N gene generates a second mRNA that is responsible for the production of the 20k isoform of HGH or HGH20k. Its smaller size is due to elimination of the first 45 nucleotides of the third exon of the mRNA and of the amino acids that correspond to positions 32-46 of the hormone, producing a protein with 176 amino acid residues (4).

This isoform comprises approximately 10% of all the GH produced in the hypophysis and although it has not been shown to be the etiological agent of any known disease, it is known that its levels are significantly higher in patients with active acromegaly and in those with anorexia nervosa (5).

The administration of exogenous HGH20k suppresses endogenous secretion of HGH22k in healthy subjects, which suggests that the regulation of secretion of both hormones is physiologically similar (6). In vitro findings suggest that both hormones can equally stimulate bone remodeling and allow anabolic effects on skeletal tissue when they are administered in vivo to laboratory animals (7).

HGHV

Several isoforms also derive from the GH gene expressed in the placenta (hGH-V) (Table 1). The most abundant mRNA from this gene in the placenta at terminus also codifies for a 22 kDa isoform. A less abundant isoform (HGHV2) originate from a species of mRNA that retains the fourth intron and due to this, it codifies for a 26 kDa protein that anchors to the membrane and which could have a local action (8). A 25 kDa protein is also derived by glycosylation of residue 140 of asparagine from the 22 kDa isoform (9, 10). Finally, two new transcripts of this gene have recently been identified: one already known that as in the case of the HGH20k also produces a 20 kDa protein, and another novel splicing variation that results in a mRNA known as hGHV3, that traduces into a 24 kDa isoform (11).

Table 1. HGH-V isoforms generated by alternative splicing and processing

Isoform	Size	Length	Characteristic
HGH-V22k	22kDa	191aa	Main isoform.
HGH-V25k	25kDa	191aa	Glycosylated version of HGH-V22.
HGH-V2	26kDa	230aa	Retains the fourth intron.
HGH-V20K	20kDa*	176aa	Deletion of aa residues 32 to 46.
HGH-V3	24kDa*	219aa	Alternate processing at level of exon 4.
*Only the mRNAs that codify each have been identified.			

During pregnancy, while hypophyseal HGHN progressively disappears from the maternal circulation until undetectable values are reached at weeks 24 to 25, HGHV progressively increases until birth, suggesting that it has a key role during human gestation (12). It has also been found that in cases of intrauterine growth restriction, circulating levels of HGHV measured between week 31 and birth are lower than those reported in normal pregnancy (13, 14, 15).

Finally, although low levels of this hormone have been associated with intrauterine growth retardation, cases of hGH-V gene deletion have also been reported, but without an apparent pathology (16).

2.3 Human chorionic somatomammotropin (HCS)

HCS is detected in maternal serum from the fourth week of gestation, increasing throughout the pregnancy in a linear fashion, and reaching high production levels of a couple of grams per day at the end of gestation. These actions result in both elevation of glucose and amino acids in the maternal circulation. These are in turn used by the fetus for his/her development. It also generates free fatty acids (by lipolytic effect), which are used as an energy source by the fetus (17, 18). Little is known about the HCS physiological role, and still is not known its action mechanism. Producing rHCS by biotechnology will help to advance these investigations.

2.4 In vitro bioassays for GHs and CSHs

As stated above except for HGH22k, the functions of the rest of hormones of the human GH family have been not completely defined. Their biological activities are being studied, classifying them into at least two general categories:

- Somatogenic activities. These involve linear bone growth and alterations in carbohydrate metabolism; effects that are in part mediated by local and hepatic generation of insulin- like growth factor-I (IGF-I). The somatogenic activity of HGHV has been studied by stimulating body weight increase in hypophysectomized rats, reporting a linear increase comparable to that produced by HGH22k (19).
- Lactogenic activities. These include stimulation of lactation and reproductive functions (20). The lactogenic activities of this hormone have been studied using a cell model (by mitogenic response to Nb2 cells) and a response that is parallel to HGH22k has been reported, although it is significantly less (19).

2.5 The human GH locus

Besides the two hGH genes (normal and variant), three HCSs complement the multigenic HGH family from the human genome and these are arranged in the following order: HGH- N, HCS-1, HCS-2, HGH-V y HCS-3 (21, 22) (Figure 2). While HCS-1 appears to be a pseudogene, HCS-2 and HCS-3 are very active in the placenta and interestingly; mature versions of the hormones that they codify are identical (23).

In the last few years, in our laboratory, all the hGH and HCS genes have been cloned and expressed in cell culture, and the factors that affect their levels of expression have been particularly studied (24).

In the same way, and using polymerase chain reaction (PCR) with consensus primers, several new genes and complementary DNAs (cDNAs) to the mRNA of numerous GHs have been isolated in our laboratory, mainly from mammals (unpublished results).

3. GROWTH HORMONE OF ANIMAL ORIGIN

3.1 Bovine growth hormone (BGH)

Bovine growth hormone (BGH) or bovine somatotropin improves the efficiency of milk production (per unit of food consumed) (25), and the production (body weight) and composition (muscle: fat ratio) of meat (26). In the case of milk cows, this permits a reduction in the number of animals needed for milk production and a subsequent savings in maintenance, feeding, water, drugs, etc. It also reduces the production of manure, and nitrogen from urine and methane (27).

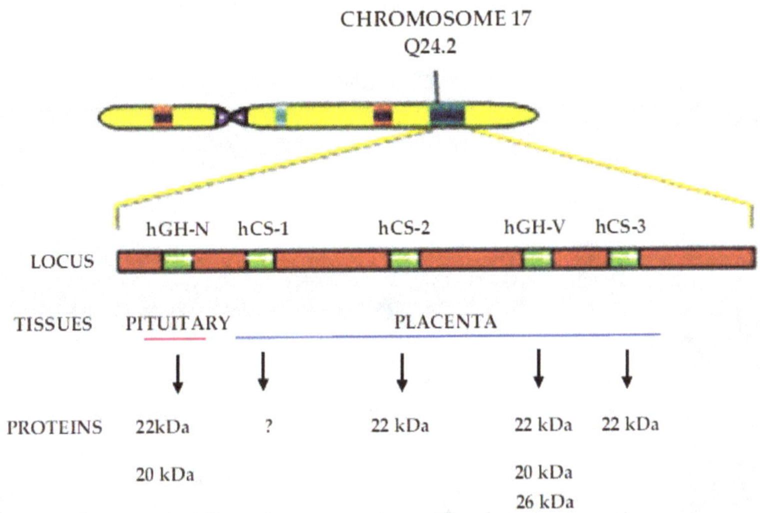

Fig. 2. HGH-HCS multigenic complex. Located on Chromosome 17, every gene is indicated; the tissue where they are expressed and the proteic isoforms that are produced are shown

Milk from cows treated with rBGH, does not differ from that of untreated cows (28, 29). The characteristics that have been evaluated include the freezing point, pH, thermal properties, susceptibility to oxygenation, and sensory characteristics, including taste; in fact all organoleptic properties are conserved. Also, differences have not been found in the properties necessary for producing cheese, including initial growth of the culture, coagulation, acidification, production and composition (29).

rBGH is administered subcutaneously and is dispensed as a long-acting suspension that is applied in a determined period of time. The taste of bovine meat and milk treated with rBGH is not altered, but the fat content is less.

3.2 Caprine growth hormone (CHGH)

For small ruminants there are studies in lactating goats in which the administration of rBGH increased milk production 23% and stimulated mammary gland growth more than in those that were frequently milked, with it being similar to prolactin (30). However, the production of recombinant CHGH, which is identical to ovine and thus can be used in both animals, had not been reported, until we achieve its expression on the methylotrophic yeast *Pichia pastoris*. (See section 7.2).

3.3 Equine growth hormone (ECGH)

With regard to horses, GH is used in the prevention of muscle wasting, in the repair of tendons and fractured bones, as well as for the treatment of anovulation in mares. Besides this, it is also used for repairing muscle tissue, to tonify and invigorate race horses, and for improving physical conditions in older horses by restoring nitrogen balance. It can also stimulate growth and early maturity in young horses, increase milk production in lactating mares and promote wound healing, especially of bone and cartilage (31, 32), as occurred in the case shown in Figure 3.

Fig. 3. Uses of equine GH. The race horse "Might and Power" (right) became the winner of the Melbourne Cup in 1997. But in 1999, a tendon from one of its hooves was severely damaged. The horse was treated with ECGH, recovered and in 2000 was able to return to horse racing (32)

3.4 Canine growth hormone (CFGH)

With regard to the dog (*Canis familiaris*), each day there is more evidence of the role that its GH (CFGH) plays in bone fracture treatment, in which the hormone helps reduce the bone restoration period (33).

It is no less important in the treatment of obesity in dogs, thanks to the metabolism activation produced by the hormone in removing fatty acids, and in general, in counteracting symptoms related to the presence or absence of the same GH. Also, since this hormone is identical to pig GH (PGH) (33), its virtues are valid for the application of CFGH in the porcine industry, where it generates leaner meat (34), which is of greater value.

3.5 Feline growth hormone (FCGH)

Although there is very little literature on cat GH (FCGH), the benefits identified in other GHs apply to this feline species, since these animals present the symptoms mentioned before for dogs, which are caused by the absence or low concentration of FCGH (dwarfism and alopecia, among others). Also, as referred to in the literature, biological tests of adipogenic activity in culture cells use cat serum (which contains FCGH) instead of bovine serum, because FCGH lacks adipogenicity (17).

Therefore, recombinant production of this GH would be useful in the mentioned tests. It is important to point out the usefulness that recombinant FCGH would have in future research on the metabolic study and role of this hormone in this and other feline species, including of course, large wild cats in captivity.

4. BIOLOGICAL POTENTIAL OF GHS

4.1 Growth hormones of human origin

Although HGH22k is widely commercialized and more functions now have been recognized to it (Table 2), the same does not occur with the other proteins and isoforms from this family; essentially the 20 kDa isoform of HGH, HGHV, also the isoform of 20 kDa of the latter (HGHV20k), and lastly, HCS. Partly because of this, many of their functions and mechanisms of action are still unknown.

Table 2. New functions atributed to HGH22k

Immunization and healing	Mental function
• Resistance to common diseases	• motional stability
• Ability to heal	• Memory
• Healing of old lesions	• General aspect and attitude
• Healing of other lesions	• Mental energy and clarity
• Ulcer treatment	

Skin and hair	Muscle strength and tone
• Skin elasticity • Skin thickness • Skin texture • Growth of new hair • Disappearance of wrinkles • Skin hydration	• Increase in energy in general • Increase muscle strength • Promotion of muscle mass gain
Sexual factors	**Circulatory system**
• Duration of an erection • Increase in libido • Potential/frequency of sexual activity • Regulation and control of the menstrual cycle • Positive effects in the reproductive system • Increase in breast-milk volume	• Improvement in circulation • Stabilization of blood pressure • Improvement in cardiac function
Bone	**Fats**
• As treatment for bone fractures • Osteoporosis treatment • Increases flexibility of the back and joints	• Increases "good" cholesterol (HDL) levels • Reduces fat
(Taken from Elian y cols., 1999), (3).	

It is believed that some of the hormone's less abundant natural variants, such as HGH20k, could retain desirable properties of the principal hormone and lack some of its other undesirable effects, such as its diabetogenic effect, which occurs with prolonged use (35).

4.2 Growth hormones of animal origin

The biotechnological potential of GHs could be enormous, since besides its use in species of the same origin, it has been demonstrated that the GHs of mammals have activity in phylogenetically lower animals.

For example, BGH and porcine GH (PGH) have been used experimentally for the treatment of hypophyseal dwarfism in dogs (36) and cats (37).

Regarding farm animals, porcine, bovine, caprine and ovine livestock have been treated with exogenous GH to improve production, since it increases food conversion efficiency, growth rate, weight gain, and milk and meat production.

What is surprising is the finding that BGH stimulates salmon growth, and even more interesting that bovine chorionic somatomammotropin (BCS) works even better (38).

5. EXPRESSION SYSTEMS FOR GROWTH HORMONES

5.1 The history of human recombinant GH

As previously mentioned, among the first cDNAs cloned and expressed in the bacteria *Escherichia coli* is precisely HGH (1). This expression system has been used since 1985 for the production of recombinant HGH by Genentech (protropin), which was later followed by Lily (humatrope), Biotech (biotropin), Novo Nordisk (norditropin), Serono (serostim), and others.

5.2 Different biotechnological hosts

Since the recombinant protein is frequently recovered from *E. coli* with undesirable modifications (extra methionine, incorrect folding, aggregated forms, etc.) and contaminated with highly pyrogenic substances, toilsome purification schemes are needed to obtain it with the desired purity, structure and biological activity. For this, subsequent efforts have focused on the search for better expression systems, with production being attempted with *Saccharomyces cerevisiae* (39), *Bacillus subtilis* (40), mammal cell cultures (41), as well as transgenic animals (42). Unfortunately, these expression systems do not offer a production level greater than that of *E. coli* and therefore in most cases they are not profitable.

In our laboratory, we succeeded in producing HGH22k in *E. coli* by fusing it with maltose binding protein (rHGH-MBP) in 1994 (unpublished results). However, due to the fact that to recover the hormone, whether from the periplasm or the cytoplasm, complicated strategies were needed, together with the limitations of the bacterial systems for folding and processing foreign proteins correctly, we proposed searching for an expression system that allows synthesizing the protein, purifying it with greater ease while retaining functionality. Thus, the evaluation of different expression systems was started in our laboratories, considering the methylotrophic yeast *Pichia pastoris* as the best (43).

5.3 Pichia pastoris as a biotechnological host for GHs

Yeasts offer the best of both prokaryotes and eukaryotes, since, in addition to performing some of the post-translational modifications that are common in superior organisms, they are easily grown in flasks and bioreactors, like bacteria, using simple and inexpensive culture media (44).

P. pastoris is a methylotrophic yeast (capable of growing in methanol as its only carbon source) that performs post-translational modifications, produces recombinant protein levels of one or two orders of magnitude above that of

Saccharomyces cerevisiae (45), is capable of secreting heterologic proteins into the culture media (where the levels of native protein are very low), and in contrast with the latter, can be cultivated at cell densities of more than 100 g/L of dry weight (46).

6. RECOMBINANT GROWTH HORMONES

In our laboratories, we identified as a scientific objective and a technological advantage, the construction and evaluation of GH protein producing *P. pastoris* strains. This as a first step in evaluating its potential in medicine as well as in animal health and productivity; searching to develop both infrastructure and experience in producing, purifying, and testing its biological activity.

Also, as previously mentioned, mammalian GHs have activity in phylogenetically inferior animals, nevertheless potentially adverse reactions to heterologic GHs can be triggered, which is why having a GH specific-species would avoid these undesirable side effects.

Regarding human hormones, we proposed constructing productive strains for the HGH22k, the HGH20k, the HGHV, and the HCS. With regard to animal GHs, we channeled our efforts into building strains to produce GHs from bovines (BGH), caprines (CHGH), ovines (OGH), equines (ECGH), canines (CFGH), porcines (PGH) and felines (FCGH); all based on the *Pichia pastoris* yeast expression system.

For this, the following experimental strategy was proposed:

- Obtain, clone, and manipulate cDNA from these hormones.
- Construct and insert into the genome of *P. pastoris* the hormones' expression cassettes.
- Develop the fermentation processes for each new strain.
- Implement the purification schemes of the recombinant hormones.
- Evaluate in vitro the bioactivity of the semipurified recombinant hormones.

As a result of this experimental work, we achieved the followings:

- Using different methodological approaches (RT-PCR, mutagenic PCR, subcloning, etc.) we cloned the cDNAs of the hormones of interest.
- Through genetic engineering manipulations, we converted the cloned cDNAs into expression cassettes capable of functioning in *Pichia pastoris.*
- The respective expression cassettes were integrated into the *Pichia pastoris* genome by homologous recombination.
- Through an inducible (with methanol) expression system, we were able to overproduce and recover from the culture media each of the respective recombinant hormones (rGHs/rHCSs).

- The data from the physicochemical and biological characterizations showed that the methodology described herein generates heterologous proteins that are identical to their natural counterparts and biologically active.

7. TECHNOLOGICAL PLATFORM FOR THE PRODUCTION OF RECOMBINANT GHS

7.1 Overall strategy

As depicted in figures 4, the following are the two stages of the overall strategy in which the work was divided:

- Construction of *P. pastoris* strains carrying the hormones′ expression cassettes producing rGHs/HCSs.
- Production and characterization of the recombinant hormones.

7.2 Construction of propagating GH cDNA plasmids (pBS-XGHs)

Oligonucleotides for GHs cDNAs amplification by PCR were designed based on consensus nucleotide sequences of GHs (mature region) of related mammals. Extra restriction sites were added on their flanks (*Xho*I and *Avr*II) to facilitate insertion of the amplicon into the expression vector. With these, each of the hormones' cDNAs was amplified from plasmids previously constructed in our laboratory carrying the respective nucleotide sequences. Each amplicon was cloned into propagating plasmid such as the pBS II KS plasmid (+) and subsequently subcloned into the yeast expression vector pPIC9 at its multicloning site, between the restriction sites *Xho*I and *Avr*II (after previous purification of the corresponding fragment and vector), thereby giving rise to each of the pPIC9-XGH expression plasmids.

In CHGH's case, which differs from BGH in a single aa residue, a different strategy was implemented. Site-directed mutagenesis was used relying on a primer to convert codon 130 of BGH cDNA into one corresponding to CHGH. A 345 bp region containing the mutated GH cDNA was thus amplified, which was cloned in pBS and later transferred into pPIC9- BGH to converted it into pPIC9-CHGH (49).

7.3 Construction of expression plasmids (pPIC9-XGHs) for each hormone

Preparative digestions of pBS-XGH and pPIC9 with the enzymes *Xho*I and *Avr*II were performed for all GHs except for CHGH. For CHGH, *Apa*I and *Xma*I (natural site) enzymes, which release a 133 bp fragment containing the mutagenized codon for CHGH, were used. This was purified and linked into the previously digested pPIC9-BGH vector in the same sites, replacing the fragment to originate the pPIC9-CHGH expressor vector The ligation reactions between pPIC9 and each cDNA were used to transform competent Ca++ cells of XL1-Blue *Escherichia coli*. PCR was used to verify that the resulting tranformants indeed carried each pPIC9-XGH, where "x" corresponds to each of the sequences of the hormone in question. The candidate clones produced by PCR with AOX1 primers for an amplicon of 1050 bp, since the expression cassette for each hormone is flanked by long regions of the AOX1 gene. While strains that were not integrated into the "cassette" gave rise to an amplicon of only 500 bp.

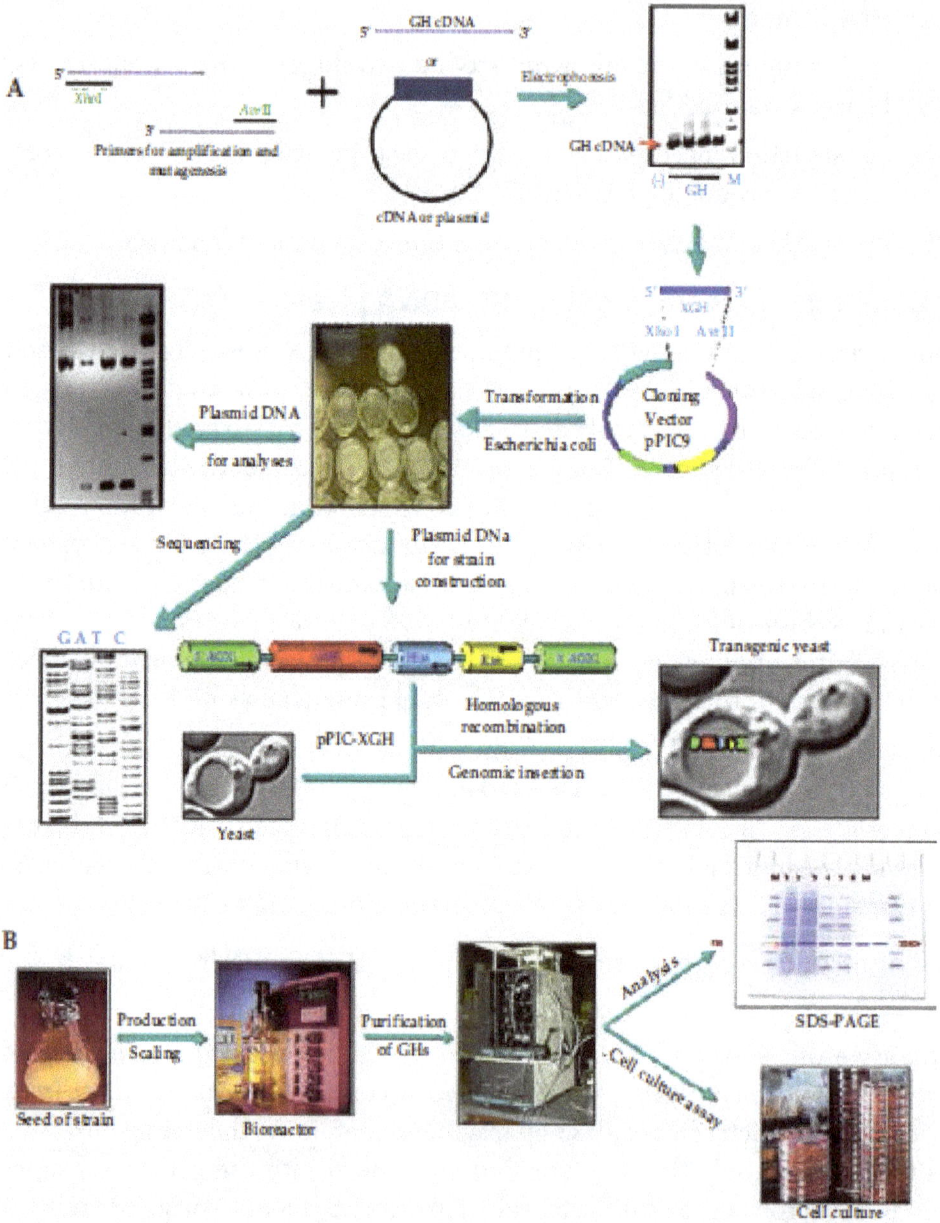

Fig. 4. General strategy for strain construction and recombinant hormones production.

(A) Genetic engineering phase. The steps followed to construct and characterize new strains of GHs and HCSs producing *Pichia pastoris* are shown. Protocols followed were based in different techniques (47, 48). (B) Biotechnology phase. The steps followed for the production and scaling, semipurification and bioassay of each of the recombinant hormones are shown

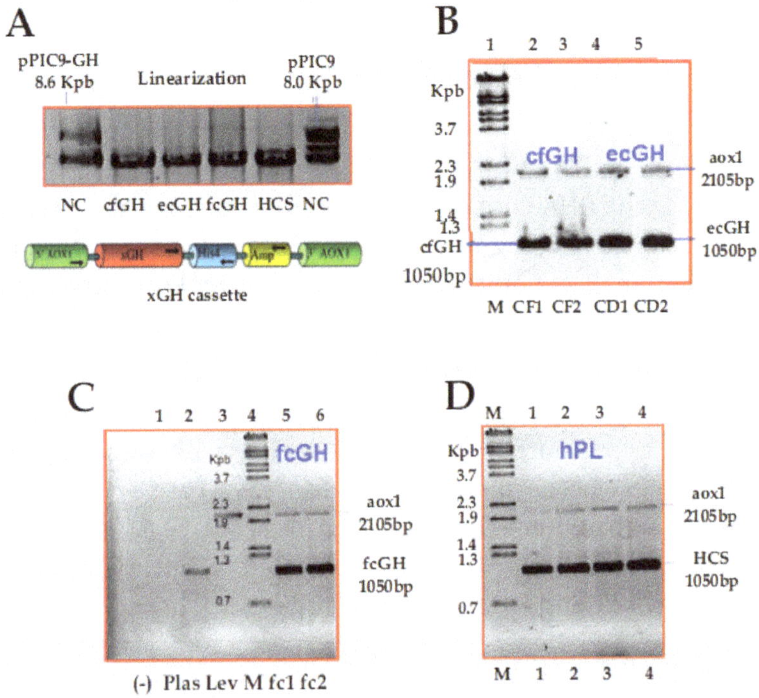

Fig. 5. Detection of the expression "cassettes" of cfGH, ecGH, fcGH and hCS in P. pastoris genome. Analysis by PCR with AOX1 primer yeast strains transfected. In each case, the 1050 bp corresponds to the expression cassette of the recombinant hormone in question, while the 2105 bp to the AOX1 gene of the yeast itself. A) The diagram shows the linearized "cassette" of XGH and the gel products (which transfected into *Pichia pastoris* integrate the "cassette" into the genome) with *Sac*I enzyme: cfGH = dog GH, ecGH = horse GH, fcGH = cat GH and HCS = human CS; in lane 1 NC-GH = uncut plasmid pPIC9 and in lane 6 NC = uncut pPIC9 plasmid. B) CF (1 and 2) = dog GH lanes 2 and 3 respectively, and CD (1 and 2) = horse GH lanes 4 and 5, respectively. C) (-) = negative PCR lane 1, Plas= amplification positive control lane 2, Lev = *Pichia pastoris* genomic DNA lane 3, M= pb marker lane 4 and fc (1 and 2) = cat GH lanes 5 and 6 respectively. D) Lanes 1 to 4 correspond to strains with the HCS "cassette". M = marker-bp □ *Bste*II. The gels correspond to 1% agarose

8. CONSTRUCTION OF GHS PRODUCING PICHIA PASTORIS STRAINS

The *Pichia pastoris* GS115 strain has a mutation in the histidinol dehydrogenase (his4) gene, which prevents it from synthesizing histidine. The class of plasmids used to transform it contains this gene (his4). The transformants are selected for their ability to restore growth in a medium lacking histidine. The plasmid vectors of the pPIC series and those constructed to express the GHs are of this class.

8.1 Insertion of GHs expression cassettes into the genome of P. pastoris

Each pPIC-XGH vector was linearized with the enzyme *SacI*, transformed into the yeast previously made competent for transformation and left exposed to the homologous regions in the yeast genome necessary for recombination.

After incubating the DNA with competent cells, transformation reactions were plated to recover clones needing no histidine to grow (HIS+ transformants). Then transformants were analyzed on their genomic DNAs by PCR using AOX1 primers to verify the presence of the transgenic hormone expressing cassette.

Verification of integration into the yeast genome of the expression cassette of the hormones was achieved in agarose gel by confirming that the amplification reaction rendered a prominent band of 1050 bp, which corresponds to the expression cassette of the hormone involved in each case and another of 2105 bp corresponding to the endogenous gene AOX1 of the yeast genome (Figure 5). In addition, each hormone "cassette" was subjected to nucleotide sequencing to verify that all they corresponded to the expected growth hormones.

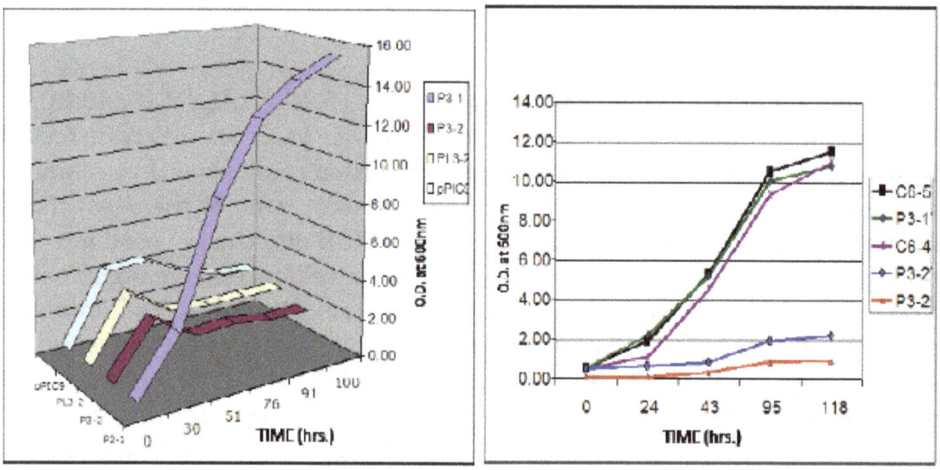

(A) (B)

Fig. 6. Mut phenotype characterization in *Pichia pastoris* strains. Growth kinetics in minimal medium using methanol as sole carbon source. Induction was started at the density of 0.5 U and ended after about 100 hrs. (A) Plot of the samples P3-1 and 2 = dog GH 1 and 2 strains, CS3 CS-2 = 2 and strain human strains pPIC9 = "mock" with the pPIC9 plasmid. (B) Graph of the C6-5T samples = horse GH, P3-1T =dog GH, C6-4T = horse GH, GH P3-Q2 = P3-dog and dog-2B = GH

8.2 Analysis of new *Pichia pastoris* strains´ phenotypes

Pichia pastoris strains were grown and the biomass was adjusted to low cell density (0.5 u at 600 nm). These were transferred to induction medium with 0.5% methanol and grown for 100 hours with the addition of methanol every 24 hours to compensate for its evaporation. Biomass growth was analyzed under methanol as the sole carbon source. The Mut+ phenotype strains metabolize methanol more rapidly, achieving significantly higher cell densities than their Muts counterparts that metabolize more slowly, appreciating a slight increase in biomass under the same fermentation conditions.

An analysis of the growth of the strains after 100-hour fermentation with 0.5% methanol identified the Mut phenotype of each strain.

After fermentation, the strains found to be Mut+ reached about 15 optical units at 600 nm, while those that were Muts did not exceed 2 units (Fig. 6). In the strains that had been built previously, their Mut phenotype was inferred when these were fermented in the bioreactor.

9. PRODUCTION AND ANALYSIS OF RECOMBINANT HORMONES IN THE FLASK

To test the fermentation of strains, a biomass was generated in a flask. This was inoculated with a colony of each strain in 25 ml of biomass producing culture medium (BMGY) (50). This was incubated at 30°C at 250 rpm for 24 to 48 hours for the first stage of growth until a biomass with an OD of 600 nm of 10 was reached.

For the second stage, which is the induction of the recombinant hormone production, yeasts were harvested by centrifugation and the packed cells were washed with 30 mL sterile water, then these were pelleted and resuspended in fresh cassette induction medium (with methanol) (BMMY) (50). The induction was maintained by adding methanol every 24 hours to a final concentration of 1% to compensate for loss by evaporation. The experiment lasted 96 hrs. Figure 7 shows the process that was followed.

When analyzing the polyacrylamide gels of proteins from the culture media of each strain, we observed that all constructions produced and directed the secretion into the medium of the recombinant hormone in question to a greater or lesser extent. For the particular case of CFGH it was observed that a strain of Muts phenotype displayed better production of recombinant protein than its counterpart Mut+. Strain of HGH-V proved to be the least productive (Fig. 8).

Figure 9 shows the gel with the results of the production of all recombinant strains generated in *Pichia pastoris*. They all produced different amounts of their respective hormone at the level of 22 kDa, except for the HGH20k, which, migrated below the rest of the recombinant hormones.

The percentage of each recombinant hormone in the culture medium was estimated by densitometry of each gel. For this we used the Gel-Doc software by BIO-RAD (Hercules, CA. EUA) and the ImageJ program (51). The results of estimation of the percentage of each hormone in relation to background proteins from *Pichia pastoris* were: HCS = 65%, CFGH = 60%, HGH22k = 30%, ECGH = 30%, BGH = 25%, FCGH = 25%, CHGH = 25%, HGH20k = 12% and HGHV = 8%.

Production kinetics was carried out for CHGH strain in a flask with a volume of 50 ml of rich medium. Samples were taken at 24, 48, 72, 96 and 120 hours of induction with methanol with restitution every 24 hours of 1% methanol. Bradford protein determination showed that the production of total protein secreted into the culture medium was 20µg/mL by densitometry and 60% represented CHGH giving us 12µg/mL of production of the recombinant hormone.

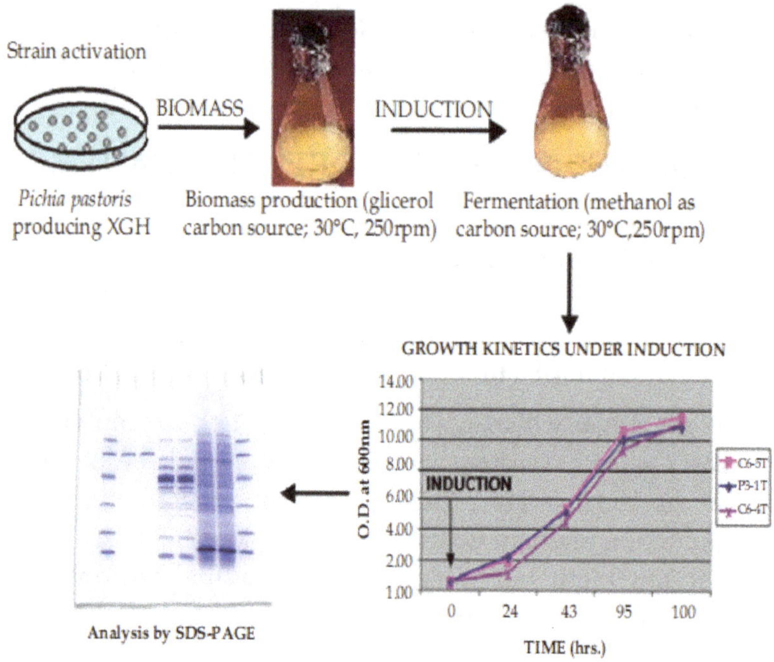

Fig. 7. Outline of the fermentation process. General procedure for the biotechnological production of recombinant hormones by fermentation of each strain. Strains were plated to activate them, incubated in liquid medium to generate a biomass flask, and the induced transgene expression by adding methanol. The culture medium was analyzed by SDS-PAGE in search of the hormones that are migrating around 22 kDa

10. PRODUCTION SCALING IN THE BIOREACTOR

When passing to a bioreactor and increasing the scale, it is possible to obtain protein concentrations 20 to 200 times greater than in flasks. In the fermentor *Pichia pastoris* reaches high cell densities greater than 100 g/L of dry weight (46).

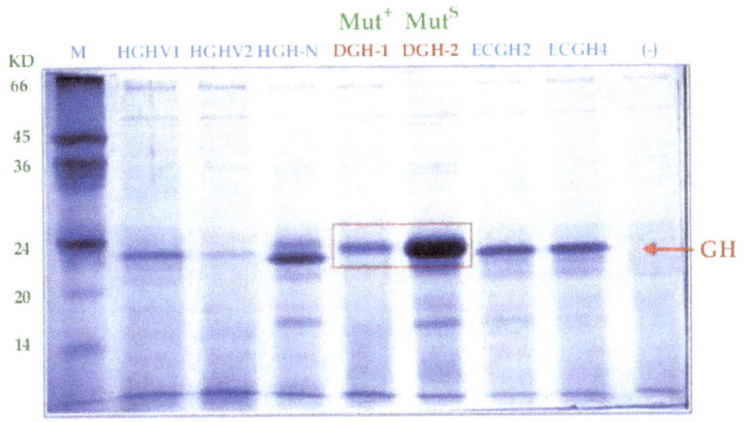

Fig. 8. Production in *Pichia pastoris* of recombinant GHs (CFGH, HGH, HGH-V and ECGH) at flask level. The bands correspond to the GHs proteins resolved at the level of 22 kDa that come from the culture media induced with methanol. The lanes are: M = molecular weight marker, HGHV1 = HGH variant strain 1; HGHV2 = HGH variant strain 2; HGH = normal pituitary HGH of 22 kDa; DGH-1 = Dog GH strain 1; DGH-2 = dog or *Canis familiaris* GH strain 2; ECGH-2 = horse GH strain 2, ECGH-4 = horse GH strain 4 and the last lane identified as (-) = negative control of PCR. Mut+= methanol utilization plus; Muts = "methanol utilization slow". Note the prominent band of the DGH-2 corresponding to the Muts phenotype, compared to the lower intensity of Mut+. The samples correspond to 500 µL concentrates of the original media. Gel corresponds to one of 15% polyacrylamide-SDS stained with Coomassie blue

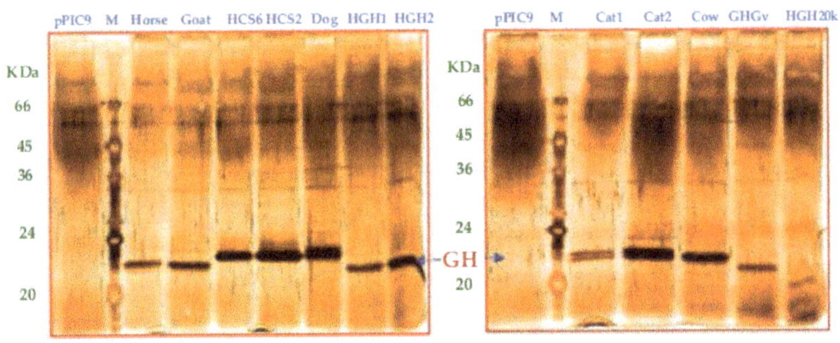

Fig. 9. Flask production in all strains yielding recombinant hormones. In all cases a prominent band is seen (except for HGH20k) at the 22 kDa level for each hormone. They are seen in their respective lane in each case indicated by their name. The lanes of the left side gel show the proteins from the culture media with recombinant hormones: horse = ECGH, goat = CHGH, HCS (6 and 2) = Human chorionic somatomammotropin clones 6 and 2, dog = CFGH and HGH (1 and 2) = cloned human GH 1 and 2. In the right gel lanes: cat (1 and 2) = GH 1 and 2 from cat; cow = BGH; HGHv = HGH placental variant and HGH20k = isoform of 20 kDa of hGH. The pPIC9 lane refers to the "mock" strain of *Pichia pastoris*. SDS-PAGE 15% gels are silver stained

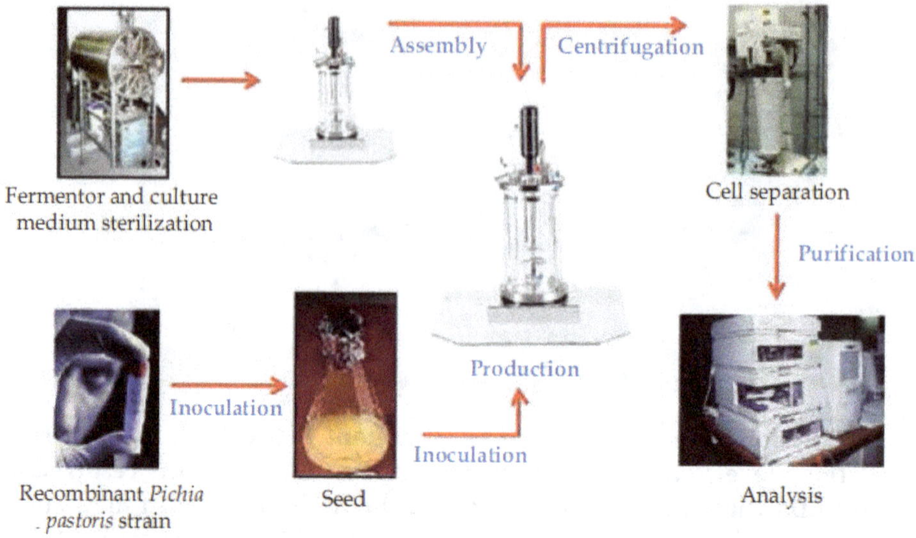

Fig. 10. Recombinant hormone production bioreactor. This illustrates the stages of the biotechnological process of production, from the preparation of the fermentor and the medium, to the analysis of proteins in the fermented culture medium

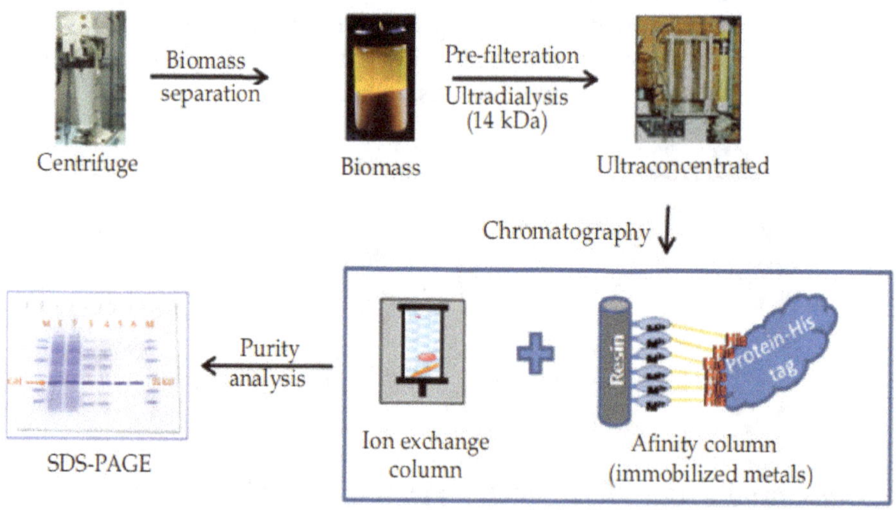

Fig. 11. "DownStream" Process for recombinant hormones. The medium was separated from the biomass by centrifugation. This was pre-filtered with a 0.45µm membrane, and ultra-dialyzed in a cutoff membrane of 10 kDa. It was then passed through an anion exchange column (FF-QS) and/or in an affinity column by immobilized metal (IMAC). Purity was analyzed in PAGE-SDS, quantified, and finally lyophilized and stored for later use

The fermentor was Bioflo 3000 (1 liter) of New Brunswick Scientific (NBSC) (NJ. EUA). The type of fermentation conducted was in fed-batch. The parameters monitored were scheduled addition of substrates to the fermentor, pH, percentage of dissolved oxygen, agitation, temperature, and aeration. The process involved three basic steps: 1) obtaining high densities of biomass, 2) induction of the cassette expression of each hormone with methanol and 3) harvest of biomass and culture medium containing the recombinant protein. Figure 10 shows the steps followed for the recombinant production of each hormone.

11. SEMIPURIFICATION OF RECOMBINANT HORMONES

Figure 11 shows the process of purification or "downstream" that was followed for each recombinant hormone produced.

Each culture medium containing the recombinant hormone was ultra-dialyzed and ultraconcentrated. The pore membrane used was 14 kDa. At the same time it was ultradyalized. The ultraconcentrate obtained was lyophilized to preserve the samples and all the powder was recovered in 50 mL plastic tubes and weighed. Total protein was quantified in each lyophilized culture medium with the Bradford method (52). Samples were stored at -20°C until use.

Each sample was prepared for loading into a chromatographic column. Five mg of the total protein from each sample was adjusted to the conditions of the loading buffer. A column was loaded with Q-Sepharose fast flow anion exchange resin; column filling with resin was carried out by gravity flow. After passing the sample, washing was done with 10 mL of loading buffer. To recover the proteins from the column, 15 mL of elution buffer (loading buffer plus NaCl) was passed with an ionic strength increased sequentially with NaCl. Total protein was measured by the Bradford method (52) for each of the fractions recovered to see what percentage of the total they represented.

The collected fractions were visualized on discontinuous polyacrylamide gel concentrations of 4-15% under denaturing conditions (SDS-PAGE), and stained with Coomassie and silver techniques (53). The fractions were subjected to lyophilization, the powder was recovered and weighted, total protein was measured by the Bradford method. In addition, using PAGE-SDS the percentage of purity of the monomer was determined with ImageJ software. The samples were stored at -20°C until used for further analysis or to determine their biological activity.

12. TESTING THE BIOLOGICAL ACTIVITY OF RECOMBINANT HORMONES

12.1 Lactogenic activity bioassay

The biological activity of hormones produced in *Pichia pastoris* recombinant was determined by their ability to promote proliferation of the Nb2 cell line, which comes from rat lymphoma (54, 55). GHs were also tested for their somatogenic activity in the adipocyte differentiation model based on the glyceraldehyde 3-phosphate dehydrogenase (GPDH) assay.

The cell-free culture medium was dialyzed and each hormone was quantified by gel densitometry. Dilutions of each recombinant hormones in the culture medium dialyzed were tested. Cell proliferation was determined by tetrazolium salt assay (MTT) (56) and was expressed as the average of three repetitions, in comparison with the positive control recombinant rat prolactin (rRPRL) and the negative control (culture medium without hormone). The concentrations tested were 0.001, 0.01, 0.1, 1, 10 and 50 nM of the following hormones, CFGH, ECGH, FCGH, HGH, HCS and RPRL. Upon completion of the testing time, we proceeded to measure the effect of hormones on cell proliferation by MTT assay (56). The activity was evaluated by color generation based on the reduction of tetrazolium salt (methyl 3- [4,5-dimethylthiazol-2-yl] -2,5-diphenyl tetrazolium) from yellow to purple forming crystals by Nb2 cell metabolism. An increase of living cells is reflected by increased metabolic activity. This increase directly correlates with the formation of absorbance monitored formazan crystals; i.e., the greater the number of cells, the greater the increase in cell metabolism with greater formazan formation and greater biological activity.

The bioassay was carried out in triplicate in a humid atmosphere with 5% CO2-95% air at 37°C. After the incubation period of 3 days, we added 10μL of MTT (to a final concentration of 0.5 mg/mL) to each well. Samples were incubated for 4 hrs under the atmospheric conditions mentioned above. After this time, we added 100 uL of formazan solubilizing solution (10% SDS in 0.01 M HCl) to each of the wells. These were left incubating overnight under the same atmospheric conditions. It was verified that the precipitate of formazan purple crystals had completely dissolved and absorbance was measured at 590 nm using an ELISA plate reader.

12.2 Somatogenic bioassay

To test the biological activity of goat GH, we assayed fibroblasts from cell line 3T3-F442A (pre-adipocytes) and with the glyceraldehyde 3-phosphate dehydrogenase (GPDH) assay as described in (18, 57). Cells were exposed to different concentrations of CHGH in the medium. The positive control was 10% fetal bovine serum (FBS) (v/v) the negative control was 0.25% FBS (v/v). The cells were incubated for 12 days, were harvested and proteins extracted; the supernatants were frozen in aliquots for subsequent tests. Specific activity of GPDH was measured in cell extracts by NADH oxidation measured in a spectrophotometer at 340 nm; 40 μg of protein was used. CHGH demonstrated biological activity in the essay. The same was doing with the others hormones showing biological activity.

13. CONTRIBUTIONS

These chapter represent the initial instances in scientific literature where the creation of the described recombinant GHs employing the Pichia pastoris methylotrophic system through biotechnology has been documented. The addition of GH cassettes, including canine-porcine, equine, feline, and caprine, to our existing strains of Pichia pastoris has expanded our collection of GH clones. These new strains, along with our previously constructed strains for GHs of 22 y 20 kDa, HCS, and bovine GH (unpublished results), have made our collection the largest known collection of GH clones in the world.

All hormones were effectively synthesized, treated, and released into the culture media. Each hormone was the predominant protein band among the proteins released by Pichia pastoris that were examined using SDS-PAGE. The Muts phenotypes were found to be the most effective in creating GHs, as observed in the case of CFGH (58) and HCS (Ascacio-Martínez and Barrera-Saldaña, unpublished data).

All of our engineered cultures successfully processed the heterologous secretion Saccharomyces cerevisiae alpha mating factor signal peptide throughout the maturation of the recombinant GHs. They were introduced into the culture media in their original and biologically active state (49, 59, 60).

The utilization of a bioreactor resulted in a significant enhancement in the yield of the recombinant proteins, with a 10 to 20-fold increase compared to the use of an Erlenmeyer flask. In this system, ionic chromatography proved to be a suitable choice for the semipurification of rGHs and HCS in all circumstances. All hormones exhibited biological activity in the Nb2 assay, indicating that human growth hormone (GH) displayed more activity compared to animal GHs. Similar results were observed in the pre-adipocyte system (3T3), indicating that Pichia pastoris is capable of synthesizing, modifying, and releasing rGHs in their active form.

Biotechnological platforms were created that enabled the transition from constructing producer clones to conducting bioassays on the semipurified recombinant protein. Using our created technology, we now have the ability to increase the production of veterinary and animal rGHs for testing purposes.

14. PERSPECTIVES: AT A GLANCE

We have created a very effective system for expressing genes and a biotechnological platform at the laboratory fermentor scale. This platform allows us to collaborate with the productive sector to make a wide range of growth hormones from both human and animal sources. The produced hormones have sufficient quantity, quality, and activity to be used in field assessments. By doing so, we can explore their complete capabilities in animal biotechnology, subsequently providing them as a viable choice in veterinary treatments and promoting cattle productivity and the well-being of competitive animals. Aside from the evident use in veterinary or livestock settings, the presence of these resources will also facilitate the identification of unforeseen biological functions for the well-being of animals.

15. REFERENCES

Goeddel D.V., Heyneker H.L., Hozumi T., Arentzen R., Itakura K., Yansura D.G., Ross M.J., Miozarri G., Crea R., Seeburg P. Direct expression in Escherichia coli of a DNA sequence coding for human growth hormone. Nature 281(5732): 544-548, (1979).

Niall H.D., Hogan M.L., Sayer R., Rosenblum I.Y., Greenwood, F.C. Sequences of pituitary and placental lactogenic and growth hormones: evolution from a primordial peptide by gene duplication. Proc. Natl. Acad. Sci. USA, 68: 866-869, (1971).

Elian, G., Jamieson, J., Gross, S. Staying Young: Growth Hormone and Other Natural Strategies to Reverse the Aging Process. Age Reversal Press. First Edition pp:120, (1999).

De Noto F, Rutter J.W., Goodman H.M. Human growth hormone DNA sequence and mRNAstructure: possible alternative splicing. Nucleic Acid Res. 9: 3719-30, (1981).

Tsushima T., Katoh Y., Miyachi Y., Chihara K., Teramoto A., Irie M., Hashimoto, Y. Serum concentrations of 20K human growth hormone in normal adults and patients with various endocrine disorders. Study Group of 20K hGH. Endocr. J. 47 Suppl: S 17-21, (2000).

Hashimoto Y., Kamioka T., Hosaka M., Mabuchi K., Mizuchi A., Shimazaki Y., Tsuno M., Tanaka T. Exogenous 20 K growth hormone (GH) suppresses endogenous 22K GH secretion in normal men. J. Clin. Endocrinol. Metab. 85(2): 601-606, (2000).

Wang D.S., Sato K., Demura H., Kato Y., Maruo N., Miyachi Y. Osteo-anabolic effects of human growth hormone with 22K- and 20K Daltons on human osteoblast-like cells.Endocr. J. 46(1): 125-132, (1999).

Cooke N.E., Ray J., Emery J.G., Liebhaber S.A. Two distinct species of human growth hormone-variant mRNA in the human placenta predict the expression of novel growth hormone proteins. J. Biol. Chem. 263: 9001-9006, (1988).

Ray J., Jones B., Liebhaber S.A., y Cooke N.E. Glycosylated human growth hormone variant. Endocrinology 125: 566-568, (1989).

Frankenne F., Scippo M., Van Beeumen J., Igout A., Hennen G. Identification of placental human growth hormone as the growth hormone-V gene expression product. J. Clin. Endocrinol. Metab. 71: 15-18, (1990).

Boguszewski C.L., Svensson P.A., Jansson T., Clark R., Carlsson M.S. Carlsson B. Cloning of two novel growth hormone transcripts expressed in human placenta. J. Clin. Endocrinol. Metab. 83(8): 2878-2885, (1988).

Frankenne F., Closset J., Gomez F., Scippo M.L., Smal J., Hennen G. The physiology of growth hormones (GHs) in pregnant women and partial characterization of the placental GH variant. J. Clin. Endocrinol. Metab. 66(6): 1171-1180, (1988).

Mirlesse V., Frankenne F., Alsat E., Poncelet M., Hennen G., Evain-Brion D. Placental growth hormone levels in normal pregnancy and in pregnancies with intrauterine growht retardation. Pediatr Res. 34: 439-442, (1993).

Chowen J.A., Evain B.D., Pozo J., Alsat E., García Segura L.M., Argente J. Decreased expression of placental growth hormone in intrauterine growth retardation. Pedriatr. Res. 39 (4 Pt 1): 736-739, (1996).

Pardi G., Marcini A.M., Cetin I. Pathophysiology of intrauterine growth retardation: rol of the placenta. Acta Pediatr. Suppl. 423: 170-172, (1997).

Rygaard K., Revol A., Esquivel-Escobedo D., Beck B.L., Barrera-Saldaña H.A. Absence of human placental lactogen and placental growth hormona (HGH-V) during pregnancy: PCR análisis of the deletion. Human Genet. 102(1): 87-92, (1998).

Morikawa M., Green H., Lewis U.J. Activity of human growth hormone and related polypeptides on the adipose convertion of 3T3 cells. Molecular and Cellular Biology 4(2): 228-231, (1984).

Juarez-Aguilar, E., Castro-Munozledo, F., Guerra-Rodriguez, N.E., Resendez-Perez, D., Martinez-Rodriguez, H.G., Barrera-Saldana, H.A., Kuri-Harcuch, W. Functional domains of human growth hormone necessary for the adipogenic activity of hGH/hPL chimeric molecules. J. Cell Sci. 112(18):3127-3135, (1999).

MacLeod J.M., Worsley I., Ray J., Friesen H.G., Liebhaber S.A., Cooke N.E. Human growth hormone-variant is a biologically active somatogen and lactogen. Endocrinol. 128(3): 1298-1302, (1991).

Cooke N.E. Prolactin: normal synthesis and regulation. En DeGroot L.J. (ed.) Endocrinology.Saunders, Philadelphia., 1: 384-407, (1989).

Chen E.Y., Liao Y.C., Smith D.H., Barrera-Saldaña H.A., Gelinas R.E., Seeburg P.H. The human growth hormone locus: nucleotide sequence, biology, and evolution. Genomics 4(4): 479-97, (1989).

Barrera-Saldaña H.A. Growth hormone and placental lactogen: biology, medicine and biotechnology. Gene 211: 11-18, (1998).

Barrera-Saldaña H.A.,Seeburg P.H., Saunders G.F. Two structurally differentgenes produce the secreted human placental lactogen hormone. J. Biol. Chem. 258: 3787- 3793, (1983).

Canizales-Espinosa, M., Martínez-Rodríguez, H.G., Vila, V., Revol, A., Castillo-Ureta, H., Jiménez-Mateo, O., Egly, J.M., Castrillo, J.L. and Barrera-Saldaña, H.A. Differential strength of transfected human growth hormone and placental lactogen gene promoters. J. Endocr. Genet. 4 (1): 25-36, (2005).

Peel C.J. Bauman D.E. Somatotropin and lactation. J. Dairy Sci. 70: 474-486, (1987).

Etherton T.D., Kensinger R.S. Endocrine regulation of fetal and postnatal meat animal growth.J Anim Sci. 59(2): 511-528, (1984).

Bauman D.E. Regulation of nutrient partitioning: homeostasis, homeorhesis and exogenous somatotropin. Keynote lecture. En: Seventh International Conference on Production Disease in Farm Animals, F.A. Kallfelz pp. 306-323, (1989).

Sun M. Market sours on milk hormone. Science 17: 246(4932): 876-877, (1989).

Juskevich J.C., Guyer C.G. Bovine growth hormone: human food safety evaluation. Science 24: 249(4971): 875-884, (1990).

Boutinaud M., Rulquin H., Keisler D.H., Djiane J., Jammes H. Use of somatic cells from goat milk for dynamic studies of gene expression in the mammary gland. J. Anim. Sci. 80(5): 1258-69, (2002).

Stewart F., Tuffnell, P.P. Cloning the cDNAfor horse growth hormone and expression in Escherichia coli. J. Mol. Endocr. 6: 189-196, (1991).

http://www.jockeysite.com/stories/e_melbournecup1.htm

Ascacio-Martínez J.A., Barrera-Saldaña, H.A. A dog growth hormone cDNA codes for mature protein identical to pig growth hormone. Gene 143: 299-300, (1994).

Evock, C.M., Etherton, T.D., Chung C.S., Ivy, R.E. Pituitary porcine growth hormone (pGH) and a recombinant pGH analog stimulate pig growth performance in a similar manner. J. Anim. Sci. 66(8): 1928-1941, (1988).

Daughaday W.H. The anterior pituitary. Williams textbook of Endocrinology. Ed. Philadelphia, WB Saunders. 568-613, (1985).

Eigenmann J.E. Diagnosis and treatment of dwarfism in a german shepherd dog. J. Am. Anim. Hosp. Assoc. 17: 798-804, (1981).

Muller G.H., Kirk R.W., Scott D.W. Small animal dermatology. Philadelphia. WB. Saunders. 4th Edition, 575-657, (1989).

Devlin R.H., Byatt J.C., Maclean E., Yesaki T.Y., Krivi G.G., Jaworski E.G., Clarke W.C. Bovine placental lactogen is a potent stimulator of growth and displays strong binding to hepatic receptor sites of coho salmon. General and Comparative Endocrinology, 95: 31-41, (1994).

Tokunaga T., Iwai S., Gomi H., Kodama K., Ohtsuka E., Ikehara M., Chisaka O., Matsubara K. Expression of synthetic human growth hormone gene in yeast. Yeast. 39: 117-120, (1985).

Franchi E., Maisano F., Testori S.A, Galli G., Toma S., Parente L., Ferra F.D., y Grandi G. A new human growth hormone production process using a recombinant Bacillus subtilisstrain. J. Biotechnology 18: 41-54, (1991).

Pavlakis G.N., Hizuka N., Gorden P., Seburg P.H., Hamer D.H. Expression of two human growth hormone genes in monkey cell infected by simian virus 40 recombinants. Proc. Natl. Acad. Sci. 78: 7398-7402, (1981).

Kerr D.E., Liang F., Bondioli K.R., Zhao H., Kreibich G., Wall R.J., Sun T.T. The bladder as a bioreactor: Urothelium production and secretion of growth hormone into urine. Nature Biotechnology 16: 75-78, (1997).

Escamilla-Treviño L.L., Viader Salvado J.M., Barrera Saldaña H., Guerrero Olazaran, M. Biosynthesis and secretion of recombinant human growth hormone. In: Pichia Pastoris. Biotechnology Letter 22: 109-114, (2000).

Romanos M.A., Scorer C.A., Clare. J.J. Foreign gene expression in yeast: A review. Yeast 8: 423-488, (1992).

Faber K.N., Harder W., Veenhuis, M. Review: Yeasts as factories for the production of foreign proteins. Yeast 11: 1131-1344, (1995).

Siegel R.S., Brierley R.A. Methylotrophic yeast Pichia pastoris produced in high-cell- density fermentation with high cell yields as vehicle for recombinant protein production. Biotechnol. Bioeng. 34: 403-404, (1989).

Ausubel, F.M., Brent, R., Kingston, R.E., Moore, D.D., Seidman, J.G., Smith, J.A., Struhl, K. Short Protocols in Molecular Biology, fourt ed., Wiley, Massachusetts, (1999).

Sambrook, J., Fristsch, E., Maniatis, T. Molecular Cloninig: A Laboratory Manual. Segunda Edición. Cols. Spring Harbor Laboratory Press. Cold Spring Harbor, (1989).

Reyes-Ruíz, J.M., Ascacio-Martínez, J.A., Barrera-Saldaña, H.A. Derivation of a growth hormone gene cassette for goat by mutagenesis of the corresponding bovine construct and its expression in Pichia pastoris. Biotechnology Letters. 28(13):1019- 25, (2006).

Invitrogen. Products for Gene Expression and Analysis. Instruction manual. Pichia Expression Kit. Protein Expression. A Manual of Methods for Expression of Recombinant Proteins in Pichia pastoris. Version L., (2000).

Rasband, W.S. ImageJ. U. S. National Institutes of Health, Bethesda, Maryland, USA, http://imagej.nih.gov/ij/, 1997-2011.

Bradford, M. (1976). A rapid and sensitive method for the quatitation of microgram quantities of protein utilizing the principle of protein-dye binding. Anal. Biochem., 72:248-254.

Merril, C.R. Gel Staining Techniques. Guide to Protein Purification: Methods in Enzymology (Methods in Enzymology Series, Vol 182). Murray P., Deutscher John N. y Abelson. pp. 477, (1990).

Tanaka, T., Shiu, R.P., Gout, P.W., Beer, C.T., Noble, R.L., Friesen, H.G. A new sensitive and specific bioassay for lactogenic hormones: measurement of prolactin and growth hormone in human serum. J. Clin. Endocrinol. Metab. 51(5):1058-1063, (1980).

Lawson, D.M., Sensui, N., Haisenleder, D.H., Gala, R.R. Rat lymphoma cell bioassay for prolactin: observations on its use and comparison with radioimmunoassay. Life Sci. 31(26):3063-3070, (1982).

Gerlier, D., Thomasset, N. Use of MTT colorimetric assay to measure cell activation. J. Immunol. Methods. 94(1-2):57-63, (1986).

Castro-Munozledo, F., Beltran-Langarica, A., Kuri-Harcuch, W., Commitment of 3T3- F442A cells to adipocyte differentiation takes place during the first 24-36 h after adipogenic stimulation: TNF-alpha inhibits commitment. Exp. Cell Res. 284, 161- 170, 2003.

Ascacio-Martínez, J.A., Barrera-Saldaña H.A. Production and secretion of biologically active recombinant canine growth hormone by Pichia pastoris. Gene. 340(2):261- 266, (2004).

Palma-Nicolás, J.P., Ascacio-Martínez, J.A., Revol-de-Mendoza A. y Barrera-Saldaña, H.A. Production of recombinant human placental variant growth hormone in Pichia pastoris. Biotechnology Letters. 27(21):1695-1700, (2005).

Treerattrakool S, Eurwilaichitr L, Udomkit A, Panyim S. Secretion of Pem-CMG, a peptide in the CHH/MIH/GIH family of Penaeus monodon, in Pichia pastoris is directed by secretion signal of the alpha-mating factor from Saccharomyces cerevisiae. J Biochem. Mol. Biol. 35(5):476-81, (2002).

08

ASSESSING THE ENVIRONMENTAL RISKS AND MANAGING CHALLENGES OF GENETICALLY ENGINEERED VIRUS-VECTORED VACCINES

1. INTRODUCTION: AN OVERVIEW

Genetically engineered or modified viruses (GMVs) are being more commonly utilized as live vaccine vectors. The potential environmental consequences of their usage must be considered in risk assessment and management procedures. Within the majority of legal frameworks, GMVs (genetically modified vaccines) are classified as GMOs (genetically modified organisms). Consequently, they must undergo an ERA (environmental risk assessment) in addition to the examination of the product's quality, safety, and efficacy before any applications for marketing authorization or clinical trials may be submitted. The Environmental Risk Assessment (ERA) is conducted to evaluate the possible hazards to public health and the environment that may result from the utilization and release of Genetically Modified Organisms (GMOs). If hazards are determined to be unacceptable, the ERA process should proceed to provide suitable risk management solutions that are capable of mitigating these risks (Anliker et al., 2010; Kühler et al, 2009).

In order to receive marketing permission in the European Union, a genetically modified variety (GMV) must fulfill the criteria and conditions outlined in the EU pharmaceutical legislation for both medicinal and veterinary purposes. Additionally, it must comply with the EU environmental regulation regarding the intentional release of genetically modified organisms (GMOs). Therefore, even though viruses are not considered living organisms, it is imperative to conduct an Environmental Risk Assessment (ERA) that follows the same approach outlined in Directive 2001/18/EC for the intentional release of Genetically Modified Organisms (GMOs) into the environment. An organism, for the purpose of an ERA, is defined as a biological entity that has the ability to replicate or transmit genetic material. This definition include viruses as well as replication-incompetent viral vectors. When submitting a marketing authorization application (MAA) for a genetically modified vaccine (GMV) to the European Medicines Agency (EMEA), it is necessary to include an Environmental Risk Assessment (ERA) that follows the guidelines outlined in Annex II of Directive 2001/18/EC and its accompanying Commission Decision 2002/623/EC. The subsequent procedure is thoroughly explained in recent literature reviews, such as the ones conducted by Anliker et al. (2010) and Kühler et al. (2009). In summary, the ERA should rely on the technical and scientific data about the GMV as specified in the Directive Annexes III and IV.

The ongoing approach differs considerably from the intentional release process of genetically modified organisms outlined in the Directive. However, challenges in creating ERAs for GMVs may occur due to the fact that Directive 2001/18/EC primarily focuses on GM plants and agricultural goods. As a result, the EMEA has created two distinct recommendations to help with the production of ERAs.

These guidelines aim to make it easier to adapt the standards and methods of the Directive for pharmaceutical goods that contain GMOs. Furthermore, the GMV ERA must be conducted in accordance with the specific regulations of each Member State in order to acquire authorization for GMV clinical and field studies. The specific requirements of each Member State may vary depending on whether the trial is classified as "contained use" under Directive 98/81/EC or as a "deliberate release" under Directive 2001/18/EC. The former is prioritizing containment measures, including the creation of physical, chemical, and biological barriers to prevent interactions between GMV and the environment. On the other hand, the deliberate release Directive relies on a comprehensive evaluation of the potential environmental hazards that may result from the release or escape of genetically modified organisms (GMOs), as well as the biosafety measures that should be implemented to avoid or reduce these risks.

The purpose of an Environmental Risk Assessment (ERA) as outlined in Directive 2001/18/EC is to systematically evaluate and analyze the potential adverse impacts of a Genetically Modified Organism (GMO) on humans, animals (both domestic and wildlife), plants, microbes, and the overall environment. This assessment is conducted on an individual basis for each instance. It is important to recognize and take into account any possible negative consequences, regardless of whether they are caused directly or indirectly, and regardless of whether they manifest immediately or with a delay. Once these consequences have been discovered, it is necessary to establish suitable strategies to reduce or eliminate them. These actions must be grounded in the understanding that the transfer of GMVs to individuals, species, or the environment in general can facilitate the future spread of GMVs. Consequently, this can lead to genetic or phenotypic alterations, rivalry with current species, or the exchange of hereditary materials between species through horizontal gene transfer. In order to comprehensively understand and assess the environmental risks linked to various potential scenarios, including the one mentioned, it is essential to possess comprehensive and specific knowledge about all potential negative impacts of a particular genetically modified variety (GMV). This is because the effectiveness and significance of the Environmental Risk Assessment (ERA), as well as the ability to manage risks efficiently, rely entirely on the capacity to anticipate, predict, and uncover potential adverse effects. This is contingent upon the society's readiness to allocate resources towards speculative and cautious scientific research, as well as recognizing and evaluating the values that consciously or unconsciously shape the decision-making process after an MAA for a GMV application.

Prior research on risk has primarily been on the functionality and intended immunological mechanisms of GMVs. However, investigations into safety concerns, specifically regarding ecological consequences, have frequently been postponed until later stages of vaccine development. At that point, implementing significant modifications to the vaccine to enhance its safety can be prohibitively expensive and time-intensive. There are situations where dangers and permanent damages may have already been set in motion. Therefore, we shall contend that risk assessment and management should not be seen as distinct activities.

Historically, risk assessment has been regarded as a "scientific" procedure, but risk management and communication have involved making value-based decisions about what is acceptable, weighing different criteria, and developing ways to deal with uncertainty. Nevertheless, risk assessments are impacted by scientific, ethical, economic, social, and political data. For example, risk assessments involve making judgments about the importance of avoiding certain consequences and the process of evaluating and describing risks. Therefore, it is essential to establish a connection between risk assessment and management techniques right from the beginning of a vaccine development project in order to fully uncover the range of environmental implications.

This article provides an overview of GMV applications and explores the associated environmental impact problems and challenges. Next, we analyze the significance and limitations of the current risk assessment framework and propose the necessity of establishing stronger links between this framework and risk management techniques. Management strategies are formulated within specific frameworks that must consider normative standards and preferences related to the relationship between humans and the natural environment. Furthermore, we will discuss how precautionary-driven research necessitates the development of hypotheses regarding the specific harmful and hazardous effects of GMV, which are influenced by both the goals of ERA and management techniques.

2. CREATION AND APPLICATIONS OF GMVS

Genetically modified viruses, derived from several taxonomic groups, are being increasingly utilized as vectors for live vaccines. Currently, the veterinary vaccines that have been licensed are primarily developed using replication-competent canarypoxvirus or herpesviruses. The European Medicines Agency (EMEA) has issued a guideline specifically for Live Recombinant Vector Vaccines for Veterinary Use.

There are 4 broad GMV application areas that may have environmental implications:

- Immunization against infectious diseases in livestock species;
- Immunization of wild life species which are reservoirs of infectious agents causing disease in humans and livestock species;
- Control of pest animal population densities by either direct lethal control operations or immuno-contraception; and
- Human vaccination programs against infections diseases or cancers.

Under some conditions, GMVs have the ability to directly cross species boundaries or undergo recombination with naturally existent viruses. The many applications can potentially result in the inadvertent or unintentional release of genetically modified viruses (GMVs) into diverse ecosystems.

Various application domains require distinct considerations and possibilities when selecting virus vectors and genetically engineering (GE) methodologies. In general, there are two strategies: The first approach involves utilizing gene-deleted viruses for homologous vaccination, specifically to induce protective immunity against the genetically modified virus (GMV) itself. The programmed deletions mostly focus on genes essential for the virus to complete its replication cycle or contribute to viral pathogenicity. In addition, genes that are not necessary may be removed to create markers that can be used to track the unintentional spread of the vaccination virus. Absence of the deleted marker gene will serve as an indication that a field virus isolate originated from a genetically modified virus (GMV). Several genetically modified vaccines (GMVs) that involve the deletion of certain genes have either been commercially released or are now undergoing advanced clinical studies for the prevention of human and livestock diseases. The majority of them are members of the herpesvirus or adenovirus families.

The second approach involves the use of recombinant virus vectors generated through transgenesis. These viruses are produced in cell cultures by introducing a plasmid containing a gene from the specific virus or bacterium that needs to be targeted immunologically, and infecting the cells with the chosen virus vector simultaneously. The plasmid construct is designed with transgenic DNA sequences that are homologous to a viral gene at both ends. Therefore, the transgene is transported and integrated into a specific location in the genome of the virus vector by the process of homologous recombination. Members of the Poxviridae and Adenoviridae families are the most often utilized vector viruses. Often, the vectors have been manipulated by the deletion of endogenous genes and the introduction of transgenes.

Currently, significant research is being conducted to develop replication-incompetent versions of the most promising vector viruses. This is feasible on the condition that the non-replicating vector can effectively produce the vaccine transgene and elicit protective immune responses at levels equivalent to those of the original replication-competent virus strain. The primary objectives are to reduce the likelihood of productive infections with the shedding of virus in the waste of individuals exposed to GMV, and to prevent attenuated or modified viruses from reverting back to full virulence due to the reversal of attenuating mutations.

GMV-vectored vaccines have been developed or are currently being developed for various diseases, such as AIDS, severe acute respiratory syndrome (SARS), Epstein-Barr virus, cytomegalovirus, West Nile virus, tuberculosis, malaria, influenza (human and equine), Rinderpest, Rift Valley fever, borreliosis, trypanosomiasis, leishmaniasis, cervical cancer, breast cancer, colorectal cancer, mesothelioma, and melanoma. The majority of the GMV vectors used are considered to be replication-incompetent and are derived from poxvirus and adenovirus families.

3. VACCINES AND VACCINATION (MODIFIED FROM TRAAVIK, 1999)

All vaccines have in common the intention to prevent disease or limit the effects of disease. Both humoral (antibody-mediated) and cellular arms of the immune system can contribute to a pathogen-specific acquired response that distinguishes specific immune protection from the innate and more general protection mediated by phagocytes (i.e. macrophages neutrophils and dendritic cells), cytokines and physical barriers. Because vaccination against a threatening disease may take place many years before exposure to the pathogen, immunological memory is a critical element. A long-lived immune response, which may be mobilized and augmented rapidly when called for, is essential.

Vaccination may have different purposes and fields of application. The most important are:

- Protection against and treatment of infectious diseases
- Protection against and treatment of cancer
- Induced infertility in domestic animals and wildlife

An ideal vaccine provides an optimal mobilization of the adaptive immune system with no unwanted side effects, and with long-lasting immunological memory. The most universal purpose is to prevent disease in individuals and prohibit transmission of disease agents between individuals. Generally, vaccination must be carried out before the individual becomes infected, but for some diseases, e.g. rabies, disease may be prevented even if vaccination takes place after infection.

Some important human and domestic animal pathogens, e.g. rabies virus, hantaviruses and a number of arboviruses, have reservoirs in free-ranging wildlife animals. Human and animal disease may then be prevented by vaccination of reservoir animals. Likewise, some free-ranging mammalian species are considered «pests» in the context of human food, animal fodder or other kinds of production. Enforced infertility following vaccination is now becoming an alternative to culling ("stamping out") for control and reduction of such pest animal populations.

Cancer cells often express surface antigens not present on their normal counterparts. Such unique antigens may provide targets for vaccines that may induce immune reactions to prevent and combat cancer cells.

Depending on the species, target-organs, epidemiological considerations etc., the vaccine delivery method and route may differ. In practical terms, the vaccine may be delivered by:

- Injection, most commonly intramuscularly or subcutaneously. The recent development of injection, so-called "gene guns" are used to propel small gold particles covered with antigen through the skin. Such procedures are often referred to as "biolistics".
- Inhalation of vaccine-containing aerosols.
- Ingestion of vaccine-containing vehicles, i.e. capsules.
- For fish: Bathing in or spraying with vaccine containing solutions
- For free-ranging animals: Baits containing vaccines that are spread out over the selected target area from airplanes or helicopters.

To a varying extent, all vaccine delivery strategies imply that vaccine-containing materials may end up in unintended locations, and hence release or escape of biologically active macromolecules (i.e. DNA or RNA), viruses or microorganisms into the environment may take place.

4. VIRUSES THAT ARE USED AS GMVS

Generally spoken, the families most widely employed as GM vaccine vectors have been members of the *Poxviridae*, *Adenoviridae* and *Herpesviridae* families. The frontrunners have been GM versions of the poxvirus genera Orthopoxvirus and Avipoxvirus. Our own experimental and sustainability oriented research have focused on these genera, and we will in section 5, 6 and 7 proceed by using such GM vectors as examples of ERA and risk management challenges.

There is now a tendency that GM viral vaccines are based on vectors that are replication- incompetent in the target species of choice. Such vectors may be selected on the basis of natural host species barriers, e.g. avipoxviruses with birds as natural hosts, and assumed replication-deficient in mammalian species. Other commonly used vectors have been made replication-incompetent by cell cultivation procedures (e.g. MVA, see below) or targeted genetic engineering, e.g. adenovirus vectors.

Veterinary vaccines against infectious diseases containing replication-competent GMVs have been on the market in the EU since the beginning of year 2000 (Kühler et al., 2009). Several vaccines for human use are currently in various stages of clinical trials, and a few of them, mostly based on adenovirus or poxvirus vectors, have already been approved for marketing.

4.1 Poxviruses as GM vaccine vectors

Our account will henceforth be directly relevant for poxvirus vectors, but at least in a general sense it will also be relevant for GMV vectors developed from other virus families.

Poxviruses are selected as potential vectors for several reasons (Liu, 2010): the extensive use of the smallpox vaccine (and the related modified vaccinia ankara), which provided knowledge of the human safety parameters; the large gene capacity for the insertion of a heterologous gene encoding the antigen of interest; the broad tropism of the virus for mammalian cells that could then result in a number of cells expressing the vaccine protein (heterologous antigen); the production of vaccine protein for a relatively short period of time (making the kinetics of the production of the heterologous protein more akin to antigen production from an acutely infecting pathogen, but less useful for gene therapy applications); and the location of the virus in the cytoplasm, thus avoiding integration risks that might occur with other alternatives, as a retroviral vector for example (Mastrangelo et al., 2000). The earliest vector for delivery of a heterologous antigen is a licensed veterinary vaccine employing a MVA (Modified Vaccinia Ankara) vector to deliver a rabies antigen (Mackowiak et al., 1999). The vaccine was developed for delivery as bait for wild animals. The success of using a vector to deliver an antigen that generated immunity sufficient to protect animals and curtail outbreaks in the wild, via oral delivery of bait (hence with imprecise dosing), in a variety of animal species, demonstrates many of the advantages of an ideal vector, in this case efficacy, safety, and even ease of distribution (oral delivery), which would not have been as facile with a non-vectored vaccine. Oral delivery may mobilize protective local immune responses on mucous membranes, a great asset for human, domestic animal and wildlife vaccines.

Various poxvirus vectors have undergone human clinical testing as HIV vaccines. In the largest clinical trial, a canarypox vector coding for HIV antigens was utilized as the first component of a prime-boost regimen in a clinical trial of an HIV vaccine involving 16,000 individuals (Rerks-Ngarm et al., 2009). In another prime-boost vaccination protocol for prophylaxis of HIV, the priming is being done with a DNA plasmid, followed by an MVA vector boost (http://clinicaltrials.gov/ct2/show/NCT00820846?term=DNA+%2B+MVA+for+HIV&rank=5).

4.2 Vaccine history, general characteristics and taxonomy of the family Poxviridae

The bicentennial celebration of the first vaccination took place 15 years ago. In 1796 Edward Jenner injected cowpox virus into the boy James Phipps, and later on challenged him with fully virulent human smallpox (variola) virus. The boy survived, and Jenner had hence protected him against one of the most dreaded human diseases of all times. The smallpox vaccination story ended in a triumphant eradication, in 1979, of variola virus and smallpox disease, due to a world-wide vaccination campaign. (Fenner et al., 1988). In recognition of Edward Jenner′s contribution, procedures that aim at protection against disease by pre- mobilization of the immune system were termed “Vaccination”, derived from the Latin word *vacca,* hence honoring the cowpoxvirus. In that context it is well worth for present day scientists to reflect on the fact that the contemporaries of Jenner rejected his findings.

The family *Poxviridae* contains a vast number of complex DNA viruses that replicate in the cytoplasm of vertebrate and invertebrate hosts. The member viruses have been sorted into two subfamilies based on their host preferences, namely, the *Chordopoxvirinae* (infecting vertebrates) and *Entomopoxvirinae* (infecting invertebrates, more specifically arthropods). The *Chordopoxvirinae* consists of eight genera with different host ranges, namely, Orthopoxvirus, Parapoxvirus, Avipoxvirus, Capripoxvirus, Leporipoxvirus, Suipoxvirus, Molluscipoxvirus and Yatapoxvirus. The *Entomopoxvirinae* is divided into three genera based on the insect host of isolation. These are designated A, B, and C entomopoxviruses, respectively. Genetic information on the entomopoxviruses is very scanty.

The distinctive characteristics of members of the family *Poxviridae* include:

- A large and complex virion (larger than any other animal virus particles) with virion- associated transcriptase used for virus-specific mRNA synthesis. The virions are very resistant to environmental degradation. This is an advantage for the shelf life of vaccines, particularly in areas with unsatisfactory cooling conditions. But persistence in the environment may also enhance the chances of transboundary movements and non- target infections.

- A large genome composed of a single linear double-stranded DNA molecule, 130-300 kbp, with a hairpin loop at each end. The genome facilitates insertion of one or more transgenes with a collective size of up to 20-30 kbp. Vaccine vectors carrying a number of vaccine transgenes may hence be constructed. Theoretically, a single vaccine may protect about a number of diseases that are prevalent in specific geographical areas or defined age groups.
- The complete viral multiplication cycle takes place in the cytoplasm of the host cell, without presence in the latter's nucleus. This is of course a major biosafety asset of poxviruses, since it precludes the insertion of viral DNA into the host genome, and the putative adverse effects that may arise by such events.

The poxviruses are diverse in their structure, host range and host specificities. Members of each of the 8 genera of the subfamily *Chordopoxvirinae* are genetically and antigenically related and share similarities in morphology and host range within the genus. For example, similarities in the restriction endonuclease maps of several Orthopoxvirus genomes, such as the 90% sequence identity of the genes of Vaccinia and Variola viruses reflect genetic relatedness of members of each genus. It is, therefore, not surprising that intra-genus cross- hybridization has been reported among these viruses.

Genetic divergence exists between members of each genus, particularly when such members come from geographically separated econiches. Indeed, Weli et al. (2004) even observed genetic heterogeneity among avipoxviruses isolated from different parts of Norway.

Sequence similarities of genomic repeats in various strains of each virus species suggests that these repeats evolved from unequal crossover events. Thus, genetic material of each virus bears a pointer to genetic recombination in its history. Further examination of the Orthopoxviruses shows evidence of genetic transpositions and deletions in the terminal hypervariable regions of the genome. The significance of these genetic interactions or phenomena in the ecology of poxviruses is at present a matter of conjecture but they makes risk assessment in the use and release of poxvirus-based GMVs absolutely necessary.

4.3 Ecological distribution of *Poxviridae*

The family *Poxviridae* contains a vast number of divergent viruses with equally divergent host specificities and host ranges. It seems justified to state that poxviruses will be found in any vertebrate species and geographical area where they are systematically looked for. However, very little is known about the characteristics and geographical distribution of most poxviruses occurring naturally in the field. This knowledge would be necessary if a meaningful risk assessment of poxvirus-based GM vaccines use or release in any target area should be undertaken. Without such knowledge there is no possibility of making meaningful ERA inclusions of adverse effects due to recombination events between GM poxvirus vaccines and naturally occurring poxviruses circulating within the actual area for application.

Many poxviruses are capable of zoonotically infecting man. It is likely that variola virus is derived from an ancient zoonotic virus that originated from a now extinct animal host species. In general, poxviruses show species specificities that range from narrow to broad, but *we still know little about the fundamental mechanisms that mediate the host tropism of individual poxviruses* (McFadden, 2005). The unpredictability in a real world situation is illustrated by a macaque outbreak caused by a European orthopoxvirus strain carried by *Rattus norvegicus* in the Netherlands (Martina et al., 2006).

Variola virus (smallpox disease) has killed more members of the human population over the span of recorded history than all other infectious diseases combined. Variola virus may never again infect humans, but there are other poxviruses that can cause serious human disease. In 2003, an outbreak of human monkeypox occurred in the mid-western United States due to the inadvertent importation of monkeypox virus in a shipment of rodents from West Africa (Reed et al., 2004). Fortunately, the strain that caused this outbreak was more benign in humans than the more pathogenic variant that is found in central Africa, which results in mortality rates of 10–15%.

Variola virus, prior to its eradication, spread all over the world. Similar worldwide distribution seems to be true of VACV and Molluscum contagiosum virus. Man was the only known host and reservoir of Variola and Molluscum contagiosum viruses. Although the reservoir host of VACV is unknown, it has been shown, under natural conditions, to exhibit a wide host range among humans, rabbits, cows and buffaloes. Other poxviruses with worldwide distribution include numerous species of Avipoxvirus. Even with the relatively limited studies so far done, certain orthopoxviruses (parapoxviruses, capripoxviruses and yatapoxviruses) are notably associated with African domestic animals and wildlife, including rodents, squirrels and monkeys, which may serve as reservoirs of infection. Knowledge of the distribution of poxviruses that infect man and with host range which includes wildlife such as rodents, squirrels and monkeys, would be important in risk assessment of poxvirus- based GE vaccines and other poxvirus-based recombinant constructs prepared for use or release in ecosystems to which these hosts belong.

Similarly, it is important to document the poxviruses carried by ectoparasites of animals in the target localities.

4.4 Poxvirus vaccine vectors

Poxviruses are being increasingly used as recombinant vectors for vaccination against numerous infectious diseases in humans, domestic animals, and wildlife (Pastoret & Vanderplasschen, 2003). For risk assessments and surveillance, information about the occurrence, distribution and ecology of poxviruses are hence important.

Poxviruses have several advantages for use as expression or vaccine vectors, including their large insertion capacity, their cytoplasmatic site of replication, their heat stability, the relatively high expression levels and proper post-translational modifications of foreign proteins as well as the ability to raise protective immune reactions locally on mucous membranes.

Vaccinia virus (VACV) strains are robust eukaryotic expression vectors that have been used for a number of different studies in biochemistry, cell biology, and immunology. Most vaccinia virus strains, such as the ones used in the smallpox eradication campaign, readily replicate in human cells and have been associated with a range of clinical complications in vaccinees. This fact is presently considered a major problem for the use of these strains as recombinant vaccines for mass vaccination. For laboratory use, in addition to mandatory biosafety level 2 procedures, proper precautions must be taken to prevent accidental exposure when managing replication-competent vaccinia strains.

4.4.1 Non-replicating poxvirus vectors

To circumvent the problems associated with "classical" vaccinia virus strains, several laboratories have been increasingly involved in efforts to develop more attenuated, host- restricted virus strains (for brief review, see McFadden, 2005). Generally, these efforts involve two related strategies: i). The isolation of chicken cell-culture adapted vaccinia virus variants that are replication-deficient in mammalian cells, for example MVA, or ii). The development of avipoxvirus platforms, such as canarypox (ALVAC) and fowlpox (TROVAC), supposed to be naturally non-permissive for mammalian cells. There is increasing evidence that such non-replicating vaccines are safer than the original vaccinia strains and are still comparably immunogenic.

4.4.1.1 MVA (Modified vaccinia virus Ankara)

Among the so-called "non-replicating poxvirus vectors", modified vaccinia Ankara

(MVA) has elicited considerable interest because of its excellent safety record in humans, its ability to mobilize different protective arms of the immune system, and the possibility to raise local mucosal membrane immunity after oral delivery. In addition to being a promising vector for the construction of poxvirus-based recombinant vaccines, one appealing feature of MVA as an expression vector is that it can be used under biosafety level 1 laboratory conditions. MVA has in common with NYVAC and other adapted VACV strains that they are *supposedly not carrying out fully productive infections in "relevant" mammalian cell types.* Such VACV strains do, however, secure efficient expression and presentation to the immune system of protein products from transgenes inserted into different locations in the viral DNA genome.

MVA was derived from a Turkish smallpox vaccine strain (Ankara) (reviewed by Drexler et al., 2004). After more than 500 passages in chicken cells, it tested defective for replication in human cells and avirulent in lab animals. From 1968–1980, MVA was inoculated into more than 100,000 individuals in Germany with no reported secondary complications and it is now considered to be a suitable platform for the next generation of safer smallpox vaccines and recombinant poxvirus vectors. Genomic mapping and sequencing studies have revealed that MVA lost nearly 30 kb of genomic information during its extended passage in chicken cells. Furthermore, it has multiple deletions and mutations compared with the parental strain. Many of these genetic alterations were in host-response genes. It is assumed that these deletions render MVA unable to carry out productive infections in human cells.

MVA has been proven very efficient in induction of mucosal immunity. This is of course utterly important since a number of human and animal disease agents use mucosal surfaces as their portal of entrance to the organism.

4.4.1.2 Avipoxvirus vectors

Like MVA, canarypox virus (ALVAC) and fowlpox (TROVAC) vaccine vectors induce antibody and cytotoxic T-cell responses, critical in the immune defense against viruses, to vectored viral antigens in a range of mammalian species (Hel et al., 2002; Srinivasan et al., 2006). Replication of avipox virus vectors is regarded abortive in mammalian cells, eliminating the safety concerns that exist for replication competent vaccinia virus vectors. The avipox viruses infect mammalian cells and produce viral proteins, with the replication block occurring at the time of viral DNA synthesis. It should, however, be noted that the experimental data backing claims for total lack of avipoxvirus replication in mammalian cells are based on studies embracing very few viral species or strains, and also very few, and mostly virus-ecologically irrelevant, mammalian cell cultures. For some avipoxvirus/mammalian cell combinations fully productive infections have been demonstrated (Weli et al., 2004).

Licensed ALVAC-vectored vaccines for dogs, cats, horses, chicken and ferrets are commercially available, and an ALVAC-vectored HIV-1 vaccine has entered phase III clinical trials. Some additional ALVAC-vectored HIV-1 and some human cancer and malaria vaccines are in clinical phase I or II trials (Weli & Tryland, 2011).

4.5 Current research on risk assessment of GMVs

There is little peer-reviewed information available that relates to ERA of GMV releases. To our knowledge research related to environmental effects is only being performed for alphaherpesviruses (Thiry et al., 2006) and poxviruses (orthopox and avipoxviruses). Such environmental biosafety-related research has been performed for a number of years in Norway, but we have no present knowledge of other research groups with a similar focus. We have concentrated on biosafety issues of the orthopoxvirus strain MVA (Modified Vaccinia Ankara), considered to be a very safe vaccine vector because of high gene expression capacity and lack of viral replication in mammalian cells (Drexler et al., 2004).

The most relevant conclusions from our studies may be summed up as follows:

- Orthopoxviruses, and hence potential recombination partners for orthopoxvirus vectored vaccines, are common in different small rodent species populations all over the country, and small rodent predator species are infected by and have antibodies to such viruses (Sandvik et al., 1998, Tryland et al., 1998).
- Recombination between an influenza-transgenic MVA and a naturally occurring orthopoxvirus is readily demonstrated in cell cultures. The recombinants may have phenotypic characteristics, some of which may point towards adverse effects, different from the parental viruses. Recombinants may be genetically unstable and "throw out" the influenza transgene. This will eliminate the most logical tag for vaccine monitoring, and will also diminish the ability of the vector to mobilize protective immune responses (Hansen et al., 2004).
- The absolute and relative permissivities for MVA multiplication and viral shedding have not been thoroughly studied. GM and unmodified MVA may, contrary to the general dogma, perform fully productive infections in highly relevant mammalian cell types, and other mammalian cell cultures that are semi-permissive to infection (Okeke et al., 2006).
- DNA sequencing revealed that orthopoxviruses can be clearly separated into geographically distinct strains, and it was inferred that these strains have distinct evolutionary histories in different rodent lineages (Hansen et al., 2009). It is an open question whether these different virus strains

have aberrant abilities to engage in recombination events with GM vaccine vectors.

- Upon sequencing of an orthopoxvirus isolated from a human clinical case, it was established that this strain was a naturally occurring hybrid between two distinct orthopoxvirus species. This is the first proof of concept for orthopoxvirus recombinations taking place under authentic ecological circumstances (Hansen et al., 2011).
- Homologous recombination between orthopoxvirus-vectored vaccine and naturally circulating orthopoxviruses, genetic instability of the transgene, accumulation of non- transgene expressing vectors or hybrid virus progenies, as well as cell line/type specific selection against the transgene are potential complications that may result if poxvirus vectored vaccines are extensively used in animals and man (Okeke et al., 2009a).
- Phenotypic characteristics of recombinants between GM and naturally occurring orthopoxviruses may be unpredictably different from any of the parental viruses (Okeke et al., 2009b).
- Contrary to common assumptions, some avipoxviruses may carry out productive infections in mammalian cells, and avipoxviruses within a restricted geographical area may be more genetically diverse than realized so far (Weli et al., 2004 and 2005).

5. FEATURES BY GMVS THAT ARE RELEVANT FOR RISK ASSESSMENT AND MANAGEMENT FRAMEWORKS

Directive 2001/18 with annexes, the EMEA guidelines and CPB with annexes give a number of leads towards GMO characteristics that may indicate hazards or adverse effects. Characteristics that must be taken into consideration (Anliker et al., 2010) include the pathogenicity, virulence, infectivity, host range, tissue tropism, replication strategy, latency/reactivation, survival and stability of the GMO. Here we will point to two important features by risk assessment and management frameworks that may cause difficulties when preparing ERA and risk management strategies for GMVs.

5.1 The distinction between chemicals and organisms

Risk perspectives used on chemical pollutions have been influential when shaping risk concepts in biosafety related to genetically modified organisms (GMOs), and hence also GMVs. We will argue that, and discuss why, this starting point is of very doubtful relevance to self-replicating organisms and molecules.

When the first genetically modified plants were approved and released the adequacy of the present framework for assessment of chemical substances i.e.

ecotoxicological assessment, for GMOs was contested (Meyer, 2011). Chemicals have a release-dependent concentration decline with a given breakdown time in the environment, while GMOs follow a different environmental routes and degradation pathways than chemical pollutants. (Trans)Genes follow the path of the host genome, possibly eventually also the path of sexually compatible and some incompatible species (through vertical and horizontal gene transfer). Thus for GMOs exposure do not necessarily predict response, and accordingly risk models (based on the premise that exposure dose predicts response) have no or little utility in predicting the environmental behaviour of released transgenes. This was illustrated for example when Hilbeck et al. and Losey et al. reported adverse effects of Bt toxins and Bt maize pollen on non-target organisms in laboratory experiments (Hilbeck et al., 1998, Losey et al., 1999). To improve ecotoxicological testing, Andow and Hilbeck (2004) have suggested that investigation of non-target effects of GMOs may be done more efficiently by employment of a combination of two models; Ecotocixology testing of chemicals and risk assessment of non-indigenous species. However, the ecotoxicological approach still remains the recommended one in risk assessment procedures of GMOs although the new EU biosafety Directive 2001/18/EC partly supports the ecological approach since it prescribes a more detailed ERA.

5.2 The distinction between viruses and organisms (modified from Traavik, 1999)

It is important to keep up the distinction between viruses and organisms. Viruses are *not* organisms. Furthermore, the differences in genome strategies and life cycles between virus families are often more fundamental than between different mammalian or plant families. Viruses multiply intracellularly in permissive host cells. One single virus particle (virion) infecting a permissive cell may give rise to billions of new particles during a short time (hours to days). The submicroscopic size of virions and the ability to spread over vast, even global, distances during short timespans are important, basic conceptions for ERA and risk management of any GMV. This is important in order to ask the relevant harm and hazard related questions, and hence to realize and conceive the risks connected with a given GM vaccine virus in a specific ecosystem and society context.

In addition to such fully productive infections, some virus/host cell combinations may result in persistent infections with virus shedding in the excreta for extended periods, while others lead to latent infection with viral DNA in a host chromosome-integrated or episomal state. Latent infections may be intermittently reactivated and accompanied by virus shedding. Integration of viral DNA into the host cell genome may by itself have harmful consequences, irrespective of viral gene expression or replication. The same virus strain may, under differently modulating conditions, display all these life cycle forms.

The host tropism, at the species-, organ- or cell type-level, is quite narrow for some viruses, while others have a much wider host-spectrum. For most viruses the molecular, genetic and epigenetic pathways determining host-cell specificity are not known in detail. Restrictions may be present at various steps during a virus multiplication cycle, from the lack of cell membrane receptors to subtle incompatibilities with host cell enzymes necessary for viral nucleic acid transcription and replication. Hence, minor genetic changes taking place during or after engineering of GM viruses may have profound effects on the host tropism and ability to spread to non-target host species.

For many virus/host cell combinations permissivity is a relative term, since it may be influenced to a considerable extent by the menu of genes expressed by the cell, and by the exact levels of gene expression. In culture, the permissivity of a given host cell may be manipulated experimentally by activation of intracellular signal transmission pathways, i.e. by hormones, growth factors, cytokines etc. Such procedures may also enhance persistent or reactivate latent infections. At the intra- as well as at the inter-species level of host animals this is illustrated by a vast variation in susceptibility for a given virus strain. Such variation may be related to sex, age, mating season, pregnancy, genetic differences, infection with other viruses or microorganisms, and environmental factors promoted by season, climatic changes or by pollution, e.g. EDCs (endocrine disrupting chemicals).

It is important to be aware of the distinction between viral infection and viral disease. An infected individual may shed virus and represent a transmission reservoir without showing clinical symptoms. Yet, other individuals within the same or other species may become clinically ill, or the viral infection may result in abortions, stillbirths, teratogenic or oncogenic effects. For persistent/latent infections, clinical symptoms may be present intermittently, only under special circumstances, or appear a long time after infection.

Different strains of the same viral species may have different virulence or pathogenicity, as well as host-cell or -species tropism. Even genetic differences at the single point mutation level may result in virus strains with aberrant phenotypic characteristics.

For GMVs it is hence conceivable that unintended phenotypic characteristics with unwanted ecological consequences may be established in addition to the intended modification(s). This may not become evident unless very comprehensive and carefully planned experiments and ecosystem surveillance/sampling programs are carried out. In many instances fully adequate studies are totally precluded by the complexity and the regular or occasional variations of the recipient ecosystem. Hence scientific uncertainty or ignorance is a state that must be accepted.

6. THE CONNECTION BETWEEN RISK ASSESSMENT AND RISK MANAGEMENT

According to Anliker et al. (2010), an ERA based on Directive 2001/18/EC should follow four general principles:

"First, the GMO should be compared to the non-modified organism from which it is derived. Second, the ERA should be carried out on a scientifically sound premise and rely on known facts supported by data derived from specific testing of the GMO-containing medicinal product including its use in previous clinical trials. If necessary, this data can be substantiated by theoretical assumptions. Third, it is necessary to perform the ERA on a case-by-case basis, since the heterogeneity of the GMO-containing medicinal products and the differences in their clinical use make it difficult to apply standardised requirements or evaluations as part of the assessment. Finally, the ERA needs to be re-evaluated if new information on the GMO or its effects on human health or the environment becomes available."

We find it an interesting and elucidating statement that "data can be substantiated by theoretical assumptions". This opens the ERA gate for creation and inclusion, as well as theoretical, mathematic and experimental modeling, of hypotheses build on "Worst case scenarios", as pointed out by Kühler et al. (2009), and also for the use of stringent Precautionary principle versions in situations and processes dominated by scientific uncertainty and ignorance (see section 9).

The case-by-case and step-by-step approaches to GMV ERAs seem to be instituted in all relevant national and international legislations, including EU Directive 2001/18/EC and the Cartagena protocol on biosafety (CPB). Any ERA should include comparative data about characteristics of the GMV and its unmodified parental virus strain. Details about the genetic modification/engineering process, including data about effects, genomic location and DNA sequence of the transgenic (vaccine gene) insert must be present. Descriptions of the vaccination regime and release/possibilities for escape of the GMV must be considered. The receiving environment/ecosystem, including possible interactions between the GMV and the environment, have to be described. Plans for monitoring, control, waste treatment and emergency response plans have to be prepared and presented, and this is, of course, tasks that overlap with risk management activities.

Anliker et al. (2010) state: "Experience gained from the release of comparable GMOs into a similar environment can be used to support the ERA". In our view this is to some extent an unwanted premise for trustworthy ERAs since it opens up for subjective interpretations of "comparable GMOs" and "similar environment". Theoretical considerations and practical experience indicate that such parameters rarely exist for engineered organisms, and are even more rare for viruses. This should become evident while reading and reflecting on the scientific ignorance, uncertainty and lack of knowledge related to the nature, characteristics and ecology of GMVs.

The ERA procedure may be scientifically and operationally divided into five steps, whereby the ERA and risk management is integrated and coordinated (described in an excellent review by Anliker et al., 2010):

- Identification of potential adverse effects;
- Evaluation of the potential consequences of each adverse effect, if it occurs;
- Evaluation of the probability (likelihood) that each identified potential adverse effect should occur;
- Estimation of the risk posed by each identified characteristic of the GMO;
- Application of risk management strategies related to marketing or deliberate release of the GMO;
- Evaluation of the overall risks of the GMO, based on the conclusions from the previous steps, taking into account the risk management strategies proposed in step 5, which were created to reduce or eliminate the risks identified in step 4.

Through the outlined procedure each potentially harmful characteristic of a GMO should be turned into risks. The total process should form a basis for consideration of the overall risk to the ecosystem, animal and human health by deliberate release or marketing of a given GMO. The ERA should conclude on whether there is a need for a risk management plan. If necessary and possible the ERA should devise appropriate risk mitigation methods. And of primary importance: The ERA should reach a conclusion as to whether the overall environmental impact is at all acceptable or not (Kühler et al., 2009).

Step 1 of the ERA procedure is focusing on the identification of GMO characteristics that may result in adverse effects. This is the decisive step for whether, and to which extent, the ERA and the risk management plans will provide biosafety and make contributions to sustainable development as well as to good health of ecosystems, animals and humans.

In our opinion the creative conception and design of "What-if?"- and Worst-case-scenario- inspired ideas, questions and working hypotheses, must be encouraged and stimulated in order to enhance the chances of high precaution levels in ERA and risk management processes. Furthermore, we will argue that these goals will only be met if independent scientists and institutions carry out the relevant intellectual processes and research projects. Finally, although national legislations, and also the Cartagena protocol, require documented scientific evidence as a basis for ERAs and risk management plans, implementation of the

Precautionary principle is always expressively required (Kûhler et al., 2009). The different versions, interpretations and possibilities for implementation of the precautionary principle are further discussed in section 9.

6.1 Relevant risk assessment and management questions for GMVs

The different virus families have their specific life cycles and host-preferences. Hence it is impossible to make risk assessment schemes that are valid for all potential virus vectors. Risk assessment must be performed on a case-by-case, step-by-step basis, taking into account the characteristics of the ecosystem into which the virus will be released, and the ability of the virus to engage in transboundary movements (Traavik, 1999; McFadden, 2005).

The most evident risk issues related to release of GMVs or unmodified viruses are the known and unknown unknowns related to (i) whether active multiplication with virus shedding in excreta takes place in target individuals, (ii) infection of non-target species, (iii) recombinations with naturally occurring virus relatives and (iii) integration of GMV DNA into host cell chromosomes. Ideally, before running the risk that any GMV becomes implanted into a species/population/new location/ecosystem a number of crucial questions should be answered (modified from Traavik, 1999), e.g.:

- Can the released virus engage in genetic recombination, or by other means acquire new genetic material? If so, will the hybrid offspring have changed their host preferences and virulence characteristics?
- Can the released virus or any hybrid or mutated offspring be shed in the excreta of intentionally or unintentionally infected individuals?
- Can the released virus or any hybrid or mutated offspring infect unintended (non- target) species?
- Can the released virus or any hybrid or mutated offspring integrate into the genomes of host cells?
- Can other viruses that are present within the ecosystem influence the infection with the released virus or its offspring?
- Can insects or migrating birds or animals function as vectors for the released virus or its offspring, to disseminate viruses out of their intended release areas?
- May climatic changes and/or xenobiotic pollutants, e.g. EDCs (endocrine disrupting chemicals), influence the virus/host animal/ecosystem interactions?
- For how long can the virus and its offspring survive outside host organisms under realistic environmental and climatic conditions?

- Are the virus and its offspring genetically stable over time?
- Can the virus or its hybrid or mutated offspring establish long-lasting, clinically mute, persistent or latent infections in naturally accessible host organisms?
- Can the virus or its offspring activate or aggravate other naturally occurring latent or persistent virus infections?

Some of these questions deal with the genetic and phenotypic characteristics of a supposedly genetically stable GMV. But the situation becomes even more complex and unpredictable if the GMV parental strain under certain conditions or circumstances is genetically unstable, giving rise to viral strains with altered characteristics (Traavik, 1999). At the present time it is strongly needed that questions and hypotheses related to the influence of climatic changes and xenobiotics, e.g. endocrine disrupting chemicals (EDCs) are included in ERAs and risk management plans.

Demanding ERA and risk management challenges connected to all the questions raised in this section are related to whether methods allowing detection of the phenomena listed, and still unlisted, have been developed, and whether the surveillance and monitoring programs that make their employment possible have been funded and operationalized.

7. BIOSAFETY IMPLICATIONS FOR ENVIRONMENTAL RISK ASSESSMENT AND MANAGEMENT FRAMEWORKS

We will argue that in order to harvest the potential benefits of any GMV, the approaches used in the Norwegian orthopoxvirus studies should be part of the regulatory risk governance frameworks all over the world (see section 4.5 as well as 6.1). In relation to ERA and risk management needs, some of the most urgent issues for targeted research will be treated in the following.

7.1 Naturally occurring relatives

With GMV use and release, it will be crucial to determine the occurrence, distribution and ecology of poxvirus relatives. In Norway rodents and other small mammals are considered to be natural reservoirs for different orthopoxviruses, and cowpoxvirus-like strains in particular (Sandvik et al., 1998). Approximately 20% of shrews and small rodents belonging to eight species carry orthopoxvirus DNA sequences in their organs, and 20% of such animals have specific serum antibodies as a sign of previous infections.

7.2 Potential recombination events and their consequences

If a GM orthopoxvirus infects an individual, animal or human, that already carries another orthopoxvirus, a recombination event, homologous or illegitimate, may follow. Foreign genes may hence be transferred from the GM donor virus to other GM poxvirus vaccines or to wild type orthopoxvirus recipient strains (Sandvik et al., 1998). The outcome may be hybrid viral progenies with unpredictable pathogenicity and altered host range. The probability and possible outcome of recombination is dependent on the characteristics of the viral vector used, and the occurrence of naturally occurring or genetically modified relatives.

In cell culture co-infection experiments with a MVA-based human influenza vaccine and a newly isolated Norwegian cowpox virus-like strain, a number of recombinant progeny virus strains were obtained (Hansen et al., 2004). One of the progeny strains displayed phenotypic characteristics different from both parents, and was genetically unstable upon cell culture passage, i.e. the influenza transgene was deleted at a high frequency. This is of paramount monitoring significance, since the inserted gene will always be the marker of choice for tracing and detection of vaccine-related effects.

Environmental release or escape of a GMV provides the opportunity for recombination with poxviruses present in target and non-target animals, including humans. In most parts of the world the occurrence and distribution of naturally occurring poxviruses are virtually unknown. But, in central-Africa it has been proposed that the vaccinia- rinderpest vaccine (RVFH) might recombine with a pox-relative, namely monkeypoxvirus. Since smallpox has been eradicated, the monkey pox is at present the most feared poxvirus (Reed, 2004), and has for instance caused virulent outbreaks in Zaire (Kaaden et al., 2002).

Under natural circumstances the probability of recombination may be low, but predictions and forecasts are made impossible by the fact that we lack knowledge about the natural occurrence and prevalence of poxviruses in all parts of the world. Furthermore, even when recombination events are rare, the consequences may be serious since one viral progeny particle may multiply into millions of identical particles in a matter of hours. In this context, it is disturbing that the present regulatory frameworks are based on a one time risk assessment of any given GMV, and does not take into account the putative consequences that may arise from successive releases of the same, or related, GMVs.

During the global smallpox eradication program (in the 1970s) transmission from vaccinated to unvaccinated persons was estimated to occur at a rate of 27 infected per million total vaccinated (Centre for Disease Control, 1991). It has been possible to isolate the vaccinia virus from domesticated animals that have been in contact with recently vaccinated humans. The potential host spectrum of vaccinia virus is very broad and includes laboratory animals, pigs, cattle, camel, and monkey species (Fenner, 1996). In the past, vaccinia virus has for instance spread from vaccinated humans to domestic animals such as cattle (through milkers) and buffaloes, before spreading within the herd. Furthermore, it has been reported that during outbreaks of buffalopox, the buffalopox virus infected unvaccinated humans (Dumbell & Richardson, 1993). In Brazil, an emerging virus with similar characteristics as a smallpox vaccine virus caused disease both in cattle and their human caretakers (Damaso et al., 2000).

GMVs made for wildlife vaccination must be thermostable and have long environmental persistence, and VACs satisfy these criteria. The vaccinia-rabies vaccine (VRG), for instance, has been reported to persist 4 months under natural conditions without a significant loss of viral infectivity (Brochier et al., 1990). Environmental stability is a prerequisite for herd immunity in target animals, but by the same token it increases the risk of non-target effects, and for spread to non-target species and other ecosystems.

VACV has been shown, in cell cultures, to easily engage in genetic recombination with other orthopoxvirus species (Ball, 1987). A high number of closely related viral species are known, and high degrees of sequence homology across species borders have been demonstrated. It may, however, be safely assumed that a high number of unknown orthopoxviruses are circulating in many different types of ecosystems and biotopes all over the world. Biological (i.e. insects, migratory birds and animals, domestic and pet animal trade, infected individuals in an incubation period etc), as well as mechanical, (i.e. automobiles, airplanes) vectors may enable further dissemination and transmission of a virus to other ecosystems. Changed tropism for particular cells or tissues may result in divergent virulence in the target, or other known, host species for the GMV or its recombinant progenies compared to the parental, unmodified poxvirus (Traavik, 1999). In addition, the GMV or its recombinant progenies may have had their host restriction programs changed, so that the viruses can transfer to and infect formerly resistant host species. Other phenotypic changes may also happen. For instance, it has been reported that GE of a poxvirus created a GMV that was more virulent than the non-modified virus. As part of a strategy to develop pest animal contraceptive vaccines, the gene encoding human interleukin-4 (IL-4) was inserted into the Ectromelia virus (mousepox) genome (Jackson et al., 2001). Expression of IL-4 was intended to curb unwanted anti-viral effects. Unexpectedly, genetically mouse pox-resistant mice infected with the IL-4 expressing virus developed symptoms of acute mouse pox accompanied by high mortality.

7.3 Non-target effects and transboundary spread

GMV administration may have non-target effects. Baits containing GMVs may be eaten both by target and non-target animals. In the second instance, viral spread may also take place by direct/indirect human/animal contact, or by installation of the GMV into a food web. A non-target GMV transfer was illustrated by the account of a woman, with a predisposing skin disease, who, having removed a VRG bait from the mouth of her dog, became infected and developed pox lesions (Rupprecht et al., 2001).

The eradication of rabies by vaccination may have indirect consequences for the ecological balance; lack of disease favours an increase in the host animal population, in this case the European red fox. Therefore, it is crucial that monitoring of GMVs in the environment is initiated with the purpose to follow-up the performed risk assessment, to map the actual environmental effects and to identify unforeseen adverse ecological effects.

GMVs or their progenies resulting from mutations and/or recombination events may achieve new phenotypic traits of importance for their ecology, spread and host preferences. The opportunities for long distance spread and cross-species transfer of mammalian viruses have increased in recent years due to enhanced contact between humans and animal reservoirs. It is, however, difficult to predict when such events will take place since the viral adaptations that are needed are multifactorial and stochastic. Recent examples of viruses that have crossed species barriers are HIV, hantaviruses, haemorrhagic fever viruses, various arboviruses (e.g. West Nile Virus), avian and porcine influenza viruses, SARS- associated coronavirus, Nipah and Hendra viruses, and monkeypox virus.

The emergence of new viral infections often follows environmental, ecological and technological changes caused by human activities (Louz et al., 2005). Such activities may lead to an increased contact between humans and livestock on one hand, and animal hosts acting as reservoirs of zoonotic viruses on the other hand. Agricultural development, an increased exploitation of environmental resources, growth and increase in the mobility of the human population and trade and transportation of food and livestock, have been identified as important factors contributing to the introduction and spread of a number of new viruses in the human population.

Against this background the intensified use of viruses and their genetically modified variants as viral gene transfer vectors for biomedical research, experimental gene therapy and as live-vector vaccines is a cause for concern (reviewed by Louz et al., 2005).

7.4 Future safety prospects

Several instances have demonstrated that undesired traits of poxvirus vectors can be altered or eliminated by targeted mutagenesis, homologous recombination, and reverse genetics techniques (Najera et al., 2006; Chakroudi et al., 2005). However, the safety advantages of these methods can only be realized if we have clearly identified the presumed properties of genetically modified viruses (GMVs) and addressed any potential negative effects in both the categories of "known unknowns" and "unknown unknowns".

8. THE NORMATIVE CHALLENGE BY THE CONCEPTS OF HARM AND UNWANTED ECOLOGICAL CONSEQUENCES

Earlier, we outlined many detrimental effects on the environment and examined their significance in relation to ERA and risk management techniques. In this discussion, we will explore the concepts and definitions associated with harms and hazards within legislative frameworks. We will assert that, both descriptively and normatively, it is necessary to employ biological, ecological, and ethical terminology to identify undesirable harm and ecological repercussions.

The endpoints of any risk assessment and risk management are invariably linked to the regulatory framework. Article 1 of the Cartagena Protocol stipulates that the primary purpose of the protocol is to save and preserve biodiversity in accordance with a cautious and preventive approach. According to the EU directive 2001/18/EC, the applicant is required to provide a notification that includes an environmental risk assessment. This assessment should consider both direct and indirect effects, immediate and delayed effects, as well as potential cumulative and long-term effects resulting from the interaction between genetically modified organisms (GMOs) and the environment. It is necessary to thoroughly characterize the negative effects on human and animal health. These factors encompass the examination of both direct consequences, such as the emergence of illness, as well as more intricate indirect consequences, such as changes in vulnerability to disease, considerations of well-being, and evaluations of harm from social and ethical perspectives. In terms of environmental damage, potential consequences encompass the impact on and interplay with all organisms in the ecosystem, and might manifest either directly or indirectly. Direct effects pertain to the biological consequences on organisms, whereas indirect effects refer to the affects on animal health, pollution of wild gene pools, or changes in ecological interactions.

The legislative framework raises numerous inquiries regarding the definition of injury, a word that encompasses both descriptive and normative aspects. Raybould highlights a significant issue in the connection between risk research and risk assessment of GM crops. He asserts that the initial step of problem formulation is heavily influenced by the interests of stakeholders who are concerned about protecting the environment and the goods and processes associated with it (Raybould, 2006). Furthermore, step 1 in environmental risk assessment is not the only step that involves normative issues. Step 5 also raises normative concerns, as pointed out by Hill (2005). Therefore, risk assessments incorporate value judgments both in terms of the outcomes that need to be prevented and the process of risk characterization. Typically, these decisions are taken prior to the initiation of the risk assessment. The subject of harm is closely linked to the concept of human welfare and the preservation of nature and biodiversity. The determination of what constitutes substantial harm to nature is contingent upon subjective value judgments that can be debated. There are clear conceptual distinctions between prioritizing the protection of human interests, known as anthropocentrism, and the preservation of ecosystems, known as biocentrism and ecocentrism (Dobson, 1998; Westra, 1998). Within an anthropocentric framework, the preservation of the environment is prioritized in order to enhance human welfare, such as for recreational purposes or as a means of acquiring new knowledge, under the assumption that ecosystems hold undiscovered information. Furthermore, biodiversity centers serve as vital reservoirs of genetic variation for future advancements in agriculture and medicine. Therefore, the desires and concerns of humans serve as strong incentives to safeguard the environment from activities that could result in significant negative impacts, such as a decrease in biodiversity, for both current and future generations.

Biocentrics and eccocentrics advocate for a shift away from the anthropocentric exploitation of the environment and towards a stronger reverence for the creatures and the environment's integrity. Biocentrics assert that it is incumbent upon us, as humans, to grant rights to species and habitats, and hence we have an obligation to protect their integrity (Regan, 1980). An ecocentric approach encompasses more than just the assessment of effects on human well-being and sustainability. It emphasizes the need to address changes in both the living and non-living components of the environment, such as the influence on soil, water, and air. Ecocentric approaches not only recognize the need of animal and environmental integrity, but also understand the significance of ecosystem integrity. Irrespective of their usefulness to humans, it is crucial to preserve and conserve biological, ecological, and genetic processes in this context. The application and release of genetically modified organisms (GMOs) can be justified when it enhances the well-being of the ecosystem or contributes to the preservation or enrichment of species variety within the community. This ideology diverges in terms of value commitments and factual views from anthropocentric GMV governance. Consequently, our attitude towards the environment and the importance we assign to it might impact the strategies and methods used in assessing and managing environmental risks.

8.1 Problem formulation processes for involvement of normative issues

Earlier, we outlined many detrimental effects on the environment and examined their significance in relation to ERA and risk management techniques. In this discussion, we will explore the concepts and definitions associated with harms and hazards within legislative frameworks. We will assert that, both descriptively and normatively, it is necessary to employ biological, ecological, and ethical terminology to identify undesirable harm and ecological repercussions.

The endpoints of any risk assessment and risk management are invariably linked to the regulatory framework. Article 1 of the Cartagena Protocol stipulates that the primary purpose of the protocol is to save and preserve biodiversity in accordance with a cautious and preventive approach. According to the EU directive 2001/18/EC, the applicant is required to provide a notification that includes an environmental risk assessment. This assessment should consider both direct and indirect effects, immediate and delayed effects, as well as potential cumulative and long-term effects resulting from the interaction between genetically modified organisms (GMOs) and the environment. It is necessary to thoroughly characterize the negative effects on human and animal health. These factors encompass the examination of both direct consequences, such as the emergence of illness, as well as more intricate indirect consequences, such as changes in vulnerability to disease, considerations of well-being, and evaluations of harm from social and ethical perspectives. In terms of environmental damage, potential consequences encompass the impact on and interplay with all organisms in the ecosystem, and might manifest either directly or indirectly. Direct effects pertain to the biological consequences on organisms, whereas indirect effects refer to the affects on animal health, pollution of wild gene pools, or changes in ecological interactions.

The legislative framework raises numerous inquiries regarding the definition of injury, a word that encompasses both descriptive and normative aspects. Raybould highlights a significant issue in the connection between risk research and risk assessment of GM crops. He asserts that the initial step of problem formulation is heavily influenced by the interests of stakeholders who are concerned about protecting the environment and the goods and processes associated with it (Raybould, 2006). Furthermore, step 1 in environmental risk assessment is not the only step that involves normative issues. Step 5 also raises normative concerns, as pointed out by Hill (2005). Therefore, risk assessments incorporate value judgments both in terms of the outcomes that need to be prevented and the process of risk characterization. Typically, these decisions are taken prior to the initiation of the risk assessment. The subject of harm is closely linked to the concept of human welfare and the preservation of nature and biodiversity. The determination of what constitutes substantial harm to nature is contingent upon subjective value judgments that can be debated. There are clear conceptual distinctions between prioritizing the protection of human interests, known as anthropocentrism, and the preservation of ecosystems, known as biocentrism and ecocentrism (Dobson, 1998; Westra, 1998). Within an anthropocentric framework, the preservation of the environment is prioritized in order to enhance human welfare, such as for recreational purposes or as a means of acquiring new knowledge, under the assumption that ecosystems hold undiscovered information. Furthermore, biodiversity centers serve as vital reservoirs of genetic variation for future advancements in agriculture and medicine. Therefore, the desires and concerns of humans serve as strong incentives to safeguard the environment from activities that could result in significant negative impacts, such as a decrease in biodiversity, for both current and future generations.

Biocentrics and eccocentrics advocate for a shift away from the anthropocentric exploitation of the environment and towards a stronger reverence for the creatures and the environment's integrity. Biocentrics assert that it is incumbent upon us, as humans, to grant rights to species and habitats, and hence we have an obligation to protect their integrity (Regan, 1980). An ecocentric approach encompasses more than just the assessment of effects on human well-being and sustainability. It emphasizes the need to address changes in both the living and non-living components of the environment, such as the influence on soil, water, and air. Ecocentric approaches not only recognize the need of animal and environmental integrity, but also understand the significance of ecosystem integrity. Irrespective of their usefulness to humans, it is crucial to preserve and conserve biological, ecological, and genetic processes in this context. The application and release of genetically modified organisms (GMOs) can be justified when it enhances the well-being of the ecosystem or contributes to the preservation or enrichment of species variety within the community. This ideology diverges in terms of value commitments and factual views from anthropocentric GMV governance. Consequently, our attitude towards the environment and the importance we assign to it might impact the strategies and methods used in assessing and managing environmental risks.

9. PRECAUTIONARY PRINCIPLE AND PRECAUTIONARY APPROACHES

We have discussed how risk assessment and management strategies are developed within particular frameworks, including normative standards and preferences regarding our relation to the natural environment and the preservation/promotion of the environment (see section 5 and 8). In such cases, decision-makers have to make decisions that will include the challenging issue of how to handle uncertainties. We consider that the uncertainties involved with GMVs entail that the precautionary principle needs to be employed (Myhr & Traavik, 2007).

The precautionary principle is a normative principle for making practical decisions under conditions of scientific uncertainty, and provides a general approach to environmental and health protection (EC, 2000). The actual content of the precautionary principle and the practical implications of its implementation in policy issues are however controversial (Foster et al., 2000). Several versions of the principle, ranging from ecocentric to anthropocentric, and from risk-adverse to risk-taking positions have been put forward (Raffensperger & Tickner, 1999). Here we would like to acknowledge the version of the principle that can be found in the 1990 Bergen Declaration on Sustainable Development:

"In order to achieve sustainable development, policies must be based on the precautionary principle. Environmental measures must anticipate, prevent and attack causes of environmental degradation. Where there are threats of serious or irreversible damage, lack of full scientific certainty should not be used as a reason for postponing measures to prevent environmental degradation."

What we appreciate with this version is the commitment to anticipate, prevent and attack causes of environmental degradation. We also acknowledge the passive voice in the following part of this version of the principle; "lack of full scientific certainty should not be used as a reason for postponing measures to prevent environmental degradation", and see that this entails lack of requirement of action. However, this version of the principle connects the importance of taking caution in innovation with the achievement of sustainable development, and is active in nature by stating that in practice its application demands anticipation and prevention of harms. For GMVs the present lack of knowledge and the uncertainties with regard to environmental effects as for example with regard to potential recombinational events, non-target effects and transboundary spread entails the necessity to initiate targeted research, in line with a precautionary approach to decision-making. In this sense, it can be argued that this version of the principle extends from being a formulaic decision-making rule to also include an approach to include scientific activity in decision- making The important elements of what represents a precautionary approach to decision-making are (Wickson et al., 2011):

- The use of scientific research that is broadly framed, interdisciplinary, able to consider indirect causal mechanisms, and contributory to a lifecycle approach to analysis.
- A recognition of the limitations of this scientific knowledge and a willingness to expose the knowledge to critical reflection and 'extended peer review', particularly so as to create transparency about embedded choices and assumptions.
- A commitment to reducing uncertainties and minimising surprises generated by ignorance through vigilance and ongoing research and monitoring.
- A transparent handling of ambiguity and indeterminacy through interdisciplinary approaches and broadly based public participation. This

handling includes the consideration and implementation of a range of socio-technical alternatives and policy options.

10. CONCISE SUMMARY

Currently, there has been no comprehensive environmental risk assessment conducted for any GMVs. Risk assessments have primarily concentrated on the unintended consequences of the vaccination in vaccinated individuals, as well as in individuals of the same species who may get infected by the virus transmitted from vaccinated persons. In this discussion, we have emphasized the importance of establishing a connection between risk assessment and management measures, as well as the necessity of adopting a cautious approach when dealing with genetically modified organisms (GMVs).

Adopting a precautionary approach to genetically modified organisms (GMOs) involves doing research to identify potential risks and develop hypotheses regarding the specific harms and hazards associated with GMOs. This research aims to analyze and manage the potential negative effects of GMOs. An inquiry into possible negative consequences and preventative measures should be started in accordance with research and policy agendas that promote comprehensive and forward-thinking approaches to support preventive efforts. We believe that problem formulation processes are an effective method for increasing the participation of experts and knowledge, as well as for identifying normative difficulties. Section 8 demonstrates that risk management endpoints rely on scientific and normative factors. The way in which nature, harm, and environmental damage are identified can impact the goals of environmental risk management. This, in turn, can be utilized to determine specific targets for protection. The determination of these endpoints is also contingent upon the goals of research in environmental risk assessment. Once the endpoints have been identified, it is possible to formulate precise hypotheses for assessing the level of risk. This allows for focused research to be conducted, using models and collecting data specifically designed to test these risk hypotheses. In section 4.5, we have provided a concise overview of the research conducted within our institution. We emphasize that these methodologies serve as a solid foundation for research endeavors in other nations. Risk-related research may encompass inquiries outlined in section 6.1. Therefore, the evaluation and control of any negative consequences must encompass consideration of the ecological context, as well as talks about standard criteria for determining outcomes. Approaches that consider this factor can ensure that the final stages of decision-making have a comprehensive foundation, encompassing both the representation and involvement of ecological aspects and ethical principles.

11. REFERENCES

Andow, D. A. & Hilbeck, A. (2004). Science-based risk assessment for nontarget effects of transgenic crops. *BioScience*, Vol. 54, pp. 637-649.

Anliker, B., Longhurst S. & Buchholz, C. J. (2010). Environmental risk assessment for medicinal products containing genetically modified organisms. *Bundesgesundheitsbl*, Vol. 53, pp. 52-57.

Ball, L. A. (1987). High-frequency homologues recombination in vaccinica virus DNA. *Journal of Virology*, Vol. 61, pp.1788-1795.

Brocheir, B., Thomas, I., Leveau, P., Pastiret, P. P., Languet, B., Chappuis, G., Desmettre, P., Blancou, J. & Artois, M. (1990). Use of vaccinia-rabies recombinant virus for the oral vaccination of foxes against rabies. *Vaccine*, Vol. 8, pp. 101-104.

CDC (Centre of Disease Control). (1991). *Vaccinia (Smallpox) Vaccine.* Recommendations of the Immunization Practices Advisory Committee. MMWR (Morbidity and mortality weekly report) (Atlanta) 40 (RR-14), pp.1-10.

Chakroudi, A., Chavan, R., Koyzr, N., Waller, E. K., Silvestri, G. & Feinberg, M. B. (2005). Vaccinia virus tropism for primary hematolymphoid cell is determined by restricted expression of a unique virus receptor. *Journal of Virology*, Vol. 79, pp. 10397-10407.

Damaso, C. R. A., Esposito, J. J., Condit, R. C. & Moussatché, N. (2000). An emergent poxvirus from humans and cattle in Rio de Janeiro State: Cantagalo virus may derive from Brazilian smallpox vaccine. *Virology*, Vol. 277, pp. 439-449.

Dobson, A. (1998). *Justice and the environment: Conceptions of environmental sustainability and theories of distributive justice*. Oxford: Oxford University Press.

Drexler, I., Staib, C. & Sutter, G. (2004). Modified vaccinia virus Ankara as antigen delivery system: how can we best use its potential? *Current Opinion in Biotechnology*, Vol. 15, pp. 506-512.

Dumbell, K. & Richardson, M. (1993). Virological investigations of specimens from buffaloes affected by buffalopox in Maharashtra State, India between 1985 and 1987. *Archives of Virology*, Vol. 128, pp. 257-267.

EC (Commission of the European Communities). (2000). *Communication on the Precautionary Principle.* (http://europa.eu.int).

Fenner, F. (1996). Poxviruses. In: *Fields virology* B. N. Fields et al. (Eds), 2673-2702. Lippincott Raven Press.

Fenner, F., Henderon, D. A., Arita, I., Jezek, Z. & Ladnyi, I. D. (1988). *Smallpox and Its Eradication*. Geneva: World Health Organization.

Foster, K. R., Vecchia, P., Repacholi, M. H. (2000). Science and the precautionary principle. *Science*, Vol. 288, pp. 979-981.

Hansen, H., Okeke, M. I., Nilssen, Ø. & Traavik. T. (2004) Recombinant viruses obtained from co-infection in vitro with a live vaccinia-vectored influenza vaccine and a naturally occurring cowpox virus display different plaque phenotypes and loss of the transgene. *Vaccine*, Vol. 23, pp. 499-506.

Hansen, H., Okeke, M. I., Nilssen, Ø. & Traavik, T. (2009). Comparison and phylogenetic analyses of orthopoxviruses isolated from cats and humans in Fennoscandia. *Archives of Virology,* Vol. 154, pp. 1293-1302.

Hansen, H., Okeke, M. I. & Traavik, T. A cowpox virus with an ectromelia virus A-type inclusion protein gene. Manuscript in preparation, 2011

Hel, Z., Nacsa, J. & Tsai, W. P. (2002). Equivalent immunogenicity of the highly attenuated poxvirus based ALUVAC-SIV and NYVAC-SIV vaccine candidates in SIVmac251- infected macaques. *Virology*, Vol. 304, pp. 125-134.

Hilbeck, A., Baumgartner, M., Fried, P. M. & Bigler, F. (1998). Effects of transgenic bacillus thuringiensis corn-fed prey on mortality and development time of immature Chrysoperla carnea (Neuroptera: Chrysopidae). *Environmental Entomology*, Vol. 27, pp. 480-487.

Hill, R. A. (2005). Conceptualizing risk assessment methodology for genetically modified organisms. *Environmental Biosafety Research*, Vol. 4, pp. 67-70.

Jackson, R. J., Ramsay, A. J., Christensen, C. D., Beaton, S., Hall, D. F. & Ramshaw, I. A. (2001). Expression of mouse interleukin-4 by a recombinat ectromelia virus suppresses cytolytic lymphocyte responses and overcomes genetic resistance to mousepox. *Journal of Virology*, Vol. 75, pp. 1205-1210.

Liu, M. A. (2010). Immunologic basis for vaccine vectors. *Immunity*, Vol. 33, pp. 504-515. Losey, J. E., Rayor, L. S. & Carter, M. E. (1999). Transgenic pollen harms monarch larvae. *Nature*, Vol. 399, pp. 214.

Kaaden, O. R., Eichborn, W. & Essbauser, S. (2002). Recent developments in the epidemiology of virus diseases. *Journal of Veterinary Medicine Series*, B 49, pp. 3-6.

Kühler, T. C., Andersson, M., Carlin, G., Johnsson, A. & Åkerblom L. (2009). Do biological medicinal products pose a risk to the environment? A current view on ecopharmacovigilance. *Drug Safety*, Vol. 32, pp. 995-1000.

Louz, D., Bergmans, H. E., Loos, B. P. & Hoeben, R. C. (2005). Cross-species transfer of viruses: implications for the use of viral vectors in biomedical research, gene therapy and as live-virus vaccines. *The Journal of Gene Medicine,* Vol. 7, pp. 1263- 1274.

Mackowiak, M., Maki, J., Motes-Kreimeyer, L., Harbin, T. & van Kampen, K. (1999). Vaccination of wildlife against rabies: Successful use of a vectored vaccine obtained by recombinant technology. *Advances in Veterinary Medicine*, Vol. 41, pp. 571-583.

Martina, B. E., van Doornum, G., Dorrestein, G. M., Niesters, H. G., Stittelaar, K. J., Wolters, M. A., van Bolhuis, H. G., & Osterhaus, A. D. (2006). Cowpoxvirus transmission from rats to monkeys. *Emerging Infectious Diseases*, Vol. 12, pp. 1005- 1007.

Mastrangelo, M. J., Eisenlohr, L. C., Gomelia, L. and Lattime, E. C. (2000). Poxvirus vectors: Orphaned and underappreciated. *The Journal of Clinical Investigations*, Vol. 105, pp. 1031-1034.

McFadden, G. (2005). Poxvirus tropism. *Nature Reviews Microbiology*, Vol. 3, pp. 201-213.

Meyer, H. (2011). Systemic risks of genetically modified crops; the need for new approaches to risk assessment. *Environmental Sciences Europe*, Vol. 23: 7.

Myhr, A. I. & Traavik, T. (2007). Poxvirus-vectored vaccines call for application of the precautionary principle. *Journal of Risk Research*, Vol. 10, pp. 503-525.

Najera, J. L., Gomez, C. E., Domingo-Gil, E., Gherardi, M. M. & Esyeban, M. (2006). Cellular and biochemical differences between two attenuated poxvirus vaccine candidates (MVA and NYVAC) and role of the C7L gene. *Journal of Virology*, Vol. 80, pp. 6033- 6047.

Nelson, K. C. & Banker M. J. (2007). *Problem formulation and options assessment handbook.* St.Paul: University of Minnesota.

Okeke, M. I., Nilssen, O. & Traavik, T. (2006). Modified Vaccinia virus Ankara multiplies in rat IEC-6 cells and limited production of mature virus occurs in other mammalian cell lines. *Journal of General Virology*, Vol. 87, pp. 21-27.

Okeke, M. I., Nilssen, Ø., Moens, U., Tryland, M. & Traavik, T. (2009a). *In vitro* host range, multiplication and virion forms of recombinant viruses obtained from co-infection in vitro with a vaccinia-vectored influenza vaccine and a naturally occurring cowpox virus. *Virology Journal*, Vol. 6, pp. 55.

Okeke, M. I., Olayiwola, A. A., Moens, U., Tryland, M., Traavik, T. & Nilssen, Ø. (2009b). Comparative sequence analysis of A-type inclusion (ATI) and P4c proteins of orthopoxviruses that produce typical and atypical ATI phenotypes. *Virus Genes*, Vol. 39, pp. 200-209.

Pastoret, P. P. & Vanderplasschen, A. (2003). Poxviruses as vaccine vectors. *Comparativ Immunology, Microbiology & Infectious Diseases*, Vol. 26, pp. 343-355.

Raffensperger, C. & Tickner, J. (1999). *Protecting public health and the environment: Implementing the Precautionary Principle*. Washington DC: Island Press.

Raybould, A. (2006). Problem formulation and hypothesis testing for environmental risk assessments of genetically modified crops. *Environmental Biosafety Research*, Vol. 5, pp. 119-125.

Reed, K. D., Melski, J. W., Graham, M. B., Regnery, R. L., Sotir, M. J., Wegner, M. V., Kazmierczak, J. J., Stratman, E. J., Li, Y., Fairley, J. A. , Swain, G. R., Olson, V. A., Sargent, E. K., Kehl, S. C., Frace, M. A., Kline, R., Foldy, S. L., Davis J. P. & Damon, 1. K. (2004). The detection of monkeypox in humans in the Western Hemisphere. *New England Journal of Medicine*, Vol. 350. pp. 342-350.

Regan, T. (1980). Animal rights, human wrongs. *Environmental Ethics*, Vol. 2, pp. 99-104. Rerks-Ngarm, S., Pitisuttithum, P., Nitayaphan, S. Kaewungwal, J., Chiu, J., Paris, R.,

Premsri., N., Namwat, C., de Souza, M., Adams, E. et al.; MOPH-TAVEG Investigators (2009). Vaccination with ALVAC and AIDSVAX to prevent HIV-1 infections in Thailand. *New England Journal of Medicine*, Vol. 361, pp. 2209-2220.

Rupprecht, C. E., Blass, L., Smith, K., Orciari, L. A., Niezgoda, M., Whitfield, S. G., Gibbons, R. V., Guerra, M. & Hanlon, C. A. (2001). Human infection due to recombinant vaccinia-rabies glycoprotein virus. *New England Journal of Medicine*, Vol. 345, pp. 582-586.

Sandvik, T., Tryland, M. Hansen., H., Mehl, R., Moens, U., Olsvik, O. & Traavik, T. (1998). Naturally occurring orthopoxviruses: potential for recombination with vaccine vectors. *Journal of Clinical Microbiology*, Vol. 36, pp. 2542-2547.

Srinivasan, V., Schnitzlein, W. M. & Tripaty, D. N. (2006). Genetic manipulation of two fowlpox virus late transcriptional regulatory elements influences their ability to direct expression of foreign genes. *Virus Research*, Vol. 116, pp. 85-90.

Thiry, E., Muylkens, B., Meurens, F., Gogev, S., Thiry, J., Vanderplasschen, A. & Schynts, F. (2006). Recombination in the alphaherpesvirus bovine herpesvirus 1. *Veterinary Microbiology*, Vol. 113, pp. 171-177.

Traavik T. (1999). *An Orphan in Science: Environmental Risks of Genetically Engineered Vaccines*. Research Report for DN 199-6 (92 pages). ISBN 82-7072-351-7. Directorate for Nature Management, Trondheim, Norway.

Tryland, M., Sandvik, T., Arnemo, J. M., Stuve, G., Olsvik, O. & Traavik, T. (1998). Antibodies against orthopoxviruses in wild carnivores from Fennoscandia. *Journal of Wildlife Diseases*, Vol. 34, pp. 443-450.

Weli, S. C., Okeke, M. I., Tryland, M., Nilssen, O. & Traavik, T. (2004). Characterization of avipoxviruses from wild birds in Norway. *Canadian Journal of Veterinary Research*, Vol. 68, pp. 140-145.

Weli, S. C., Nilssen, O. & Traavik, T. (2005). Avipoxvirus multiplication in a mammalian cell line. *Virus Research*, Vol. 109, pp. 39-49.

Weli, S. C. & Tryland, M. (2011). Avipoxviruses: infection biology and their use as vaccine vectors. *Virology Journal*, Vol. 8, pp. 49.

Westra, L. (1998). Biotechnology and transgenic in agriculture and aquaculture; the perspectives from ecosystem integrity. *Environmental Values*, Vol. 7, pp. 79-96.

Wickson, F., Gillund, F. & Myhr, A.I. (2010). Treating Nanoparticles with Precaution: The Importance of Recognising Qualitative Uncertainty in Scientific Risk Assessment. In: *Nano goes macro, Social Perspectives on Nanoscience and Nanotechnology*. K. Kjølberg & F. Wickson (Eds.), 445-473. Pan Stanford Publishing.

09 THE ETHICAL IMPLICATIONS OF GENETIC ENGINEERING

1. INTRODUCTION: AN OVERVIEW

Throughout the course of the last 12,000 years, people have progressively acquired a more profound comprehension and mastery of the phenomenon of life. Early key developments were agriculture, encompassing both plant and animal husbandry. Medicine has played a significant role in managing life by combating diseases and, more recently, by utilizing technologies to regulate and alter fertility. Physics and chemistry offer the means to study biological processes at a molecular and atomic scale. The genetic research of the late 20th century and 21st century has brought about significant breakthroughs in our comprehension of life and our capacity to govern and modify it for our intended purposes. Modern scientific research is currently focused on emerging biotechnologies.

Theories and technologies such as evolutionary theory, Mendel's laws of inheritance, the discovery of DNA, the mapping of the human genome, genetic engineering (GE) of organisms, gene therapy, synthetic biology, cloning, stem cell therapies, epigenetics, and life extension research offer valuable insights into the nature of life and enable the manipulation of various aspects of life. This knowledge is challenging and reshaping our understanding of life and the essence of being human, as well as our place within the natural hierarchy. Table 1 is a concise compilation of significant achievements in humanity's comprehension and manipulation of life, along with some loosely related perspectives on the world.

Genetic technology possesses the capacity to alter both biological and societal reality. The development and deployment of this technology have significant implications for people, other species, and the planetary environment. These implications can be subject to moral scrutiny, prompting concerns such as: what are the probable social and moral ramifications? Is this considered progress? Do these repercussions have positive or negative outcomes? Does the potential benefit outweigh the potential harm? For whom is this intended? What is the level of fairness in the consequences? What is the level of accessibility or avoidability? How can various social and biophysical settings influence the moral standing of individuals? Another pertinent inquiry is whether the favorable outcomes achieved through the utilization of genetic technology can be achieved by alternative technologies that may have a lower likelihood of undesirable repercussions. These problems illustrate that the field of genetic science, as well as science as a whole, is closely connected to moral reasoning, moral behavior, and technological foresight.

This chapter will examine the societal and ethical concerns related to several components of emerging genetic technologies, with a specific emphasis on genetic engineering. The relationship between these technologies and moral problems will be examined in light of the requirement of sustainability. The discussion will explore and compare the public's moral sentiments towards genetic engineering with those held by the scientific community. Lastly, the chapter will end with an examination of the privileges and obligations of scientists within society.

Table 1. Selected milestones in the understanding and control of life, approximate dates and associated worldviews

Biological Milestones	Approx. date	Associated worldview
Agriculture - plant and animal husbandry	10,000BC -	Animistic/magical/mythological
Ancient medicine (e.g.,Imhotep, Hippocrates, Galen)	2500BC - 180AD	Animistic/magical/mythological /religious/Ptolemaic
Medieval medicine (e.g., Avicenna, Ibn an-Nafis, Paracelsus)	1000-1500AD	Religious/Ptolemaic
Renaissance medicine (e.g., Vesalius to Jenner)	1500-1800	Religious/Copernican/scientific
Darwin's Theory of Evolutionary	1860	
Mendel's Laws of inheritance	1865	
Pasteur invents vaccines	1880	
Morgan' discovery of the chromosomes	1915	
Fleming invents antibiotics	1928	Religious/Copernican/scientific/ modernist
Watson and Crick discover DNA	1953	
Fertility control – oral contraceptive, in vitro fertilisation	1960	Copernican/scientific/modernist
Genetic engineering	1971	
Tissue engineering	1987	Copernican/scientific/post-modernist
Gene therapy	(1970) 1990	
Epigentics	1990	
Animal cloning	1996	
Stem cells therapy	1998	
Life extension	2000	
Synthetic biology	2000	

2. THE RISE OF GENETIC SCIENCE

Darwin's Theory of Evolution completely revised our notions of the nature of life and its origins. Species were no longer created individually by God, nor once 'created', were they fixed and immutable. No longer were we a unique and special creature, made in the image of a miraculous supernatural creator, rather, it became apparent that humans were one of approximately ten millions species inhabiting earth, evolving to fit selection pressures in a similar fashion to the other animals on the planet. Gregor Mendel's laws of inheritance statistically demonstrated that characteristics could be passed on from one generation to the next. The discovery, in the early twentieth century by Thomas Hunt Morgan, of chromosomes and the genetic diversity engendered by sexual reproduction, and the mid century discovery of DNA by Crick and Watson provided a causal mechanism for inheritance and a molecular level mechanism for Darwinian natural selection. Technology has enabled the genomes of organisms to be 'read' and compared, showing that humans share more than 98% of our genes in common with the chimpanzee (Jones, 2006), giving us new insights into our biological and moral position within nature.

The Human Genome Project (HGP) achieved three major goals. First, it sequenced the order of all the 2.9 billion base pairs in the genome. Second, it developed maps locating genes for major section of all our chromosomes. Third, it produced 'linkage maps' enabling inherited traits to be tracked over generations. Francis Collins, the director of the HGP described the results and meaning of the HGP as:

> *It's a history book - a narrative of the journey of our species through time. It's a shop manual, with an incredibly detailed blueprint for building every human cell. And it's a transformative textbook of medicine, with insights that will give health care providers immense new powers to treat, prevent and cure disease.*
>
> (Cited by National Human Genome Research Institute, 2009)

As the relationship between genes and individual health and behaviour becomes more apparent, moral questions arise as to who may have access to an individual's genome, and what will they be able to do with this information. As significant a milestone as it is, sequencing of the genome merely marks a beginning. It will take many decades (and massive computer power) to understand how the approximately 20,000 genes in the human genome interact with one another to produce over two hundred thousand different proteins. A great deal is not currently understood about how the genome works. Long held theories continue to be questioned. For example, contrary to the last hundred years of scientific belief, Mendel's Laws have recently been challenged. Although still believed to be fundamentally correct, it has been claimed that Mendel's Laws are not absolute and exceptions occur (Lolle, Victor, Young, & Pruitt, 2005). Likewise, the idea of inherited acquired characteristics was for a long time considered biological and scientific heresy, but the received scientific dogma has been challenged by the new science of epigenetics (Jablonka & Raz, 2009; Kaati, Bygren, & Edvinsson, 2002; Lumey, 1992). Similarly, a dozen years ago, with perhaps a little scientific arrogance, molecular biologists designated long stretches of organisms' genomes as "junk DNA" claiming that these non-coding segments served no purpose. However, it is logically obvious that human lack of knowledge about the function of elements of nature does not mean they lack function.

Recently, research has shown important roles for junk DNA (Nowacki, et al., 2009), demonstrating the hubris of the junk DNA assumption. Indeed, it now appears that junk DNA plays a vital role in evolution (in particular enabling fast genetic adaptation to changing environmental circumstances) and will be crucial for the refining of GE techniques and for gene therapy (Feng, Naiman, & Cooper, 2009; Vinces, Legendre, Caldara, Hagihara, & Verstrepen, 2009). New evidence also suggests that the rDNA repeats known as "junk DNA" are essential for repairing the DNA damage caused by factors such as UV light (Ide, Miyazaki, Maki, & Kobayashi, 2010). The use of technologies with powerful potential to affect the physical and social worlds, without a good understanding of the science involved, has the potential for unexpected and unforeseen negative social and moral impacts.

3. DEVELOPMENTS IN GENETIC SCIENCE AND MORAL QUESTIONS

3.1 Genetic engineering

The breeding of promising individuals over generations in order to create desirable phenotypic characteristics in plants and animals has long been practiced in horticulture and animal husbandry. This is a relatively slow process with progressive changes made over many generations, not by nature or natural selection, but by human intervention in the evolutionary progress of the species. Racehorses, domestic cattle, show dogs and the staple grains are prime examples of centuries and even millennia of breeding to slowly bend nature to the aesthetic tastes and teleological desires of humans.

In the past forty years, with the discovery of recombinant DNA, humans have gained the power to make changes to an organism's genome in a single generation. Genetic engineering (GE) involves the chemical addition or deletion of a specific gene from an organism's genome in order to bring about a desired change in the organism's phenotype. With this process organisms can have current characteristics enhanced or removed and even entirely new characteristics, not evident in the organism's species, added. Thus, a gene from one species (or a synthetic analogue of the gene), may be spliced into the genome of the same or a different species, or even an organism from a different biological kingdom, giving the new GE organism phenotypic characteristics from the donor species (Small, 2004a).

In this way GE can create organisms with desired attributes much more quickly than traditional breeding (i.e., in a single generation). This amounts to a speeding up of evolution in a direction decided by humans. This also differs from normal evolution and animal and plant husbandry in that the new organism does not co-evolve, in little steps, over time with the other organisms in its environment. Instead an evolutionary leap is engineered within a single generation. Another difference between GE and selective breeding is that organisms can be created that could not possibly have come about naturally, as organisms generally cannot breed with others from different species or kingdoms. Proponents see great hope for the common good of humanity in GE technology, and often claim that the technology will be necessary to produce enough food to feed the future population (Borlaug, 1997; Fedoroff, et al., 2010; Ortiz, 1998).

While GE offers the potential to further bend nature to our desires, critical commentators express concern about negative extrinsic moral impacts. These include the potential to develop dangerous organisms, the impossibility of reversibility once such organisms are loose in the environment, and the potential for negative impacts on humans, other animals and the environment (Antoniou, 1996; Fox, 1999; Ho, 2000; Rifkin, 1998; Straughan, 1995b). Others criticise the technology from an intrinsic moral perspective; creating life is the province of 'God' or nature – human attempts to usurp the role of God or nature are seen as acts of hubris – against God or disrespectful to nature (Appleby, 1999; Straughan, 1995a).

Currently GE is being used to engineer micro-organisms and bacteria (particularly for the production of medicines such as insulin, factor 9 clotting agent, human growth hormone, etc.), plants and animals for food production, production of medicines, industrial production and phytoremediation. An example of a potential GE food animal is the 'eco- friendly' GE pig, engineered to contain bacteria which help pigs remove phosphate from their food, thus stopping it from passing through into the environment, where it causes harm to life in streams and rivers (Golovan, et al., 2001). Pigs have also been genetically engineered to contain human genes, so that their organs will be less susceptible to immune system rejection when used for xenotransplantation (White, Langford, Cozzi, & Young, 1995); the replacement of failing human organs with those from animals.

Advocates of GE claim that the technology is safe. In 2008 GE crops were grown on 300 million acres worldwide. GE crops have been consumed for over 13 years without any incident, it is claimed. Furthermore, production has increased and so have farmers' profits, while pesticide and herbicide use have been reduced and the use of the no-till method of agriculture (helpful for reducing soil erosion) increased (Fedoroff, et al., 2010). However, so far the principal use of GE in food crops has been to engineer insect resistance (bt crops) or to make the crops resistant to a specific herbicide used to eliminate weeds from fields of growing crops – a major beneficiary being the company selling the proprietary herbicide and seeds (one and the same company – Monsanto). On the positive side, the herbicide for which resistance is engineered (Roundup or glyphosate) is relatively environmentally benign and the whole process eliminates the need for further applications of less environmentally benign herbicides.

One possibility presented by GE is the enhancement of nutritional qualities of crops, as for example, the much heralded golden rice. Golden rice has been engineered to contain extra beta-carotene which converts to vitamin A when consumed by humans. Many people in developing countries, where rice is the primary staple, suffer from vitamin A deficiency (Tang, Qin, Dolnikowski, Russell, & Grusak, 2009). Foods with genetically enhanced health qualities or with healthy additives are referred to as functional foods and the science of developing them and studying the relationship between food plant genes, health and the individual human genome is called nutrigenomics. Of course, the societal benefits of functional foods will be dependent upon the public's acceptance of GE food.

Genetic engineering for medical purposes is considerably more acceptable to the general public than GE of food crops (Small, Parminter, & Fisher, 2005). Proponents hope that numerous medicines will be able to be grown in GE plants and/or GE animals and produced more cheaply than through current techniques. A biotech company, SemBioSys, has submitted an Investigational New Drug application for safflower-produced recombinant human insulin to the U.S. FDA (SemBioSys, 2008). Edible vaccines (e.g., potatoes, tomatoes, bananas etc) are being developed for a range of diseases (e.g., cholera, measles, malaria, hepatitis B, type 1 diabetes etc) and are proposed as a logistically simpler resolution of the problem of getting vaccines to those in need in developing countries (Chowdhury & Bagasra, 2007; Levi, 2000). However, it remains unclear how vaccine dosages would be controlled and how accepting the public will be of the conflation of food and medicine. Nonetheless, biotech and pharmaceutical companies have high hopes for rich profit streams from genetically enhanced medical foods and functional foods.

GE animals have been used as 'bioreactors' to produce medicines and industrial products. Cows, sheep and goats have been genetically engineered to produce human proteins in their milk for medical purposes (Wells, 2010). Silk worms have been genetically engineered to produce a form of the human protein collagen which scientists hope to harvest for applications such as artificial skin and wound dressings (Tomita, et al., 2003). The industrial sector also contains many potential applications for GE technology in terms of new methods of producing currently available materials, new materials with desirable qualities, and the production of chemicals and biofuels. For example, spider silk is stronger than steel and as resilient as kevlar, but it is very expensive to produce. Scientist have placed artificial versions of silk genes in various plants (potatoes, tobacco) and animals (goats) and, using this technology, hope to be able to mass produce silk protein for the development of new biodegradable 'super-materials' (Scheller, Guhrs, Grosse, & Conrad, 2001). Gene engineered viruses have even been used to manufacture a 'green battery' which the authors claim is capable of powering an iPod three times as long as current iPod batteries (Lee, et al., 2009).

However, some GE animals seem largely for human entertainment, for example, the first GE pet commercially available in the U.S. was a florescent red zebrafish called a GloFish (GloFish.com, 2010). A company called Lifestyle Pets has marketed a genetically engineered hypoallergenic cat. Given the history of animal breeding for traits of interest to humans, further such applications seem highly probable. Indeed, GE pets suggest mythological sized possibilities; anyone for a pet gryphon? Chimeras are indeed possible using genetic technologies, with a number of research projects having already created them (however, a gryphon might be a bit of a stretch). Of particular concern to some is the possibility of human-animal chimeras (Robert & Baylis, 2003). Robert and Baylis imagine a fusion between a chimp and a human. They suggest that there might be confusion over the status of such a creature and that it might lead to social disorder. However, Savulescu (2003) argues that there might be good reasons to create human chimeras. He suggests medical reasons (e.g., to confer resistance to specific diseases such as AIDS), to delay aging, or to enhance human capabilities.

Clearly, a range of ethical questions are opened by the creation of chimeras. Undoubtedly, there will be a range of different responses to these questions. Another question some ethicists have raised regarding GE animals concerns respect for the *telos* of the animal. Telos refers to the "genetically based drives or instincts that, if frustrated, would result in a significant compromise to the welfare of an animal" (Thompson, 2010, p. 817). Some ethicists claim that it may be morally acceptable to alter an animal's telos using GE so long as it enhances wellbeing (Rollin, 1998), while others have argued that it is not (Fiester, 2008).

3.2 Gene therapy

Humans could be considered as another potential target for GE. Genetic problems in humans can potentially be treated using genetic engineering (GE) and an associated technique called gene therapy. Genetic engineering (GE) is performed on eggs or embryos, while gene therapy is a method that can be employed to modify the genome of germ-line cells (such as eggs or sperm) or the somatic cells of specific organs in a growing or fully grown organism, either in a laboratory setting or within the organism itself. The modifications introduced to somatic cells using gene therapy are not passed down to future generations of the organism

(Gene Therapy Net, 2010). Gene therapy employs a vector, often a disabled virus, to introduce the desired gene into target cells. The field of genetic engineering has achieved successful modifications in the germ-line of marmoset monkeys, as demonstrated by Sasaki et al. in 2009. Modifications made to germ-line cells using gene therapy are heritable by the organism's offspring. By employing these methods, it is possible to treat and eradicate inherited disorders from the reproductive cells, therefore potentially eradicating the disease from an entire species (Gene Therapy Net, 2010).

Although there were some initial obstacles (Neimark, 2009), gene therapy is now showing great potential, as seen by several recent successful experiments. Gene therapy has effectively treated certain medical conditions, including 'bubble boy syndrome', a degenerative visual disorder known as 'Leber congenital amaurosis', and a blood cancer called 'EBV lymphoma' (Neimark, 2009). The efficacy of the gene therapy medication ProSavin in treating a primate model of Parkinson's Disease has been demonstrated, and it is currently undergoing clinical trials with human participants (Jarraya, et al., 2009). There is great optimism regarding the potential of gene therapy, often known as GE, to effectively treat various life-threatening genetic disorders such as cystic fibrosis, hemophilia, Tay-Sachs, muscular dystrophy, multiple sclerosis, and diabetes. Gene therapy can also be employed to stimulate the immune system's response to infectious diseases and malignancies, as well as to deceive the body into regenerating new tissue for the purpose of healing wounds, restoring damaged hearts, and revitalizing arthritic joints (Neimark, 2009).

3.3 Genetic enhancement

The application of gene therapy has successfully conferred color vision to monkeys that are normally color blind, as demonstrated by Mancuso et al. (2009). This finding implies that comparable upgrades to human sensory capabilities may be attainable. Therefore, in addition to therapeutic human genetic engineering, there is also the potential to utilize genetic engineering to augment people. Considering the significant financial and social status associated with sports, as well as the inherent inclination towards hostility in human nature and politics, one naturally thinks of very skilled athletes and warriors. Cattle with doubled muscular mass, such as the Belgium Blue breed, exhibit a 20% increase in muscle mass due to a mutation in the bovine MSTN gene, which codes for myostatin (Grobet, et al., 1997). Researchers have successfully generated genetically engineered mice with a double muscled phenotype by knocking down the myostatin gene (McPherron, Lawler, & Lee, 1997). These mice possess muscles that are 2-3 times more substantial in weight compared to regular mice. Although highly uncommon in humans, there have been documented cases of at least two children who had this genetic mutation spontaneously, which results in remarkable levels of strength and speed (Associated Press, 2007; Schuelke, et al., 2004). This indicates a potential gene target that could be used to enhance the performance of athletes and warriors.

A number of genes in mice have been identified that, when modified through genetic engineering, enhance brain function by promoting the formation of nerve fibers, improving problem-solving abilities, and enhancing memory (Routtenberg, Cantallops, Zaffuto, Serrano, & Namgung, 2000). An estimated 40-80% of the diversity in human intellect is thought to be due to hereditary causes. A comprehensive analysis of the entire genome, which included 634 pairs of siblings, revealed the presence of two specific areas on chromosomes 2 and 6 that are responsible for the differences in IQ levels, as reported by Posthuma et al. in 2005. These genes provide insights on enhancing human intellectual capacity or mitigating the impact of disorders including Parkinson's, Alzheimer's, autism, and dyslexia. Therapeutic gene editing (GE), which aims to cure diseases, is a topic of significant moral controversy. However, the use of gene editing for enhancement purposes raises even more moral concerns for the majority of individuals. According to Small (2006), almost 80% of the general people in New Zealand, as well as a similar amount of scientists in the country, expressed significant disagreement or disagreement with the idea of genetically modifying individuals to improve their talents. Nevertheless, certain ethicists contend that if we possess the capability to augment human beings and as long as this augmentation contributes to the overall welfare of humanity, it is our ethical duty to engage in such enhancement (Savulescu, 2005).

Savulescu (2005) references gene therapy research that transformed lethargic primates into industrious individuals and promiscuous rodents into monogamous ones. Undoubtedly, there might be disagreement on the definition of an enhancement, and it is often difficult to distinguish between therapy and enhancement. Another ethical and societal dilemma with augmentation is the issue of distribution and access. Is it limited to only those who have the financial means? Does it serve as another method to enhance the positional advantage of the affluent over the impoverished? To what degree will individuals opt to modify their genetic makeup or the genetic makeup of their offspring? Will those who are affluent and have access to genetic enhancement technologies evolve into a distinct human subspecies compared to those who are economically disadvantaged, hence exacerbating genetic inequality? The act of enhancement gives rise to social and moral concerns regarding principles of justice and fairness. A more radical yet less imminent danger is if humans may undergo genetic enhancement and diverge into numerous distinct species that are incapable of interbreeding. What conflicts could arise between different species that have evolved from homo sapiens? Enhancement raises complex social and moral inquiries.

Considering the excessive pride in disregarding junk DNA and our current little comprehension of gene interactions that lead to the production of various proteins, human contemplation of genetic upgrades may likewise be misguided. The endeavors we do may indeed inflict more harm upon our species than benefit. Only time will reveal the truth. Undoubtedly, the acquisition of future knowledge has the potential to eradicate the perilous state of ignorance. Nevertheless, achieving a comprehensive understanding of the human genome may need a considerable amount of time. This is due to the immense number of possible combinations and interactions within the genome, which consists of around 2.9 billion 'letters'.

3.4 Eugenics or conscious evolution?

Another issue that concerns some people regarding genetic therapy and enhancement is the historically sinister shadow of eugenics. Eugenics "requires that natural selection be replaced by intentional human control" (Hansen, Janz, & Sobsey, 2008, p. S105). Somewhat perversely, human medical control is currently unintentionally replacing aspects of natural selection. Advanced medical interventions enable individuals who would have died before procreation to contribute to the human gene pool. Our moral concern for the health and wellbeing of individual members of our species, unfit under conditions of natural selection, unintentionally or incidentally negatively impacts the overall fitness of the human species.

One thing is clear, new medical and genetic technologies gives us the power to consciously, (or unconsciously), manipulate the evolution of our species. Should we do it just because we can? Or is it too much of an intellectual, moral and ontological minefield? What might be the *unintentional effects* of such tinkering with human nature? If we do it, on what basis should it be done? Should individuals be free to design their own offspring and, if so, what degree of freedom should be allowed? Would it, for example, be acceptable for deaf parents to have their children genetically engineered to be deaf? What are the rights of the genetically engineered offspring in this regard? Who makes the decision and by what authority?

Another possibility is that conscious human evolution might be 'centrally controlled' so that a uniformity remains about our species rather than allowing us the freedom to diverge into a range of new and different species which may have quite disparate capabilities and perhaps even needs. How much 'central control' would be acceptable? Will there come a time when specific genetic enhancements are enforced for the good of the species? Designer babies, conscious evolution and eugenics are complex moral topics raised by current scientific and technological development in the field of genetic science.

3.5 Synthetic biology

"Synthetic biology aims to design and build new biological parts and systems or to modify existing ones to carry out novel tasks" (Parliamentary Office of Science and Technology, 2008, p. 1). Synthetic biology offers prospects for novel methods to produce food, drugs, chemicals or energy, environmental biosensors, and new therapeutic techniques. Engineering principles may be used to build standardised interchangeable segments of DNA for use as biological building blocks to make new, or alter existing, organisms. DNA sequences (potentially even whole genomes) can be designed on computers and manufactured in chemical laboratories. It is possible to construct the genome of a medium- size virus in about three weeks (Parliamentary Office of Science and Technology, 2008).

The potential for the *malevolent* use of synthetic biology is clear. In 2002, researchers from the University of New York constructed the poliovirus by following a recipe downloaded from the Internet using synthetic gene sequences sourced from a mail-order supply firm. Their purpose was to show how easily terrorists could create deadly biological weapons (Cello, Paul, & Wimmer, 2002). It is conceivable, that in the future, entirely new forms of life might be created in the laboratory using the techniques of synthetic biology (Chopra & Kamma, 2006). Synthetic biology attracts the 'do it yourself' brigade', including groups with such exciting names as: OpenWetWare (OpenWetWare, 2009), DIYbio (DIYbio, 2010), Biopunk (Biopunk.org, n.d.), and Biohack (Bishop, 2008). Synthetic biologists in California are about to launch an open source biological production facility called BIOFAB (International Open Facility Advancing Biotechnology). BIOFAB "aims to boost the ease of bioengineering with "biological parts" that are shared resources, standardized and reliable enough that they can be switched in and out of a genome like electronic parts in a radio" (Katsnelson, 2010).

J. Craig Venter, who led the private research project to decipher the human genome in competition with the publically funded project, has been working on creating the world's first synthetic organism since 1995. He envisaged creating organisms that have the ability to manufacture biofuels and other useful compounds. He claims a "new design phase of biology" is about to begin (quoted by Grant, 2008). On May 21, 2010, Venter's team published details of the creation of a synthetic genome that began replicating and producing proteins (Gibson, et al., 2010). Although many useful organisms may be created, the possibility of dangerous ones being created either accidentally or deliberately, is also very real. The Oxford ethicist, Savulescu, reflecting on the achievement of Venter's team said:

> *Venter is creaking open the most profound door in humanity's history, potentially peeking into its destiny. He is going towards the role of a god: creating artificial life that could never have existed naturally. The potential is in the far future, but real and significant: dealing with pollution, new energy sources, new forms of communication. But the risks are unparalleled. (quoted by Henderson, 2010)*

Savulescu continued:

> *We need new standards of safety evaluation for this kind of radical research and protections from military or terrorist misuse and abuse. These could be used in the future to make the most powerful bioweapons imaginable. The challenge is to eat the fruit without the worm. (quoted by BBC News, 2010)*

Biosecurity, biosafety, intellectual property, stakeholder engagement and involvement, unforeseen harmful consequences, human malevolent use of the technology and technological governance are among the serious ethical issues facing synthetic biology.

3.6 Life extension

Impressive work is currently being conducted in the area of life extension (de Grey, 2005; Finkel, 2003). The average life span in the developed countries has been steadily increasing and current projections are that someone born today could potentially live much longer than the current maximum natural full term of around 120 years. It has long been known that calorie restriction (CR) diets can improve the health (in particular reduction of cancers and increased immunity) and extend the life of most mammals by up to 40% (Weindruch, Walford, Fligiel, & Guthrie, 1986). Recently a molecular level epigenetic mechanism for the effects of CR has been proposed (Li, Liu, & Tollefsbol, 2009), glucose restriction produced increased expression of hTERT (human telomerase reverse transcriptase). This provides targets for drugs for life extension.

Telomerase is an enzyme that is responsible for the formation of the telomere DNA sequence. This sequence forms a cap on the ends of chromosomes. Telomeres are responsible for maintaining genomic stability and regulating cellular division. As somatic cells divide the telomere sequences get shorter and shorter, limiting cells to a fixed number of divisions (Harley, Futcher, & Greider, 1990). Cellular senescence and eventually death occur when telomeres reach a critical value. Heritability of telomeres is strong with studies estimating it at 40-80% (Codd, et al., 2010). This is thought to be an important component of aging at a cellular level. Cancer cells preserve their telomeres no matter how many times they divide (i.e., they are immortal). Cancer cells have increased telomerase activity, thus suggesting a possible mechanism for increasing the longevity of normal cells.

A number of genes have been shown to be related to the aging process, the health of individuals throughout their lifespan, and the overall length of lifespan. By manipulating specific genes in various organisms scientist have been able to greatly increase their lifespan. Adding an additional sir-2.1 gene to the genome of *C. elegans* (a nematode worm), led to a 50% increase in lifespan (Tissenbaum & Guarente, 2001). Decreasing the activity of the daf-2 gene led to a doubling of the lifespan of *C. elegans* (Kenyon, Chang, Gensch, Rudner, & Tabtiang, 1993). Single genes which have significant lifespan effects in fruit flies have also been discovered. Mutants fruit flies, with reduced activity (down regulation) of the mth gene (the methuselah gene), have a 35% increase in average lifespan and increased resistance to stressors such as starvation and heat (Lin, Seroude, & Benzer, 1998). Reduced expression of the Indy (I'm not dead yet) gene doubles the fly's lifespan with no noticeable negative side effects (Rogina, Reenan, Nilsen, & Helfand, 2000).

Helfand & Inouye (2003, p. 276) claim that "There is great conservation between different organisms suggesting that what is learned in one model system will be true for others." Research with healthy centenarians (and their very old - 91+ siblings), using genetic linkage analysis, found a region on chromosome 4, that contains between 100 and 500 genes, associated with extra long healthy lives (Puca, et al., 2001). In the past few years a handful of these genes have been identified as important to the aging process (Rucz, 2008). Genetic engineering, gene therapies and stem cell therapies are expected to play a significant role in life extension in terms of rejuvenation of aging or diseased cells and organs and elimination of genetic disorders. Significantly increasing human life spans raises moral problems of justice and fairness in a world still undergoing exponential human population growth and facing depletion of the resources necessary for survival.

3.7 Patenting genes

An important ethical, legal and economic issue with implications for the practice of science, the dissemination of scientific knowledge, and medical practice, is the question of patenting genes and gene sequences (Schacht, 2006). According to an article in Nature Biotech, in the year 2000, over 6000 gene patents had been granted with over 1000 of these specifically related to human genes and more than 20,000 gene patents were pending at that time (Grisham, 2000). While commercial biotech companies are strongly in favour of being able to patent genes, claiming that it is necessary in order to fund research and innovation (Schacht, 2006), many scientists and scientific bodies are opposed to it, claiming that patenting slows the progress of science by restricting open access and use of the genes in further research.

Physicians and patients also claim that patenting of genes restricts patients' access to medical care (Andrews, 2002; Leonard, 2002; Wadman, 2010). Others are strongly opposed to patenting genes on the principle that 'no one should be able to patent life' and that genes are products of nature and are merely discovered by humans and are, therefore, not patentable, as inventions are (Ho, 2000; Wadman, 2010). Some object that the 'patenting of life' "turns organisms, including human parts, into saleable commodities" (Ho, 2000, p. 30).

The topic of gene patenting remains controversial and undecided. It will continue to pose legal, social, moral and economic issues for some time to come.

3.8 Genetic engineering and technological convergence

There are a range of powerful new technologies under development that in convergence will enhance both the power and availability of genetic engineering and other cell based technologies. Two of particular importance are nanotechnology and information technology. Computer processor power has undergone exponential growth since 1965 when Gordon Moore first stated the principle that has become known as Moore's Law; the number of transistors on a chip doubles every two years. This law is still holding true. Also in 1965, I. J. Good proposed a concept that has become known as 'the technological singularity'. This is the point in time when computers become more intelligent than humans, and hence better able than humans to develop their own intellectual capacity, eventually giving rise to super intelligent machines, fostering a near infinite rate of knowledge innovation (Good, 1965). Some futurists consider the technological singularity will occur in the not so distant future. Vinge (1993) predicts 2030 while Kurzweil (2005) predicts 2045. Computers with this level of processing power may enable an understanding and prediction of the interaction of genes leading to a vastly enhanced understanding of genetics and increasing the potential use and power of genetic engineering and related technologies.

Similarly, in the rapidly advancing area of nanotechnology, new tools are being developed that will enhance our understanding and ability to control molecular and cellular behaviour. Nanotechnology will play an important role in the field of tissue engineering and organ regeneration through the generation of biomaterial scaffolding to maintain and regulate cell behaviour such as apoptosis, proliferation and differentiation (Chen, Mrksich, Huang, Whitesides, & Ingber, 1997; McBeath, Pirone, Nelson, Bhadriraju, & Chen, 2004). Small and Jollands (2006) argued that convergence will make these technologies, including genetic engineering, more accessible to an ever widening population. Such accessibility will enhance the ability of the 'do it yourself' brigade to genetically engineer and 'biohack' life. It will also make it increasingly easier for terrorists and others with malevolent intent to use GE technology to cause great harm.

3.9 Genetic engineering and some potential planetary threats

Sustainability has become an ecological, social and moral imperative of our times (Lubchenco, 1998; The World Commission on Environment and Development, 1987). There is currently a range of potential threats to the sustainability of human and other life on earth (Bostrom, 2002; Brown, 2008; Rees, 2003). These threats are related to the increasing number of humans inhabiting Earth and our technological power to effect and impact nature (Small & Jollands, 2006). Main threats include resource depletion, deforestation, land and soil depletion, species extinction, pollution, peak oil, and climate change (Brown, 2008; Rees, 2003). Human behaviour, in particular fossil fuel pollution, is causing climate change, global warming, ice melt, rising sea levels and ocean acidification (Intergovernmental Panel on Climate Change, 2007). These changes will make it increasingly difficult to produce enough food to feed the increasing human population (Fedoroff, et al., 2010; The Royal Society, 2009). Despite the negative impacts of fossil fuel use, without oil it is estimated that the planet can only feed 2-3 billion people (Youngquist, 1999).

The optimistic potential of GE crop production is the development of conventional breeding varieties of crops that are resistant to heat, salinity, drought, disease, pests and toxic heavy metals (The Royal Society, 2009). GE may also be able to increase the effectiveness of plants to extract necessary minerals for growth (such as nitrogen and phosphorus) enabling them to be grown on poorer quality soils using less fertiliser (West, 2010). The Royal Society claimed that the world needs genetically engineered crops to minimise environmental impacts and increase food yields to meet the challenge of feeding another 2.3 billion people by 2050 (The Royal Society, 2009). Another recent paper in the journal *Science* made similar claims (Fedoroff, et al., 2010). Noting that many important crops have sharp declines in production (20-30%) once the temperature exceeds 30 degrees Celsius, they claimed global warming will drastically reduce production in tropical and sub-tropical zones by the mid 21st Century, causing food scarcity.

Fedoroff et al. (2010) claimed that a radical rethink of agriculture is required. They argued for the development of crops which are heat, salt and drought tolerant and which do not require the current high levels of chemicals and fertilisers. Genetic engineering is their preferred radical strategy. They stated that GE will be necessary in order to produce crops at current production levels, let alone the production levels required in 2050 (Fedoroff, et al., 2010). Given the likelihood of significant future temperature increases (Anderson & Bows, 2008; J. Hansen, et al., 2008; Intergovernmental Panel on Climate Change, 2007), water shortages (Brown, 2008; Gleick, 2003; Vorosmarty, et al., 2004), salinity and degraded soil conditions in many of the world's major growing regions (Lal, 2007; Pimentel & Sparks, 2000), genetic engineering of crops is certainly an appropriate research strategy.

However, Fedoroff, et al's. (2010) arguments may be queried on several accounts. First, they made their claims about the *necessity* of GE without considering possible viable alternatives. Second, they ignored the impact of fossil fuel depletion on food production (Pimentel & Pimentel, 1996; Pimentel, Pimentel, & Karpenstein-Machan, 1999; Youngquist, 1999). This is a factor which would further add to their pessimism about actually being able to achieve the goal of feeding the planet's population in 2050. Currently available GE crops are heavily dependent on oil products for their success. Third, they omitted to point out that, so far, no commercial GE plant crops have the particular attributes that they claimed will be beneficial (however, research is being conducted to this end and GE crops may well have these attributes in the future). Fourth, they made their case sound more favourable by failing to note that the beneficial attributes they proposed are attributes that increase the fitness of plants and, therefore, increase the probability that horizontal gene transfer to weedy relatives will also make problem weeds fitter.

Further, it is clear that, even if GE technology did contribute to enable production of adequate supplies of food to feed the population of 2050, there is no guarantee that economic and political action would not thwart these good intentions. Indeed, that is the very reason that, despite the currently adequate food supply, over a billion people still suffer from hunger and thousands starve to death each day (May, 1999). Although it is no reason to stop this line of GE research, there is abundant evidence that having a technological solution to a problem is no guarantee that humans will implement it. If people have no land to produce food themselves, or no money to buy it, they will likely starve, irrespective of any biotechnological bounty. No matter how useful the technology potentially is, the crucial issue is how people will choose to use it. Currently, economics trumps morals.

This well known political fact is probably a major reason why approximately 60% of both the New Zealand public and New Zealand scientists are either sceptical or uncertain about the claim posited by some scientists' that 'GE will help solve the world's food problems'. Regarding this statement a survey of the New Zealand public (n=860) found 9% strongly agreed, 29% agreed, 22% neither agreed nor disagreed, 21% disagreed, 11% strongly disagreed and 7% did not know (Small, 2009). This survey was conducted in May 2005 at the same time as a survey of NZ scientists (n=733) which also asked the same question. A very similar level of scepticism was also evident amongst NZ scientists; only 6% strongly agreed, 33% agreed, 24% neither agreed nor disagreed, 21% disagreed, 10% disagreed strongly and 6% did not know (Small & Botha, 2006). Nonetheless, GE has the potential to help address several of the planetary threats noted above (including food production). Synthetic genetic organisms could be used to produce fuel to help replace petroleum with bioethanol, butanol and other such products (Sticklen, 2008). Genetically engineered plants could also be used for phytoremediation of polluted soils (Cherian & Oliveira, 2005). Some have suggested that genetically engineered trees may be able to help sequester greater amounts of carbon from the atmosphere – helping to reduce a major cause of climate change and global warming (Jansson, Wullschleger, Kalluri, & Tuskan, 2010). While it would clearly be better to not create these problems in the first place, rather than relying on a 'technological fix' being developed in the future, given the current problems and their magnitude and urgency, it would be foolish not to continue with GE research that may help us address these issues.

Table 2. (continues on next page) Some current applications of GE and some potential benefits and harms (note: these are meant to be illustrative rather than exhaustive)

Area of application	Tech/product	Potential benefits	Potential harms
Food		Increased production Less pesticides and herbicides Less fertilisers No till Agriculture (soil conservation) Environmentally resilient crops Crops with enhanced nutritional value	**Extrinsic** Resistant pests (evolve) Super weeds (outcrossing and escape) Irreversibility Single generation evolutionary impacts Conflation of food and medicine Lack of knowledge Accidental or incidental negative impacts on humans, animals, and environment **Intrinsic and emotional** Playing God Disrespectful to nature Morally/spiritually wrong Emotional yuk factor
	GE animals	Increased production Healthier meat More resilient animals (less medicines, increased environmental tolerance)	**Extrinsic** Reduced species diversity Single generation evolutionary impacts Conflation of food and medicine Lack of knowledge Accidental or incidental negative impacts on humans, animals, and environment **Intrinsic and emotional** Playing God Disrespectful to nature Disrespectful to animal telos Morally/spiritually wrong Emotional yuk factor

Area of application	Tech/product	Potential benefits	Potential harms
Medicine	Therapy Medicines derived from GE micro-organisms, plants, animals. Gene therapy Stems cells Tissue engineering	New medicines for curing illness, and injury Organ replacement Elimination of some diseases Increased life expectancy	**Extrinsic** Outcrossing (and/or escape) Irreversibility Lack of knowledge Accidental or incidental negative impacts on humans, animals, and environment Zoonotic disease (e.g. from xenotransplantations) Overpopulation Malevolent actions (GE virus developed as weapon) **Intrinsic and emotional** Same as for GE food animals

3.10 Genetic engineering and the potential for harm

Despite having a large range of intentionally positive applications, it is also clear that GE and gene therapy could create harm in the world through *accidental* unforeseen and unintended side effects, *incidental* effects – known side effects associated with positive intentional effects, or from *malevolent* intent (Small & Jollands, 2006). For example, scientists experimenting with the mouse pox virus (a mouse analogue of smallpox in humans) accidentally succeeded in making the virus much more virulent and deadly –killing even mice vaccinated against the disease (Jackson, et al., 2001).

The potential to alter human smallpox or to combine genes from different diseases to create super diseases for the purpose of warfare or terror is clear and very real (*malevolent* intent). Indeed, it is reported the former Soviet Union succeeded in using recombinant DNA techniques to combine features of smallpox and Ebola (Katz, 2001). As Sir Martin Rees noted regarding biological and chemical weapons "A few adherents of a death-seeking cult, or even a single embittered individual, could unleash an attack" (Rees, 2003, pp. 48- 49). The possibility of creating lethal pathogens that target specific groups in society based on gene markers specific to that group also exists (Katz, 2001). Given the current racial and religious fanaticism demonstrated by some groups in society this is an extremely alarming possibility.

Table 2. (continues on next page) Some current applications of GE and some potential benefits and harms (note: these are meant to be illustrative rather than exhaustive)

Area of application	Tech/product	Potential benefits	Potential harms
Medicine	Therapy Medicines derived from GE micro- organisms, plants, animals. Gene therapy Stems cells Tissue engineering	New medicines for curing illness, and injury Organ replacement Elimination of some diseases Increased life expectancy	**Extrinsic** Outcrossing (and/or escape) Irreversibility Lack of knowledge Accidental or incidental negative impacts on humans, animals, and environment Zoonotic disease (e.g. from xenotransplantations) Overpopulation Malevolent actions (GE virus developed as weapon) **Intrinsic and emotional** Same as for GE food animals
	Enhancement Somatic and germ- line therapy (enhanced physical, social mental capabilities, life extension) Chimeras	Enhanced human (and non-human) capabilities Increased human resilience Disease elimination Promotion of human wellbeing Much increased life expectancy	**Extrinsic** Super warriors Eugenics Lack of knowledge Accidental or incidental negative impacts on humans, Fairness/justice Autonomy Species divergence Potential enforcement Overpopulation **Intrinsic and emotional** Playing God Disrespect to nature Disrespectful to human telos Morally/spiritually wrong Emotional yuk factor

Area of application	Tech/product	Potential benefits	Potential harms
Industry	GE Pets GE plants, animals, micro-organisms for manufacturing Chemicals and materials Energy and fuels Synthetic biology Bioinformatics Biomimetics	Pets with reduce allergic potential New and existing chemicals and materials with a range of new or enhanced properties Mitigation of peak oil New production methods and processes	**Extrinsic** Outcrossing or escape Dangerous organisms Irreversibility Competition between food and fuel for land and water Lack of knowledge Accidental or incidental negative impacts on humans, animals, and environment Malevolent bioweapons **Intrinsic and emotional** Same as for GE food crops
Area of application	Tech/product	Potential benefits	Potential harms
Ecosystem services	Phytoremediation Trees with enhanced carbon absorption	Remediation of pollution and toxic sites Climate change mitigation	**Extrinsic** Outcrossing or escape Irreversibility Lack of knowledge Unforeseen or incidental negative impacts on humans, animals, and environment Accidental or incidental negative impacts on humans, animals, and environment Malevolent application as bioweapons **Intrinsic and emotional** Playing God Disrespectful to nature Morally/spiritually wrong

Another a worrying trend is 'do it yourself' (DIY) bioengineering. Reportedly, home hobbyists (often without training in the field) are conducting GE and synthetic biology experiments from information and products found on the Internet and equipment constructed in home labs. Such experiments are conducted without regulation or control. These biohackers, as they call themselves, claim that the future Bill Gates of biotech could be developing a vaccine for cancer in their garage (Wohlsen, 2008). A worrying scenario is that synthetic organisms escaping from uncontrolled home GE labs could cause outbreaks of dangerous diseases and serious environmental damage. Table 2 presents some selected applications of GE technologies along with some potential benefits and some potential harms.

3.11 Challenges to the development and use of GE technologies and products

3.11.1 New Zealand public perceptions of GE and GE products

In the preceding sections, I have considered some potential positive applications of genetic engineering, potential problems with some of these applications, and some potential negative applications of the technology. Of course, the examples considered are only a fraction of the potential applications, the potential benefits and the potential harms. However, from these examples, it is clear that GE technology may result in both benefit and harm. Next, I consider the challenges facing the adoption of GE technologies with applications that ostensibly are 'for good'. Even for technologies with no harmful uses or side effects, in order for them to provide benefit there must be significant degrees of acceptance by society. People have to want and use the products that provide benefit. If people do not want the products then their development may be a very expensive waste of time and resources, which could be better spent elsewhere.

A recent study conducted with a random sample the of New Zealand public (*N* = 1008) showed complex public attitudes to GE. Small (2009) asked respondents whether they had ever consumed GE food products and whether they had ever used GE medicines. The results are presented in Table 3.

Table 3. Percent of respondents consuming GE food and using GE medicine (*N* = 1008) Most notable is that nearly 70% of respondents did not know whether they had consumed GE food or used GE medicines. Respondents were next asked their level of support for GE food and GE medicine. Results are presented in Table 4.

Technology	Yes (%)	No (%)	Don't know (%)
Consumed GE food	20.2	12.1	67.7
Used GE medicine	6.4	23.8	69.8

Table 4. Percent of respondents supporting GE food and GE medicine (*N* = 1008)

Level of support	GE food (%)	GE medicine %
Totally support it	7.3	26.2
Support it in some circumstance	61.2	55.0
Totally oppose it	20.8	9.4
Don't know	10.6	9.4

GE medicine is supported more strongly than GE food, with more respondents finding it acceptable than unacceptable. However, more respondents found GE food unacceptable than acceptable. Of note is the high percent of respondents who are prepared to support both GE food and GE medicine "under some circumstances" This indicates that although they have concerns about GE food and medicine they perceive that under some circumstances it may be justified. Also supporting this contention, 60% of the sample agreed that GE applications need to be considered on a case-by-case basis rather than totally supporting or totally opposing all applications of the technology. For 67% of the respondents their primary concern about GE was potential negative consequences for humans, animals, and the environment, only 6.5% considered their primary concern was that GE is in principle unethical, disrespects nature or is against God, while 26.1% did not have any major concerns about GE technology. Small (2009) also reported that respondents were neutral regarding "GE helping to cure disease" but sceptical about other proposed benefits such as "helping to feed the world", being "environmentally friendly" and being "of more benefit than harm".

3.11.2 The practice of genetic science

The technoethicist Mario Bunge held the position that although pure science "is intrinsically valuable, technology can be valuable, worthless, or evil, according to the ends it is made to serve. Consequently technology must be subjected to moral and social controls" (Bunge, 1977, p. 106). This is a teleological ethical argument. People may reject new technologies, including genetic engineering, for two main ethical reasons (Appleby, 1999; Small, et al., 2005; Straughan, 1995a, 1995b). They may reject it for deontological (intrinsic) reasons (i.e., they consider the technology bad in itself, or an affront to God or nature) or for teleological (extrinsic) reasons (i.e., they believe the consequences of the use of the technology will be bad e.g., negative impacts on humans, animals or the environment). However, generally people tend to use both deontological and teleological reasoning when considering moral issues. When these two ethical perspectives are at odds with one another the individual may weigh the degree of perceived good and bad to reach a 'balanced' conclusion. Understanding public attitudes and moral perspectives may have important social consequences for the development and promulgation of powerful new technologies. The theoretical assumption of post-normal science (Funtowicz & Ravetz, 1993) and the methods of empirical ethics (Borry, Schotsmans, & Dierickx, 2007) may help provide understandings of public perceptions appropriate to policy development regarding genetic engineering.

However, even applications of GE that are developed for good purposes may produce harm. Small and Jollands (2006) identified two primary ways in which technologies developed for good purposes may lead to harm. The first is through *accidental* causes, that is, unforeseen and unintended consequences, that are premised by a lack of knowledge or a lack of precaution regarding the use and potential consequences of the technology. The second is through *incidental* causes or coincidental effects of the technology. In this case there is awareness that the technology has potential harmful side effects, but nonetheless, we choose to use it for the perceived benefits (e.g., for many years leaded petrol was one such technological product). This highlights the importance of two related approaches to harm reduction associated with the development of powerful modern technologies. The first is foresight and the second is the precautionary principle.

Jonas (1985) argued forcefully that, in order to respond to the techno-crises threatening nature and humans, a new ethic of scientific responsibility was necessary. Jonas' (1985) insight, and a contention of the current work, is that modern technology changes the landscape of ethics. In the past, without modern technologies, the effects of human actions were proximally located in time and space and limited by their degree of control of energy and matter, as were their consequent impacts on human life and other conscious beings. Under such circumstances our moral responsibility need extend only as far as the effects of our actions. Now that humans possess sufficient power over nature as to affect the global conditions for human and non-human life, the far-off future and even the physical destiny of the planet, the framework of former ethics is no longer valid.

Jonas (1985, p. x) claimed that to discharge this new moral responsibility "we must lengthen our *foresight* with *a scientific futurology*", by which he meant using scientific knowledge about cause and effect relationships to extrapolate and attempt to predict future states associated with technological development. Bradshaw and Bekoff (2001) and Small and Jollands (2006) made the further claim that any such endeavour must take account of the psychology of human nature, as technological impacts on nature and humans arise from the application of technology *by humans*. Jonas acknowledged that, due to insufficiency of our predictive knowledge, foresight will always be uncertain and incomplete. Therefore, given the magnitude of what is at stake, he proposed that we should adopt a "pragmatic rule to give the prophesy of doom priority over the prophesy of bliss" (p. x). Foresight is a logical prerequisite for teleological ethical reasoning. Both Bunge and Jonas, like more recent technoethicists (e.g., Luppicini, 2008; Moor, 2005), claimed that, because of their role in developing technology, scientists have an extra responsibility (and perhaps accountability) regarding the social and moral impacts of technology.

The precautionary principle is another proposed response to the risks and potential negative impacts of technology on nature and humans. It may be the only possible effective response to existential risk (Bostrom, 2002). One widely accepted formulation of the precautionary principle states: "when an activity raises threats of harm to human health or the environment, precautionary measures should be taken even if some cause and effect relationships are not fully established scientifically" (Raffensperger and Tickner, 1999, cited in Kriebal, et al., 2001, p. 872). The precautionary principle is proposed as a guideline for decision-making. It has four main components: "taking preventative action in the face of uncertainty; shifting the burden of proof to the proponents of an activity; exploring a wide range of alternatives to possible harmful actions; and increasing public participation in decision making" (Kriebal, et al., 2001, p. 872). The term 'precautionary principle' is an English translation of the German word *Vorsorgeprinzip* which might also be translated as 'foresight principle', a translation which focuses on anticipatory action rather than the slightly more negative reactive focus of the English word 'precaution'.

The precautionary principle is, therefore, a teleological ethical approach. As noted earlier, genetic knowledge is still undergoing rapid discovery and revision. Suzuki (2001) pointed out that, considered today, the leading ideas of genetics in 1961 seem naïve and absurd. He made the further plausible claim that many of our current scientific ideas will ultimately be wrong, irrelevant or unimportant. From this, he concluded that the wise approach to technological development is to take a precautionary approach rather than rush to apply the latest ideas. However, such an approach does not necessarily come easily to scientists engaged in leading edge research. Scientists tend to have their focus on the good that could be accomplished from their discoveries (Small, 2011) Indeed, top experts tend to be overly optimistic about the technologies they are developing, often underestimating the risk associated with their field of work and underestimating realisation and diffusion problems (National Science Board, 1977; Rollin, 1996; Ticky, 2004). Less specialised experts were found to be less optimistic than top experts (Ticky, 2004).

Small and Jollands (2006) also identified a third way in which technology may cause harm to nature and humans. This cause of technological harm they called *malevolent.* In this case, the technology is developed specifically to cause harm to humans or the environment. Thus, the 'telos' of such technologies is evil. Nuclear weapons and other weapons of mass destruction are examples of technologies whose telos is evil. Although gene technologies hold fantastic potential for public good and the benefit of humanity, they also hold fantastic potential for harm. In a worst case scenario, through accidental, incidental or malevolent causes, the result could be the demise of life on Earth (Bostrom, 2002; Joy, 2000; Rees, 2003). Much depends upon how and what humans choose to use powerful technologies for. As these technologies become increasingly accessible and available for use by all echelons of society, what humans choose to use these technologies for will be determined by the full range and extent of human nature and human behaviour, from the altruistic to the malevolent. In keeping with the advice of Jonas (1985) to give priority to the prophesy of doom over the prophesy of bliss, a problem that will have to be solved, sooner rather than later, is how to stop the potential malevolent applications of genetic technologies (some of which pose existential threats).

4. RIGHTS AND RESPONSIBILITIES OF SCIENCE IN SOCIETY

In the case of genetic engineering it is reasonably clear that some applications of the technology are almost universally acceptable to society (i.e., production of GE insulin raises almost no resistance), and medical applications are generally viewed with reasonable favour. Others, such as GE food raise greater deontological and teleological moral concern for society, but may not be rejected outright and are likely to be considered acceptable under some circumstances (Small, 2005; Small, et al., 2005). Human technology and actions are changing the ecological balance of the planet with some potential catastrophic outcomes (Lubchenco, 1998). Perhaps, if as suggested by Federoff (2010) and The Royal Society (2009), GE technology will be able to help us address these issues, society will consider the circumstances appropriate. However, there are other GE technologies or potential products which society has greater moral trepidation about. The genetic engineering of animals is less acceptable than the genetic engineering of plants or crops, while the use of genetic engineering to enhance human potential is currently very unacceptable to society. Although, to my knowledge, there is no empirical data regarding the public acceptability of genetically engineered weapons of mass destruction, I would suggest that such use of the technology would be considered morally reprehensible by any right thinking person.

A question that the analytical and empirical considerations discussed in this chapter raise is: What are the rights and responsibilities of science and scientists with respect to the wider society in regard to the development of controversial science and technologies such as genetic engineering? A common argument made by scientists working in the field of GE, when considering public criticism of the technology, is that the public are arguing from an emotional rather than a rational perspective and, therefore, their arguments should be dismissed. (As an aside, scientists too might have an emotional stake in the issue – attachment to years of education, passion for their field of science, justification of their own research endeavour and life path, and the social status and financial security provided by their career, etc.).

However, Small (2004b) argued that emotions are just as relevant to humans as rationality. Individuals grow up within a particular culture and are trained to adopt the culture's explicit moral values and assimilate its tacit ones from an early age. These deeply embedded values comprise the core of an individual's self identity and provide a lens through which they examine and evaluate the world and construct personal meaning. People have strong emotional attachments to their core values. To be human is to be both rational and emotional. To deny either of these components is to be less than human. Without our emotions our moral sensibilities would be severely diminish. If we do not respect people's emotional experience then we are not respecting them. When it comes to understanding public acceptance of a technology and respecting people's core values, it is essential that science consider how the technology will impact on people's emotional, moral, cultural and spiritual sensibilities, as well as their rational reasoning (again emphasising the need to foresight the societal implications of new technologies from an understanding of human nature and behaviour). Responsible science is obligated to acknowledge, respect and appropriately incorporate the cultural, spiritual, and moral values of society. This may be an essential requirement for science to gain and maintain the trust and co-operation of society.

Nonetheless, when considering the rights of science in society, Small (2004b) argued that moral beliefs about what is acceptable or unacceptable differ between cultures and societies, and between groups within a society. While some moral values enjoy almost unanimous support (e.g., murder is wrong) others may be more controversial (e.g., abortion is wrong). Similarly, cultures change and evolve over time, along with their moral values. Practices that were once commonplace, and consistent with the moral values of the time, are no longer acceptable in modern societies (or are becoming less so). Slavery, child labour, child marriages, drink driving, environmental pollution, and the death penalty are clear examples. Similarly, practices and values once unacceptable to society, have over time, become more morally acceptable, sometimes even enshrined in changed law. Clear examples include religious freedom, freedom of speech, birth control, homosexuality, divorce, and children born outside traditional marriage (Svensson & Wood, 2003).

Science and ethics are inseparably interconnected. Scientific discoveries and other forms of new information have the potential to influence and alter the moral, spiritual, and cultural beliefs and practices of society. The influence of science has been essential in shaping the progress of contemporary Western culture and the formation of modern values. Two prominent historical instances were Galileo's demonstration of the Copernican heliocentric model and Darwin's theory of evolution. These examples are highly valuable since they demonstrate the impact of research on shaping contemporary values. Moreover, they highlight the significance of scientists challenging prevailing scientific views as well as spiritual, cultural, and moral norms. By proving the heliocentric model, Galileo displaced mankind from the geocentric center of the cosmos, establishing that the Earth orbits around the sun. Darwin's idea of evolution furthered this paradigm shift by displacing humans from the metaphysical core of the cosmos. These scientific theories revolutionized our comprehension of our position within the natural world. These consequences pertain to the ethical standing of humans, non-human life, and the environment (Small, 2004b).

Galileo and Darwin not only questioned the prevailing scientific knowledge of their era, but also challenged the cultural, spiritual, and moral conventions. Both theories elicited moral and religious indignation upon their initial introduction. Today, they are widely accepted as the prevailing knowledge or beliefs. In retrospect, it is evident that the beliefs held by Galileo and Darwin were more advanced than those of the majority of society. These individuals are significant personalities in human history because they demonstrated the bravery to question not only the prevailing scientific beliefs of their era, but also the cultural, spiritual, and moral norms of their period. A crucial requirement for scientific advancement is that the assertions and hypotheses of science are susceptible to scrutiny and modification in response to fresh evidence. I suggest that a crucial factor in the development of cultural, spiritual, and moral values is their willingness to be questioned and examined in light of new information and other perspectives, including advancements in scientific understanding.

5. CONCISE SUMMARY

I have contended that it is imperative for the scientific community to recognize, esteem, and integrate the cultural, spiritual, and moral convictions and principles of society. I have also contended that given the fact that these ideas are not fixed and can evolve, it is equally crucial for science to have the liberty to question the cultural, spiritual, and moral principles of society. Galileo and Darwin's examples demonstrate that questioning values and beliefs is a crucial means by which science and culture evolve and advance. Undoubtedly, it is a crucial duty of science to society to raise such concerns. The challenge thus lies in achieving a suitable equilibrium between these two sometimes conflicting goals. Not all scientific challenges to societal norms will ultimately be embraced. Science and logic are highly effective methods of acquiring knowledge about the universe, but they may not always possess superior understanding compared to culture, religion, emotion, and morality in every circumstance. The reasons presented above endorse the adoption of a post-normal science approach to technological advancement. This approach emphasizes the importance of transparency, discussion, and continuous interaction between the scientific community and society regarding the trajectory of scientific progress and its alignment with society's cultural, spiritual, and moral values. Through practicing openness, transparency, and active participation in public discourse and deliberation over contentious or pioneering scientific investigations, the scientific community not only exhibits social accountability but also potentially assumes a pioneering role in the cultural, spiritual, and ethical advancement of society.

According to Jonas (1985), a crucial aspect of this conversation is the scientific anticipation of the possible biophysical, social, and moral consequences of emerging technologies. Scientists can assist the public in comprehending the consequences of emerging technology in relation to their personal values and the material possessions they cherish. Bunge (1977) asserts that scientists should bear responsibility and be held accountable for the applications and societal consequences of their technological innovations. Undoubtedly, scientists receive recognition and admiration for the advantages resulting from their technical innovations. Consequently, it follows that they should also be held responsible and accountable for the negative consequences that arise. Nevertheless, within this particular framework, there seems to be a noticeable imbalance between the allocation of credit and blame. This asymmetry arises from two facts. Firstly, scientists lack the ability to exert control over the choices made by others regarding the utilization of the technologies they create. Secondly, as Jonas highlights, even while there is a moral obligation to anticipate and evaluate the consequences and effects of new technologies, the process of foresight will inevitably be flawed. Therefore, when it comes to the advancement of influential emerging technologies like genetic engineering, which could permanently transform both human and non-human life, as well as the ecosystems of our planet, it is crucial for the scientific community to adhere to the precautionary principle.

Engaging in debate and dialogue between science and society is crucial for both the scientific community and the public. However, it is important to note that this may not always lead to the most precise understanding of public attitudes and values, or provide insight into the current trends in public norms. Debate and dialogue often attract those who lean towards extreme opinions, resulting in the majority remaining uninvolved and the nuances of their stances going unnoticed. Empirical ethics, a relatively recent field (Borry, Schotsmans, & Dierickx, 1995, 2005; Borry, et al., 2007), integrates normative ethics and descriptive ethics. Descriptive ethics is grounded in the social sciences, specifically sociology, anthropology, psychology, and epidemiology. It employs both qualitative and quantitative empirical methodologies. Empirical ethics seeks to integrate the descriptive ethics perspectives of different publics with the analytical insights of normative ethics in order to provide morally appropriate decisions that are relevant to the specific context (Borry, et al., 2007). Post-normal science and empirical ethics offer rigorous methodologies to enhance transparency on the social and moral ramifications of genetic engineering.

6. REFERENCES

Anderson, K., & Bows, A. (2008). Reframing the climate change challenge in light of post- 2000 emission trends. *Philosophical Transactions of the Royal Society A: Mathematical, Physical and Engineering Sciences, 366*(1882), 3863-3882. doi: 10.1098/rsta.2008.0138

Andrews, L. B. (2002). Genes and patent policy: rethinking intellectual property rights. *Nat Rev Genet, 3*(10), 803-808. doi: 10.1038/nrg909

Antoniou, M. (1996). Genetic pollution. *Nutritional Therapy Today, December 1996. 6,* 8-11.

Appleby, M. C. (1999). Tower of Babel: Variation in ethical approaches, concepts of welfare and attitudes to genetic manipulation. *Animal Welfare, 8,* 381-390.

Associated Press. (2007). Rare condition gives toddler super strength Retrieved 15 March, 2010, from http://www.ctv.ca/servlet/ArticleNews/story/CTVNews/20070530/strong_tod dler_070530/20070530

BBC News. (2010, 20 May). 'Artificial life' breakthrough announced by scientists Retrieved 23 May, 2010, from http://news.bbc.co.uk/2/hi/science_and_environment/10134341.stm

Biopunk.org. (n.d.). Biopunk.org Retrieved 26 January, 2010, from http://www.biopunk.org/

Bishop, B. (2008, 28 January). The open biohacking project / Kit Retrieved 26 January, 2010, from http://biohack.sourceforge.net/

Borlaug, N. (1997). Feeding a world of 10 billion people: The miracle ahead. *Plant Tissue Culture and Biotechnology, 3,* 119-127.

Borry, P., Schotsmans, P., & Dierickx, K. (1995). Empirical research in bioethical journals. *Journal of Medical Ethics, 32,* 240-245.

Borry, P., Schotsmans, P., & Dierickx, K. (2005). The birth of the empirical turn in bioethics. *Bioethics, 19*(1), 49-71.

Borry, P., Schotsmans, P., & Dierickx, K. (2007). Bioethics and its methodology: The rise of empirical contributions. In K. D. Chris Gastmans, Herman Nys and Paul Schotsmans (Ed.), *New Pathways for European Bioethics.* Antwerpen: Intersentia.

Bostrom, N. (2002). Existential risks: Analyzing human extinction scenarios and related hazards. *Journal of Evolution and Technology, 9*(1). Retrieved from http://www.jetpress.org/volume9/risks.pdf

Brown, L. R. (2008). *Plan B 3.0: Mobilizing to save civilization.* New York: W. W. Norton & Company. Bunge, M. (1977). Towards a technoethics. *Monist, 60*(1), 96-107.

Cello, J., Paul, A. V., & Wimmer, E. (2002). Chemical synthesis of poliovirus cDNA: Generation of infectious virus in the absence of natural template. *Science, 297*(5583), 1016-1018. doi: 10.1126/science.1072266

Chen, C. S., Mrksich, M., Huang, S., Whitesides, G. M., & Ingber, D. E. (1997). Geometric control of cell life and death. *Science, 276*(5317), 1425-1428. doi: 10.1126/science.276.5317.1425

Cherian, S., & Oliveira, M. M. (2005). Transgenic plants in phytoremediation: recent advances and new possibilities. *Environmental Science Technology, 39*(24), 9277-9290.

Chopra, P., & Kamma, A. (2006). Engineering life through synthetic biology. *In Silico Biology, 6*(5), 401-410.

Chowdhury, K., & Bagasra, O. (2007). An edible vaccine for malaria using transgenic tomatoes of varying sizes, shapes and colors to carry different antigens. *Medical Hypotheses, 68*(1), 22-30. doi: 10.1016/j.mehy.2006.04.079

Codd, V., Mangino, M., van der Harst, P., Braund, P. S., Kaiser, M., Beveridge, A. J., et al. (2010). Common variants near TERC are associated with mean telomere length. *Nat Genet, advance online publication*. doi: 10.1038/ng.532

de Grey, A. D. N. J. (2005). A strategy for postponing aging indefinitely. *Studies in Health Technology and Informatics 118*, 209-219.

DIYbio. (2010). DIYbio Retrieved January 26, 2010, from http://diybio.org/

Fedoroff, N. V., Battisti, D. S., Beachy, R. N., Cooper, P. J. M., Fischhoff, D. A., Hodges, C. N., et al. (2010). Radically rethinking agriculture for the 21st century. *Science, 327*(5967), 833-834. doi: 10.1126/science.1186834

Feng, J., Naiman, D. Q., & Cooper, B. (2009). Coding DNA repeated throughout intergenic regions of the Arabidopsis thaliana genome: evolutionary footprints of RNA silencing. *Mol. BioSyst., 5*, 1679 - 1687. doi: 10.1039/b903031j

Fiester, A. (2008). Justifying a presumption of restraint in animal biotechnology research. *American Journal of Bioethics, 8*(6), 36-44. doi: 10.1080/15265160802248138

Finkel, T. (2003). Ageing: A toast to long life. *Nature, 425*(6954), 132-133. doi: 10.1038/425132a

Fox, M. W. (1999). *Beyond evolution: The genetically altered future of plants, animals, the earth... and humans*. New York: Lyons Press.

Funtowicz, S. O., & Ravetz, J. R. (1993). Science for the post-normal age. *Futures*(September), 739-755.

Gene Therapy Net. (2010). What is gene therapy? Retrieved 16th March, 2010, from http://www.genetherapynet.com/what-is-gene-therapy.html

Gibson, D. G., Glass, J. I., Lartigue, C., Noskov, V. N., Chuang, R.-Y., Algire, M. A., et al. (2010). Creation of a bacterial cell controlled by a chemically synthesized genome. *Science*, science.1190719. doi: 10.1126/science.1190719

Gleick, P. H. (2003). Global freshwater resources: Soft-path solutions for the 21st century. *Science, 302*(5650), 1524 - 1528. doi: 10.1126/science.1089967

GloFish.com. (2010). GloFish: Experience the glo Retrieved 8 February, 2010, from http://www.glofish.com/

Golovan, S. P., Meidinger, R. G., Ajakaiye, A., Cottrill, M., Wiederkehr, M. Z., Barney, D. J., et al. (2001). Pigs expressing salivary phytase produce low-phosphorus manure. *Nat Biotech, 19*(8), 741-745. doi: 10.1038/90788

Good, I. J. (1965). Speculations concerning the first ultraintelligent machine. *Advances in Computers, 6*, 31-88.

Grant, B. (2008). Artifical life, a step closer. *TheScientist*. Retrieved from http://www.the- scientist.com/blog/display/54212/

Grisham, J. (2000). New rules for gene patents. *Nat Biotech, 18*(9), 921-921. doi: 10.1038/79381 Grobet, L., Royo Martin, L. J., Poncelet, D., Pirottin, D., Brouwers, B., Riquet, J., et al. (1997).

Hansen, J., Sato, M., Kharecha, P., Beerling, D., Berner, R., Masson-Delmotte, V., et al. (2008). Target atmospheric CO2: Where should humanity aim? *Open Atmos Sci J, 2*, 217- 231. doi: 10.2174/1874282300802010217

Hansen, N. E., Janz, H. L., & Sobsey, D. J. (2008). 21st century eugenics? *The Lancet, 372*(Supplement 1), S104-S107. doi: 10.1016/S0140-6736(08)61889-9

Harley, C. B., Futcher, A. B., & Greider, C. W. (1990). Telomeres shorten during ageing of human fibroblasts. *Nature, 345*(6274), 458-460. doi: 10.1038/345458a0

Helfand, S. L., & Inouye, S. K. (2003). Aging, life span, genetics and the fruit fly. *Clinical Neuroscience Research, 2*(5-6), 270-278. doi: 10.1016/S1566-2772(03)00003-3

Henderson, M. (2010, 21 May). Scientists create artificial life in laboratory Retrieved 23 May, 2010, from http://www.timesonline.co.uk/tol/news/science/biology_evolution/article71322 99.ece

Ho, M.-W. (2000). *Genetic engineering: Dream or nightmare?: Turning the tide on the brave new world of bad science and big business* (2nd ed.). New York: Continuum.

Ide, S., Miyazaki, T., Maki, H., & Kobayashi, T. (2010). Abundance of ribosomal RNA gene copies maintains genome integrity. *Science, 327*(5966), 693-696. doi: 10.1126/science.1179044

Intergovernmental Panel on Climate Change (Producer). (2007, 24 August 2009). Climate change 2007: Fourth assessment report of the intergovernmental panel on climate change. Retrieved from http://www.ipcc.ch/

Jablonka, E., & Raz, G. (2009). Transgenerational epigenetic inheritance: Prevalence, mechanisms, and implications for the study of heredity and evolution. *Quarterly Review of Biology, 84*(2), 131-176. doi: 10.1086/598822

Jackson, R. J., Ramsay, A. J., Christensen, C. D., Beaton, S., Hall, D. F., & Ramshaw, I. A. (2001). Expression of mouse interleukin-4 by a recombinant ectromelia virus suppresses cytolytic lymphocyte responses and overcomes genetic resistance to mousepox. *J. Virol., 75*(3), 1205-1210. doi: 10.1128/jvi.75.3.1205-1210.2001

Jansson, C., Wullschleger, S. D., Kalluri, U. C., & Tuskan, G. A. (2010). Phytosequestration: Carbon biosequestration by plants and the prospects of genetic engineering. *BioScience, 60*(9), 685-696. doi: 10.1525/bio.2010.60.9.6

Jarraya, B., Boulet, S., Scott Ralph, G., Jan, C., Bonvento, G., Azzouz, M., et al. (2009). Dopamine gene therapy for Parkinson's Disease in a nonhuman primate without associated dyskinesia. *Science Translational Medicine, 1*(2), 2ra4-2ra4. doi: 10.1126/scitranslmed.3000130

Jonas, H. (1985). *The imperative of responsibility: In search of an ethics for the technological age*. Chicago: The University of Chicago.

Jones, S. (2006). Why creationism is wrong and evolution is right. London: The Royal Society.

Kaati, G., Bygren, L. O., & Edvinsson, S. (2002). Cardiovascular and diabetes mortality determined by nutrition during parents' and grandparents' slow growth period. *European Journal of Human Genetics, 10*(11), 682-688. doi:10.1038/sj.ejhg.5200859

Katsnelson, A. (2010). DNA factory launches. *The Scientist*. Retrieved from http://www.the-scientist.com/blog/display/57090/

Katz, R. (2001). Biological weapons: A national security problem that requires a public health response. Princeton: Office of Population Research Princeton University.

Kenyon, C., Chang, J., Gensch, E., Rudner, A., & Tabtiang, R. (1993). A C. elegans mutant that lives twice as long as wild type. *Nature, 366*(6454), 461-464. doi: 10.1038/366461a0

Kurzweil, R. (2005). *The singularity is near*. New York: Viking.

Lal, R. (2007). World soils and global issues. *Soil and Tillage Research, 97*(1), 1-4. doi: 10.1016/j.still.2007.04.002

Lee, Y. J., Yi, H., Kim, W.-J., Kang, K., Yun, D. S., Strano, M. S., et al. (2009). Fabricating genetically engineered high-power lithium-ion batteries using multiple virus genes. *Science, 324*(5930), 1051-1055. doi: 10.1126/science.1171541

Leonard, D. G. B. (2002). Medical practice and gene patents: A personal perspective. *Academic Medicine, 77*(12, Part 2), 1388-1391.

Levi, G. (2000). Vaccine cornucopia: Transgenic vaccines in plants: new hope for global vaccination? *EMBO reports 1*(5), 378-380. doi: doi:10.1093/embo-reports/kvd103

Li, Y., Liu, L., & Tollefsbol, T. O. (2009). Glucose restriction can extend normal cell lifespan and impair precancerous cell growth through epigenetic control of hTERT and p16 expression. *FASEB J.*, Published online. doi: 10.1096/fj.09-149328

Lin, Y.-J., Seroude, L., & Benzer, S. (1998). Extended life-span and stress resistance in the drosophila mutant methuselah. *Science, 282*(5390), 943-946. doi: 10.1126/science.282.5390.943

Lolle, S. J., Victor, J. L., Young, J. M., & Pruitt, R. E. (2005). Genome-wide non-mendelian inheritance of extra-genomic information in Arabidopsis. *Nature, 434*(7032), 505- 509. doi: 10.1038/nature03380

Lubchenco, J. (1998). Entering the century of the environment: A new social contract for science. *Science, 279*(5350), 491-497. doi: 10.1126/science.279.5350.491

Lumey, L. H. (1992). Decreased birthweights in infants after maternal in utero exposure to the Dutch famine of 1944-1945. *Paediatric and Perinatal Epidemiology, 6*(2), 240-253. doi: 10.1111/j.1365-3016.1992.tb00764.x

Luppicini, R. (2008). The emerging field of technoethics. In R. Luppicini & R. Adell (Eds.), *Handbook of research on technoethics* (pp. 1-18). Hersey: Idea Group Publishing.

Mancuso, K., Hauswirth, W. W., Li, Q., Connor, T. B., Kuchenbecker, J. A., Mauck, M. C., et al. (2009). Gene therapy for red - green colour blindness in adult primates. *Nature, 461*(7265), 784-787. doi: 10.1038/nature08401

May, R. (1999). Melding heart and head Retrieved 10 March, 2010, from http://www.unep.org/OurPlanet/imgversn/111/may.html

McBeath, R., Pirone, D. M., Nelson, C. M., Bhadriraju, K., & Chen, C. S. (2004). Cell shape, cytoskeletal tension, and RhoA regulate stem cell lineage commitment. *Developmental Cell, 6*(4), 483-495.

McPherron, A. C., Lawler, A. M., & Lee, S.-J. (1997). Regulation of skeletal muscle mass in mice by a new TGF-p superfamily member. *Nature, 387*(6628), 83-90. doi: 10.1038/387083a0

Moor, J. (2005). Why we need better ethics for emerging technologies. *Ethics and Information Technology, 7*(3), 111-119. doi: 10.1007/s10676-006-0008-0

National Human Genome Research Institute. (2009, 27 October). What was the human genome project? Retrieved 15 December, 2009, from http://www.genome.gov/12011238

National Science Board. (1977). Science at the bicentennial. A report from the research community (pp. 154). Washington, DC: National Science Foundation.

Neimark, J. (2009). The second coming of gene therapy. *Discovery,* (September). Retrieved from http://discovermagazine.com/2009/sep/02-second-coming-of-gene- therapy/article_view?b_start:int=0&-C=

Nowacki, M., Higgins, B. P., Maquilan, G. M., Swart, E. C., Doak, T. G., & Landweber, L. F. (2009). A functional role for transposases in a large eukaryotic genome. *Science, 324*(5929), 935-938. doi: 10.1126/science.1170023

OpenWetWare. (2009, 3 March 2009). OpenWetWare Retrieved 26 January, 2010, from http://openwetware.org/wiki/Main_Page

Ortiz, R. (1998). Critical role of plant biotechnology for the genetic improvement of food crops: Perspectives for the next millenium. *Electronic Journal of Biotechnology, 1*(3). Retrieved from http://www.ejb.org/content/vol1/issue3/full/7/

Parliamentary Office of Science and Technology. (2008, January). Synthetic biology. *Postnote, 298.*

Pimentel, D., & Pimentel, M. (Eds.). (1996). *Food, energy and society* (Revised ed.). Niwot: University Press of Colorado.

Pimentel, D., Pimentel, M., & Karpenstein-Machan, M. (1999). Energy use in agriculture: An overview. *Agricultural Engineering International: The CIGR EJournal, 1*, Retrieved 21 January, 2010 from http://www.cigrjournal.org/index.php/Ejounral/article/view/1044.

Pimentel, D., & Sparks, D. L. (2000). Soil as an endangered ecosystem. *BioScience, 50*(11), 947- 947. doi: 10.1641/0006-3568(2000)050[0947:saaee]2.0.co;2

Posthuma, D., Luciano, M., de Geus, E. J. C., Wright, M. J., Slagboom, P. E., Montgomery, G. W., et al. (2005). A genomewide scan for intelligence identifies quantitative trait loci on 2q and 6p. *Am J Hum Genet. 2005, 77*(2), 318-326.

Puca, A. A., Daly, M. J., Brewster, S. J., Matise, T. C., Barrett, J., Shea-Drinkwater, M., et al. (2001). A genome-wide scan for linkage to human exceptional longevity identifies a locus on chromosome 4. *Proceedings of the National Academy of*

Sciences of the United States of America, 98(18), 10505-10508. doi: 10.1073/pnas.181337598

Rees, M. (2003). *Our final century?* London: William Heineman. Rifkin, J. (1998). *The biotech century*. London: Victor Gollanz.

Robert, J. S., & Baylis, F. (2003). Crossing species boundaries. *The American Journal of Bioethics, 3*(3), 1-13.

Rogina, B., Reenan, R. A., Nilsen, S. P., & Helfand, S. L. (2000). Extended life-span conferred by cotransporter gene mutations in drosophila. *Science, 290*(5499), 2137-2140. doi: 10.1126/science.290.5499.2137

Rollin, B. E. (1996). Bad ethics, good ethics and the genetic engineering of animals in agriculture. *J. Anim. Sci., 74*(3), 535-541.

Rollin, B. E. (1998). On telos and genetic engineering. In A. Hollands & A. Johnson (Eds.), *Animal biotechnology and ethics* (pp. 156-187). London: Chapman and Hall.

Routtenberg, A., Cantallops, I., Zaffuto, S., Serrano, P., & Namgung, U. (2000). Enhanced learning after genetic overexpression of a brain growth protein. *Proceedings of the National Academy of Sciences of the United States of America, 97*(13), 7657-7662.

Rucz, K. (2008). Longevity genes. *Hungarian Medical Journal, 2*(4), 499-507. doi: 10.1556/HMJ.2.2008.28335

Sasaki, E., Suemizu, H., Shimada, A., Hanazawa, K., Oiwa, R., Kamioka, M., et al. (2009). Generation of transgenic non-human primates with germline transmission. *Nature, 459*(7246), 523-527. doi: 10.1038/nature08090

Savulescu, J. (2003). Human-animal transgenesis and chimeras might be an expression of our humanity. *The American Journal of Bioethics, 3*(3), 22-25.

Savulescu, J. (2005). New breeds of humans: the moral obligation to enhance. *Reproductive BioMedicine Online, 10*(Supp 1), 36-39. Retrieved from www.rbmonline.com.article/1643

Schacht, W. H. (2006). Gene patents: A brief overview of intellectual property issues. Washington: Congressional Research Service - Report for Congress.

Scheller, J., Guhrs, K.-H., Grosse, F., & Conrad, U. (2001). Production of spider silk proteins in tobacco and potato. *Nat Biotech, 19*(6), 573-577. doi: 10.1038/89335

Schuelke, M., Wagner, K. R., Stolz, L. E., Hubner, C., Riebel, T., Komen, W., et al. (2004). Myostatin mutation associated with gross muscle hypertrophy in a child. *N Engl J Med, 350*(26), 2682-2688. doi: 10.1056/NEJMoa040933

SemBioSys. (2008). Insulin Retrieved 4 August, 2008, from http://www.sembiosys.com/pdf/SBS-1723-Product-FS(Insulin).pdf

Small, B. (2004a). Emotion and evolution in science and ethics. *Reflections on the use of human genes in other organisms: Ethical, spiritual, and cultural dimensions.* Wellington: Toi te Taiao - the Bioethics Council.

Small, B. (2004b). *Responsibilities and rights of science in society: the case of placing human genes in other organisms.* Paper presented at the Technologies, Publics and Power: The Terrain of the 6th Framework in New Zealand and Beyond Conference, Akaroa, New Zealand,1-5 February 2004.

Small, B. (2005). *Genetic engineering: New Zealand public attitudes 2001, 2003 and 2005.* Paper presented at the Talking Biotechnology: Reflecting on science in society, Wellington, New Zealand.

Small, B. (2006, 10-12 January). *GE and medicine: A comparison of NZ public and scientists' attitudes.* Paper presented at the New Zealand Bioethics Conference, University of Otago, Dunedin.

Small, B. (2009). Interim Report: New Zealander's Perceptions of Genetic Engineering Including a Comparison with Surveys Conducted in 2001, 2003, and 2005. Hamilton: AgResearch Ltd.

Small, B., & Botha, N. (2006, 23-29 July). *Scientists' moral attitudes and beliefs about genetic engineering* Paper presented at the XVI International Sociology Association Congress, Durbin, South Africa.

Small, B., & Jollands, N. (2006). Technology and ecological economics: Promethean technology, Pandorian potential. *Ecological Economics, 56*(3), 343-358.

Small, B., Parminter, T. G., & Fisher, M. W. (2005). Understanding public responses to genetic engineering through exploring intentions to purchase a hypothetical functional food derived from genetically modified dairy cattle. *New Zealand Journal of Agricultural Research, 48*, 391-400.

Sticklen, M. B. (2008). Plant genetic engineering for biofuel production: towards affordable cellulosic ethanol. *Nature Reviews Genetics, 9*(6), 433-443.

Straughan, R. (1995a). Ethics, morality and crop biotechnology. 1. Intrinsic concerns. *Outlook on Agriculture, 24*(3), 187-192.

Straughan, R. (1995b). Ethics, morality and crop biotechnology. 2. Extrinsic concerns about consequences. *Outlook on Agriculture, 24*(4), 233-240.

Suzuki, D. (2001). Introduction: A geneticist's reflections on the new genetics. In R. Hindmarsh & G. Lawrence (Eds.), *Altered genes II: The Future.* Carlton North, Victoria Australia: Scribe Publications Pty Ltd.

Svensson, G., & Wood, G. (2003). The dynamics of business ethics: a function of time and culture - cases and models. *Management Decision, 41*(4), 350-361.

Tang, G., Qin, J., Dolnikowski, G. G., Russell, R. M., & Grusak, M. A. (2009). Golden rice is an effective source of vitamin A. *Am J Clin Nutr, 89*(6), 1776-1783. doi: 10.3945/ajcn.2008.27119

The Royal Society. (2009). Reaping the benefits: Science and the sustainable intensification of global agriculture. London: The Royal Society.

The World Commission on Environment and Development. (1987). *Our Common Future (The Brundtland Report).* Oxford: Oxford University Press.

Thompson, P. B. (2010). Why using genetics to address welfare may not be a good idea. *Poultry Science, 89*, 814-821. doi: 10.3382/ps.2009-00307

Ticky, G. (2004). The over-optimism among experts in assessment and foresight. *Technological Forecasting & Social Change, 71*, 341-363.

Tissenbaum, H. A., & Guarente, L. (2001). Increased dosage of a sir-2 gene extends lifespan in Caenorhabditis elegans. *Nature, 410*(6825), 227-230. doi: 10.1038/35065638

Tomita, M., Munetsuna, H., Sato, T., Adachi, T., Hino, R., Hayashi, M., et al. (2003). Transgenic silkworms produce recombinant human type III procollagen in cocoons. *Nat Biotech, 21*(1), 52-56. doi: 10.1038/nbt771

Vinces, M. D., Legendre, M., Caldara, M., Hagihara, M., & Verstrepen, K. J. (2009). Unstable tandem repeats in promoters confer transcriptional evolvability. *Science, 324*(5931), 1213-1216. doi: 10.1126/science.1170097

Vinge, V. (1993). The coming technological singularity: How to survive in the post-human era. *Vision-21: Interdisciplinary Science & Engineering in the Era of CyberSpace Symposium.* Retrieved from http://www-rohan.sdsu.edu/faculty/vinge/misc/singularity.html

Vorosmarty, C., Lettenmaier, D., Leveque, C., Meybeck, M., Pahl-Wostl, C., Alcano, J., et al. (2004). Human transformation of the global water system. *EOS, 85*(48), 509-513.

Wadman, M. (2010). Breast cancer gene patents judged invalid: Court ruling may spell bad news for biotech industry. *Nature, 30 March, doi:10.1038/news.2010.160.*

Weindruch, R., Walford, R. L., Fligiel, S., & Guthrie, D. (1986). The retardation of aging in mice by dietary restriction: Longevity, cancer, immunity and lifetime energy intake. *J. Nutr., 116*(4), 641-654.

Wells, D. J. (2010). Genetically modified aninals and pharmacological research. In F. Cunningham, J. Elliot & P. Lees (Eds.), *Comparative and veterinary pharmacology* (Vol. 199, pp. 213-226): Springer.

West, A. (2010). *Approaching the limits: Feeding 9+ billion humans whilst sustaining civilisation.* Paper presented at the West Oxford Farming Conference, Oxford. http://gw/noticeboard/Pages/Speeches-2010.aspx

White, D. J. G., Langford, G. A., Cozzi, E., & Young, V. K. (1995). Production of pigs transgenic for human DAF: A strategy for xenotransplantation. *Xenotransplantation, 2*(3), 213-217. doi: 10.1111/j.1399-3089.1995.tb00097.x

Wohlsen, M. (2008, 26 Dec). Hobbyists try genetic engineering at home: Critics worry that amateurs could unleash an environmental or medical disaster Retrieved 21 Jan, 2010, from http://www.msnbc.msn.com/id/28390773

Youngquist, W. (1999). The post-petroleum paradigm -- and population. *Population and Environment: A Journal of Interdisciplinary Studies 20*(4).

INDEX

A

Abscisic 190

Absorbance 232

Acceptor 51, 62

Adenovirus 245

Adsorption 78

Agarose 163

Alcohol 113, 116, 131, 141

Aldehyde 24, 113

Angiosperm 116

Antigens 244

Antisense 124, 136, 142, 155

Asexual 177

Asparagine 189

B

Bacteria 11, 16, 24, 57, 221, 277

Binary 42, 46-48, 50, 53, 58-59, 61-62, 67

Bioassay 232, 238

Bioethics 301, 307-308

Biology 3, 32-33, 35-36, 38, 131-132, 138, 169-170, 173-174, 200, 203-204, 206-207, 209, 235, 283, 302

Biomass 109, 227, 231

Bioreactor 228

Biosafety 145

Bioscience 67

Breeding 147, 149-150, 205, 276-277

Buffalopox 260

Buffer 96, 160, 231

C

Catalytic 99, 117

Cellulose 110, 118

C-fragment 58

C-fusion 50

Chemistry 130, 137, 140

Chicken 250

Chimeric 162

Chloride 79

Chromatin 22

Chromosome 4, 146

Climate 287, 304

Coding 4, 15, 17, 20, 157

Codons 7, 10, 13, 15

Coenzyme 142

Community 299-300

Computers 286

C-terminus 55

Cucumber 150

Cytoplasm 7-8, 156

D

Deletions 250

Depletion 287

Domains 75

E

Economics 308

Ecosystems 260, 263

Elements 17, 19, 21-22

Embryos 127, 177-190, 195-196, 198-201, 206, 209

Energy 5, 54, 71-72, 306

Enzymes 5, 16, 71, 75, 90, 92, 112, 114-117, 119, 223

Enzymology 238

Eugenics 282

Eukaryotes 3-7, 10, 19-20

Evolution 36, 93-95, 103, 277, 282, 298, 304

F

Family 112, 116, 213-214, 216, 246

Fibers 118

Floral 169

Forage 134

Formation 117, 180, 198, 232

Fossil 287

Fragments 4, 55-56, 58, 98, 158

Fusions 47-48

G

Gametes 177

Genera 86, 246

Generation 277

Genetics 33-34, 38, 65, 172-173, 200, 205, 308

Genome 11-13, 41-44, 116, 152-153, 156-157, 222, 225-226, 247, 253, 275-276, 282-283, 303, 306

Genotypes 183-185

Germ-line 280

Glutamine 189

Glycine 200-201, 203-205, 207-212

Glycol 206

Growth 25-26, 109, 128-130, 135, 140, 181-182, 187-188, 190, 195, 213-214, 216-221, 226-227, 233-238

H

Half-life 78

Hazards 239

Herbicide 278

Histone 37

Homologous 243

Hormones, 208, 213-216, 222-223, 226-229, 231-233

Hybrid 252, 257

Hydrogen 91

Hydrolysis 8, 72

Hypocotyls 123

Hypophysis 214

I

Immunity 152, 155, 250

Incubation 80, 95

Introns 4

K

Kanamycin 162

L

Legumes 178, 180
Letters 35
Lifespan 285
Lignins 110, 133
Liposomes 164
Luciferase 2, 23-24, 27, 31, 33, 35, 71, 74-96, 98-108
Luciferins 72

M

Maltose 187-189, 191
Markers 155
Metabolism 112, 130, 133, 232
Methanol 227-229
Methionine 189
Microtools 163
Morphology 185
Mosaic 150-151, 168, 173
Mutants 80-81, 88-90, 92-95, 98
Mutations 13, 81, 97

N

N-terminal 76
Nucleotide 223
Nucleus 156

O

Oryzae 147
Oxidation 71-72

P

Papaya 148, 175
Pathogen 150-151
Pepper 148
Peroxidase 127, 131
Peroxisome 59-60
Phenolate 73
Physiology 63, 65-67, 131-132, 135-136, 138-142, 169, 201, 204, 208, 210
Placenta 215, 235
Plantlets 123
Plants 24, 41-42, 44, 47, 58, 60, 62, 67, 111, 115, 118, 122-123, 127-130, 132, 139, 142, 146-152, 155, 158-160, 162, 164, 168-174, 177, 185, 190, 196-198, 200-201, 210-211, 213, 240, 288, 302
Plasma 58, 158
Plasmids 42, 49, 153, 170, 197, 223
Pollution 287
Polymerase 4-6, 17, 33, 37, 152
Progeny 198, 259
Prolactin 238
Proteins 25, 44, 50, 54-58, 60, 96, 106, 148, 157, 182, 213, 229, 233
Protocols 159, 182-185, 187

R

Ribosomes 7, 35

S

Salinity 287
Serine 90
Signals 58
Sodium 118
Sorbitol 188-189
Soybean 148, 169, 177-178, 180-189, 195-198, 200-212
Strains 151, 197, 222, 225-227, 233, 249-251, 254, 258-259

Sucrose 186, 188-189, 191, 193, 206, 209

Sugarcane 148, 150

Surrogate 1-2

Symbols 96

Symptoms 219, 254

Systems 9, 24, 72, 221

T

Telomerase 284

Thaliana 68, 133

Ti-plasmid 43-44, 153

Tobacco 131, 139-140, 148, 151, 164, 168, 170, 173

Transgene 17, 22, 37-38, 158, 162, 172, 197, 243, 252

V

Vaccines 239, 242-246, 248-251, 278

Vectors 26, 38, 41-42, 44-48, 50, 54-60, 62, 64-67, 149, 153, 155, 239, 242-243, 245, 249-250, 252, 257, 261, 269-270

Virology 34, 39, 173, 268-270

Virulence 153

Vitamins 191, 193

X

Xenogeneic 16

Z

Zinnia 142

Zygotic 177-182, 187, 197-198

www.ingramcontent.com/pod-product-compliance
Lightning Source LLC
LaVergne TN
LVHW062222200926
848484LV00006B/1452